State Feedback Control and Kalman Filtering with MATLAB/Simulink Tutorials

State Feedback Control and Kalman Filtering with MATLAB/Simulink Tutorials

Liuping Wang and Robin Ping Guan

WILEY

IEEE PRESS

This edition first published 2023

Registered Office
John Wiley & Sons Ltd, The Atrium, Southern Gate, Chichester, West Sussex, PO19 8SQ, UK

Editorial Office
The Atrium, Southern Gate, Chichester, West Sussex, PO19 8SQ, UK

For details of our global editorial offices, customer services, and more information about Wiley products visit us at www.wiley.com.

Wiley also publishes its books in a variety of electronic formats and by print-on-demand. Some content that appears in standard print versions of this book may not be available in other formats.

Library of Congress Cataloging-in-Publication Data is Applied for:

Hardback ISBN: 9781119694632

Cover Design: Wiley
Cover Image: © fotoslaz/Shutterstock.com

Set in 9.5/12.5pt STIXTwoText by Straive, Chennai, India
Printed and bound by CPI Group (UK) Ltd, Croydon, CR0 4YY

C9781119694632_260922

To

Jianshe

Contents

Author Biography

Liuping Wang, PhD, is a Professor of Control Engineering at RMIT University, Australia. She obtained her PhD from the Department of Control Engineering at the University of Sheffield, UK. Professor Wang gained substantial process control experience by working in the Chemical Engineering Department at the University of Toronto, Canada, and the Center for Integrated Dynamics at the University of Newcastle, Australia. She is the author of five books in systems and control.

Robin Ping Guan, PhD, obtained his Masters degree in Electrical Engineering from the University of Melbourne in 2014 and his PhD degree from RMIT University, Australia in 2019. He is a research fellow in RMIT University and is the co-author of the book on state feedback control and Kalman filter.

Preface

Between PID Control and Model Predictive Control

Control systems are usually designed using one of the two types of mathematical models: transfer function model and state space model. The PID controller is an example of transfer function-based design and is very popular in control system applications because of its simplicity in the design, analysis and implementation. However, the PID controller suffers from performance loss (mild or severe), when it is applied to higher order systems or systems with interactive dynamics. Among the second category of control systems designed using state space models are the model predictive controllers[1]. The model predictive controllers are widely used in industrial applications and are suitable for systems with complex dynamics and interactions between many input and output variables. Based on optimization techniques in real-time, the model predictive controllers have the advantage of maintaining optimality for control systems in the presence of operational constraints. As a result, process efficiency and profit margins may improve. On the other hand, the deployment of online optimization techniques with operational constraints causes complexity in real-time computation and implementation, and in an industrial environment it also leads to higher cost of maintenance and commission of such a complex control system.

When we consider the two major groups of control systems, there is an opportunity for a compromise. This compromise exists between the classical PID controllers and the modern model predictive controllers. The design formulation using a state space model for the description of system dynamics retains some of the key features of model predictive control, and avoiding online optimization solutions reduces the cost of real-time computation, implementation and maintenance of control systems in applications. This book is to present learning materials for such a class of control systems. There are two key aspects of the associated

1 Model predictive controllers are also designed using transfer function models and step response models

control systems that may be attractive for practical applications. The first aspect is the design formulation using state space models, which naturally handles higher order systems or systems with interactive dynamics. The second aspect is the vector-based control system implementation that has control signal limits imposed together with an anti-windup mechanism, which is exceedingly simple and effective in real-time computations.

The Book Structure

This book consists of three parts. Part I and Part II sequentially present the continuous-time and discrete-time state feedback control systems. Part III presents the learning materials for the Kalman filter. Because the Kalman filter has the ability to perform state estimation for linear time-varying systems, nonlinear systems and systems with multi-rate sampled data and missing measurement, it offers endless opportunities for extending the state feedback control systems to various challenging applications.

Part I: Continuous-time State Feedback Control

One of the readers of this book (John) has had many decades of work experience in the field of industrial control and asked us the question why we go beyond PID control. To answer this question, the book begins by constructing a simple analytical example to show the limitations of PID control for systems with interactive dynamics. Upon completion of this detailed case study, we move on to present the fundamental theory in state feedback control (see Chapter 1), including pole-assignment controller, linear quadratic regulator, observer and state estimate feedback control.

Disturbance rejection and reference tracking are two primary objectives for feedback control. In Chapter 2, we introduce two types of practical multivariable controllers for tracking constant reference signals and rejecting low frequency disturbances. In this chapter, we have applied the controllers to three industrial systems with design and simulation: drum boiler control, sugar mill control and drive train control of a wind turbine. MATLAB tutorials are developed in a step-by-step manner for the control system design and the closed-loop simulation. We have also constructed an example to show the scenario of integrator windup and the role of anti-windup mechanism.

Part II: Discrete-time State Feedback Control

The continuous-time control systems are designed in continuous-time and are implemented in the digital environment through discretization of the control law. They are limited to systems that have a fast sampling rate. Furthermore, many complex or new systems lack a continuous-time model from the start and face the potential complications of missing measurements in a multi-rate sampled environment. For these regards, the discrete-time system design and analysis offer a natural solution.

Part II of this book begins by presenting the introductory materials for discrete-time systems (see Chapter 3). This chapter is prepared for some of us who are not familiar with the basic concepts and analysis tools in discrete-time systems.

In parallel with the state feedback control in the continuous-time, Chapter 4 presents the theory for discrete-time state feedback control, including discrete-time observer, discrete-time linear quadratic regulator (DLQR) and DLQR with a prescribed degree of stability.

In discrete-time, the reference signals and disturbance signals can be conveniently described by a discrete-time model. According to the internal model principle (Francis and Wonham (1976)), the desired properties for reference tracking and disturbance rejection of a particular signal can be achieved in a stable closed-loop control system if the signal generator is embedded into the control system. Based on these observations and the internal model principle, in Chapters 5 and 6 we develop the learning materials for discrete-time control system design, simulation and implementation where a disturbance model is used to describe either the reference signal or the disturbance signal. In the design framework, the external signals may be commonly encountered such as step or ramp signals, or a complex signal, and the control system design and implementation remain the same. This generalization is especially useful because the tasks for reference tracking and disturbance rejection of complex signals in control system design and simulation are simplified.

Among these two chapters, Chapter 5 introduces the idea of disturbance observer and the associated control system design to compensate the effects of disturbance via observer design. In contrast, Chapter 6 is centered around the controller design for reference tracking and disturbance rejection. Repetitive control systems, which will track a periodic reference signal, are a natural extension of both control systems. For both chapters, MATLAB tutorials are developed in a step-by-step manner for the control system design and closed-loop simulation. The examples include three applications: a quadruple tank, a heating furnace and a robotic arm with complex moving trajectories.

Part III: Kalman Filtering

The Kalman filter has always been fundamental in the control engineering field for the past six decades. Its importance has increased significantly in the recent decades as the Kalman filter is widely used in the navigation and localization of autonomous aerial and ground vehicles, and in the even wider applications of robotics and the modern manufacturing industry. Understanding the Kalman filter and having the capability to design and implement it are desirable and even necessary for an engineer. In Part III of this book we present the Kalman filter and the approaches associated with the implementation of the Kalman filter in a low-cost computational device.

In the literature, there are many ways to derive the Kalman filter. In Chapter 7, we choose to derive the Kalman filter using an intuitive approach from a control engineer's point of view. Also, from this point of view, we address the commonly encountered issues such as sensor bias, load disturbance, multi-rate sampled data and missing measurement data in state estimation. In this chapter, we give a detailed presentation on the extended Kalman filter for state estimation of nonlinear systems. Additionally, MATLAB tutorials are given in a step-by-step manner to show the computations of the Kalman filter, the Kalman filter with multi-rate sampled data and the extended Kalman filter.

Considering the challenges of implementing the Kalman filter in a low-cost computational device, Chapter 8 discusses the sequential Kalman filter to avoid matrix inversion and the Kalman filter with UDU^T factorization for improving the filter's numerical accuracy when it is implemented for real-time applications. MATLAB tutorials are given to show the implementation of the sequential Kalman filter and the Kalman filter with UDU^T factorization.

Key Features of this Book

Systems and control engineering underpins many engineering disciplines. It becomes increasingly important for an engineer to have the skills of control system design, simulation and implementation. The goal of this book is to facilitate the learning activities of undergraduate and graduate students, industrial researchers and engineers in any engineering field. To achieve this goal, we combine the control system theory with various applications together with MATLAB/Simulink tutorials.

- **Learning by examples.** There are 68 examples and 9 major case studies in the book. The major case studies are associated with practical applications. The examples and case studies contain the illustration of the theory, numerical solutions, and closed-loop simulations.
- **Learning by practice.** There are 23 MATLAB/Simulink tutorials given to show the design, simulation and implementation of the control systems and the Kalman filter. These tutorials are presented in a step-by-step manner so that the learner can follow the process to create their own solutions and closed-loop simulations. Once having understood the closed-loop simulations, we can convert the MATLAB/Simulink simulation programs into C-code for real-time control system implementation.

For each section, there is a set of questions for us to consider. At the end of each chapter, there is a set of problems for us to practice the design and simulation of the control systems.

The Audience

For the past 10 years, we have developed the book materials for classroom teaching of an undergraduate/graduate course across the entire engineering school in RMIT University, Australia. We have also had the privilege to teach short and intensive graduate courses based on the book materials in several other universities overseas with the delivery mode either in-person or through the internet. The courses are typically designed using the combination of lectures and MATLAB tutorials with exercises. The students in all the classes have reported excellent learning outcomes and positive experiences after taking these control system subjects.

An undergraduate course can be developed based on the book materials from Part I in continuous-time control systems. An advanced undergraduate or a graduate course can be developed entirely based on the book materials from Part II and Part III.

We hope that the undergraduate and graduate students in any engineering field will find this book particularly helpful due to a good amount of theory presented in a combination with applications. Because the book is practical and application oriented, we hope that the industrial researchers and engineers will find this book useful for their control system design, simulation and implementation. Finally, this book is written in a way that is enjoyable to read.

Melbourne, Australia

Liuping Wang
Robin Ping Guan

Acknowledgments

We would like to express our gratitude to Professor Graham Goodwin at the University of Newcastle in Australia and Professor Robin Evans at the University of Melbourne in Australia for their help and support over the years. We wish to thank Professor Eric Rogers and Professor Chris Freeman at the University of Southampton, UK, for the collaborative work on repetitive predictive control systems. For discussions, we wish to thank Professor Kevin Moore at the Colorado School of Mines in USA, Professor Tore Hagglund at the Lund Institute of Technology in Sweden, Professor Alex Penlidis at the University of Waterloo in Canada. We wish to thank Dr Arash Vahidnia at RMIT University in Australia for discussions on the topic of drive train control of a wind-turbine.

For valuable comments towards improvement of this book, we wish to thank Professor Antonio Visioli at the University of Brescia in Italy, Dr Minh Tran and Mr Chuong Nguyen in RMIT Vietnam, Dr John Tsing, who had worked in Measurex Corp. USA as a process control engineer and was an adjunct professor at San Jose State University, USA. We wish to thank Dr Junaid Saeed at RMIT University for drawing the block diagrams of drum boiler, quadruple tank and hot air balloon.

List of Symbols and Acronyms

Symbols

A	State matrix of state-space model
B	Input-to-state matrix of state-space model
C	State-to-output matrix of state-space model
$C(s)$	Laplace transfer function of controller
$C(z)$	z-transfer function of controller
$D(q^{-1})$	Disturbance model
$A_{cl}(s)$	Closed-loop polynomial
$A_{cl}(s)^d$	Desired closed-loop polynomial
Δt	Sampling interval
$D_o(s)$	Laplace transform for output disturbance
$D_i(s)$	Laplace transform for input disturbance
$D_m(s)$	Laplace transform for measurement noise
$G(s)$	Transfer function model
J	Performance index for optimization
J_N	Performance index for optimization with finite horizon
K	Feedback control gain vector, also Kalman filter gain
K_{ob}	Observer gain vector
$\kappa(A)$	Condition number of A matrix
q^{-i}	Backward shift operator, $q^{-i}[f(k)] = f(k-i)$
q^i	Forward shift operator, $q^i[f(k)] = f(k+i)$
$Q,\ R$	Pair of weighting matrices in the cost function of linear quadratic regulator, also the covariance matrices for the process noise and measurement noise in the Kalman filter
$r(\cdot)$	Reference signal
$S(s)$, $S(z)$	Sensitivity function in continuous-time and discrete-time
$S_i(s)$, $S_i(z)$	Input disturbance sensitivity function in continuous-time and discrete-time

σ^2	Noise variance
$T(s)$, $T(z)$	Complementary sensitivity function in continuous-time and discrete-time
τ_{cl}	Closed-loop time constant
$u(\cdot)$	Control signal
u^{min}, u^{max}	Minimum and maximum limits for u
$x(\cdot)$	State vector
$\hat{x}(\cdot)$	Estimated state vector
$\hat{x}(k)^{-}$, $P(k)^{-}$	*A priori* state estimate and *a priori* covariance
$\hat{x}(k)^{+}$, $P(k)^{+}$	*A posteriori* state estimate and *a posteriori* covariance
$y(\cdot)$	output signal
0_m	Zero matrix with dimension $m \times m$
$0_{p\times q}$	Zero matrix with dimension $p \times q$
I_m	Identity matrix with dimension $m \times m$

Acronyms

DLQR	Discrete-time linear quadratic regulator
FFT	Fast Fourier transform
KF	Kalman filter
LQR	Linear quadratic regulator
NMSS	Non-minimal state space

About the Companion Website

This book is accompanied by a companion website:

www.wiley.com/go/feedback19

From the website you can find the following online materials:

- MATLAB files
- Lecture Slides

Part I

Continuous-time State Feedback Control

1

State Feedback Controller and Observer Design

1.1 Introduction

State feedback control systems open up a different landscape to control system design for complex systems that have a higher order or have many input and output variables. For those of us who are familiar with the classical control systems designed using a Laplace transfer function model, the continuous-time control systems using a state space model will be a natural extension. Therefore, it is easier to learn and master the design steps in the continuous-time.

The chapter begins with a case study on the effectiveness of PID control of a system with interactions (see Section 1.2). This is to answer the question why to go beyond PID control because this question is urgent for some of us who work in the field of process control and use PID control as bread and butter for daily operations. In Section 1.3, the basic idea about state feedback control is illustrated using a simple analytical example and is followed by the introduction of controllability with an explanation what it means when a system loses its controllability. The concept of closed-loop eigenvalues (or poles) is fundamental in state feedback control. By choosing a set of desired closed-loop eigenvalues, a state feedback controller is designed (see Section 1.4). With a similarity transformation, this section takes us in a step-by-step manner for the design of a pole-assignment controller. However, the pole-assignment controllers have limitations when the systems have multiple inputs as illustrated in Section 1.5. For a control system with multiple inputs, the Linear Quadratic Regulator (LQR) offers a new effective design tool in which optimization underpins the design methodology (see Section 1.5). State estimation using an observer has always been an integrated part of state feedback control because it solves the fundamental problem of how to implement a control system when the state variables are not measured. Section 1.6 introduces observer design and implementation for state feedback control. It also explains the concept of observability with a simple analytical example and duality between observer and controller. The final section of this chapter (see Section 1.7) introduces the state

State Feedback Control and Kalman Filtering with MATLAB/Simulink Tutorials, First Edition.
Liuping Wang and Robin Ping Guan.

Companion website: www.wiley.com/go/feedback19

estimate feedback control system, which is to combine the state feedback control system with the observer. Additionally, the closed-loop stability of such a system is ensured by the stability of the state feedback control system and the observer error system.

1.2 Motivation for Going Beyond PID Control

PID control has been instrumental for the success of wide control engineering applications. Without doubt, the majority of control engineering applications used PID controllers. Before we begin our journey in studying state space control system design methods, we need to answer the question why we want to go beyond PID control. Because the PID controllers have a limited complexity, it is not difficult to imagine that when a system has complex dynamics, a controller beyond a PID structure could provide performance improvement. However, it is harder to imagine that if a system has multi-input and multi-output, even though it has a first order or a second order structure for the subsystems, a controller beyond PID structures could also provide performance improvement, as long as there are interactions between its inputs and outputs. For this purpose, we will explore the limitations of PID control systems in the context for systems with multiple input and output. When using PID controllers for a multi-input and multi-output system, the control system is completely decentralized. For those who wish to further study decentralized control systems, we refer the books by Skogestad and Postlethwaite [2007] and Goodwin et al. [2000].

The limitations of PID control system will be examined using a very simple example with respect to two aspects: (i) closed-loop stability and dynamic performance, and (ii) steady-state performance under the assumption that the closed-loop system is stable.

System under study

We consider a system with two inputs and two outputs described by the following transfer function:

$$\begin{bmatrix} Y_1(s) \\ Y_2(s) \end{bmatrix} = \begin{bmatrix} \frac{1}{s+1} & \frac{\gamma_1}{s+1} \\ \frac{\gamma_2}{s+2} & \frac{1}{s+2} \end{bmatrix} \begin{bmatrix} U_1(s) \\ U_2(s) \end{bmatrix}, \tag{1.1}$$

where $\gamma_1(\neq 0)$ and $\gamma_2(\neq 0)$ are constants. From this transfer function model, there are interactions between the inputs and outputs. The interactions are reflected by the following relationships:

$$Y_1(s) = \frac{1}{s+1}U_1(s) + \frac{\gamma_1}{s+1}U_2(s) \tag{1.2}$$

$$Y_2(s) = \frac{\gamma_2}{s+2}U_1(s) + \frac{1}{s+2}U_2(s). \tag{1.3}$$

Clearly when $\gamma_1 \neq 0$ and $\gamma_2 \neq 0$, the input signal U_2 will affect the output Y_1, and the same is true for the effect of output Y_2 by the input signal U_1.

Simple integral controllers

Because PID controllers are designed for single input and single output systems, in their design, we will neglect the interactions shown in (1.2) and (1.3) by regarding the constants $\gamma_1 = \gamma_2 = 0$. This is the essential assumption in the design of PID controllers, where the interactions are regarded as a form of disturbance. Thus, we will consider the following two independent systems:

$$Y_1(s) = \frac{1}{s+1} U_1(s) \tag{1.4}$$

$$Y_2(s) = \frac{1}{s+2} U_2(s). \tag{1.5}$$

These are two first order systems for which we can use two PI controllers to achieve a desired closed-loop performance (see for example, Wang [2020]). In order to illustrate the basic ideas and keep the solution simple, we will consider two integral controllers $C_1(s)$ and $C_2(s)$, where

$$C_1(s) = \frac{k_1}{s}; \quad C_2(s) = \frac{k_2}{s}.$$

The closed-loop characteristic equations for the two systems are

$$s^2 + s + k_1 = 0 \tag{1.6}$$

$$s^2 + 2s + k_2 = 0, \tag{1.7}$$

which lead to the closed-loop poles as

$$\frac{-1 \pm \sqrt{1 - 4k_1}}{2}; \quad \frac{-2 \pm \sqrt{4 - 4k_2}}{2}.$$

Therefore, as long as we choose $k_1 > 0$ and $k_2 > 0$, the two closed-loop systems will be stable with poles strictly on the left-half of the complex plane. Note that the stability statement is true if there is no interaction in the system, meaning that $\gamma_1 = \gamma_2 = 0$.

Effect of interactions

Now, we explore what happens when there is an interaction between the two subsystems where $\gamma_1 \neq 0$ and $\gamma_2 \neq 0$. When the two integral controllers $C_1(s)$ and $C_2(s)$ are used for the original system (1.1), we obtain the following matrix fraction models:

$$G(s) = \begin{bmatrix} s+1 & 0 \\ 0 & s+2 \end{bmatrix}^{-1} \begin{bmatrix} 1 & \gamma_1 \\ \gamma_2 & 1 \end{bmatrix} \tag{1.8}$$

$$G(s)C(s) = \begin{bmatrix} \frac{1}{s+1} & \frac{\gamma_1}{s+1} \\ \frac{\gamma_2}{s+2} & \frac{1}{s+2} \end{bmatrix} \begin{bmatrix} \frac{k_1}{s} & 0 \\ 0 & \frac{k_2}{s} \end{bmatrix} = \begin{bmatrix} \frac{k_1}{s(s+1)} & \frac{k_2\gamma_1}{s(s+1)} \\ \frac{k_1\gamma_2}{s(s+2)} & \frac{k_2}{s(s+2)} \end{bmatrix} \tag{1.9}$$

$$= \begin{bmatrix} s(s+1) & 0 \\ 0 & s(s+2) \end{bmatrix}^{-1} \begin{bmatrix} k_1 & k_2\gamma_1 \\ k_1\gamma_2 & k_2 \end{bmatrix} \tag{1.10}$$

$$I + G(s)C(s) = \begin{bmatrix} \frac{s(s+1)+k_1}{s(s+1)} & \frac{k_2\gamma_1}{s(s+1)} \\ \frac{k_1\gamma_2}{s(s+2)} & \frac{s(s+2)+k_2}{s(s+2)} \end{bmatrix}$$

$$= \begin{bmatrix} s(s+1) & 0 \\ 0 & s(s+2) \end{bmatrix}^{-1} \begin{bmatrix} s(s+1)+k_1 & k_2\gamma_1 \\ k_1\gamma_2 & s(s+2)+k_2 \end{bmatrix}. \tag{1.11}$$

Figure 1.1 illustrates the closed-loop control system in terms of reference signal ($R(s)$), input disturbance signal ($D_i(s)$), output disturbance signal ($D_o(s)$) and measurement noise ($D_m(s)$).

From this figure, the output response $Y(s)$ to the reference signal $R(s)$ is governed by

$$Y(s) = (I + G(s)C(s))^{-1}G(s)C(s)R(s) = T(s)R(s), \tag{1.12}$$

where I is identity matrix with the dimension 2×2, $T(s)$ is called complementary sensitivity function. The output response $Y(s)$ to the output disturbance $D_o(s)$ is governed by

$$Y(s) = (I + G(s)C(s))^{-1}D_o(s) = S(s)D_o(s), \tag{1.13}$$

where $S(s)$ is called sensitivity function. The output response $Y(s)$ to the input disturbance $D_i(s)$ is governed by

$$Y(s) = (I + G(s)C(s))^{-1}G(s)D_i(s) = S_i(s)D_i(s), \tag{1.14}$$

where $S_i(s)$ is called input disturbance sensitivity function.

The next step in the investigation of the control system is to compute the closed-loop transfer functions for reference tracking and disturbance rejection when we use the two integral controllers for the multi-input and multi-output system.

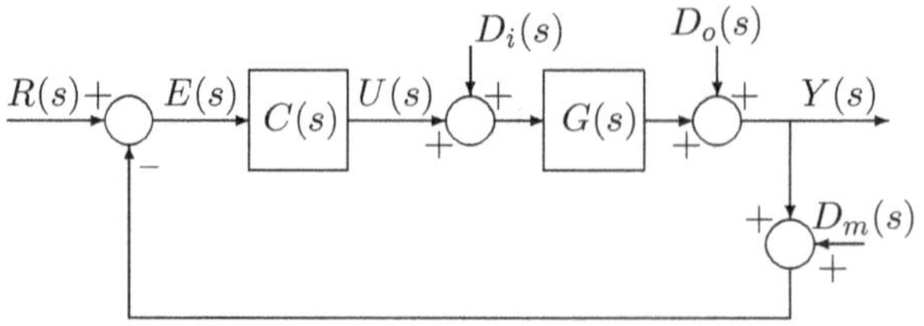

Figure 1.1 Closed-loop control system.

By using (1.10) and (1.11), we can verify that

$$T(s) = \begin{bmatrix} s(s+1)+k_1 & k_2\gamma_1 \\ k_1\gamma_2 & s(s+2)+k_2 \end{bmatrix}^{-1} \begin{bmatrix} k_1 & k_2\gamma_1 \\ k_1\gamma_2 & k_2 \end{bmatrix}$$

$$= \frac{1}{A_{cl}(s)} \begin{bmatrix} s(s+2)+k_2 & -k_2\gamma_1 \\ -k_1\gamma_2 & s(s+1)+k_1 \end{bmatrix} \begin{bmatrix} k_1 & k_2\gamma_1 \\ k_1\gamma_2 & k_2 \end{bmatrix}, \tag{1.15}$$

where $A_{cl}(s) = (s(s+1)+k_1)(s(s+2)+k_2) - k_1k_2\gamma_1\gamma_2$, and we have used the matrix inversion equality:

$$\begin{bmatrix} a_{11} & a_{12} \\ a_{21} & a_{22} \end{bmatrix}^{-1} = \frac{1}{a_{11}a_{22} - a_{12}a_{21}} \begin{bmatrix} a_{22} & -a_{12} \\ -a_{21} & a_{11} \end{bmatrix}. \tag{1.16}$$

Similarly, we can verify that the sensitivity function

$$S(s) = \begin{bmatrix} s(s+1)+k_1 & k_2\gamma_1 \\ k_1\gamma_2 & s(s+2)+k_2 \end{bmatrix}^{-1} \begin{bmatrix} s(s+1) & 0 \\ 0 & s(s+2) \end{bmatrix}$$

$$= \frac{1}{A_{cl}(s)} \begin{bmatrix} s(s+2)+k_2 & -k_2\gamma_1 \\ -k_1\gamma_2 & s(s+1)+k_1 \end{bmatrix} \begin{bmatrix} s(s+1) & 0 \\ 0 & s(s+2) \end{bmatrix} \tag{1.17}$$

and the input disturbance sensitivity function

$$S_i(s) = \frac{1}{A_{cl}(s)} \begin{bmatrix} s(s+2)+k_2 & -k_2\gamma_1 \\ -k_1\gamma_2 & s(s+1)+k_1 \end{bmatrix} \begin{bmatrix} s & 0 \\ 0 & s \end{bmatrix} \begin{bmatrix} 1 & \gamma_1 \\ \gamma_2 & 1 \end{bmatrix}. \tag{1.18}$$

Upon obtaining the closed-loop transfer functions, closed-loop stability and performance can be analyzed. There are two topics specifically concerned in the analysis: (i) closed-loop stability and dynamic performance, and (ii) steady-state performance for reference tracking and disturbance rejection under the assumption that the closed-loop system is stable.

(i) Closed-loop stability and dynamic performance

From the closed-loop transfer functions (1.15)–(1.18) that govern the output responses to the reference, output disturbance, and input disturbance, we see that the closed-loop poles for the system with interactions are determined by the solutions of the characteristic equation:

$$A_{cl}(s) = (s(s+1)+k_1)(s(s+2)+k_2) - k_1k_2\gamma_1\gamma_2 = 0. \tag{1.19}$$

There are four poles for this system. We have two cases listed as below.

Case A. If either $\gamma_1 = 0$ or $\gamma_2 = 0$, then the closed-loop poles consist of the two sets of the poles for the two single-input and single-output systems, which are

$$\frac{-1 \pm \sqrt{1-4k_1}}{2}; \quad \frac{-2 \pm \sqrt{4-4k_2}}{2}.$$

Therefore, as long as we choose $k_1 > 0$ and $k_2 > 0$, the closed-loop system is guaranteed to be stable with poles strictly on the left-half of the complex plane. This is the ideal case.

Case B. If $\gamma_1 \neq 0$ and $\gamma_2 \neq 0$, then the closed-loop poles are no longer the same as those we originally chose in Case A. The closed-loop poles are unknown if the parameters γ_1 and γ_2 are unknown. In fact, the closed-loop poles are dependent on the values of k_1, k_2, γ_1, and γ_2. In order to guarantee the closed-loop system to be stable, a necessary condition for this system is that the constant term

$$k_1k_2(1 - \gamma_1\gamma_2) > 0, \tag{1.20}$$

so that the closed-loop poles obtained through the closed-loop characteristic equation (1.19) are strictly on the left-half of the complex plane. The second necessary condition is not obvious, but can be determined through Routh-Hurwitz stability criterion (see Goodwin et al. [2000], Wang [2020]), which is

$$(6 + k_1 + 2k_2)(2k_1 + k_2) > 9k_1k_2(1 - \gamma_1\gamma_2). \tag{1.21}$$

If both (1.20) and (1.21) are satisfied for the parameters γ_1 and γ_2 with the given integral controller gains k_1 and k_2, then the closed-loop control system is stable. We choose $k_1 = k_2 = 1$ to investigate the variations of closed-loop poles as a function of $\gamma_1\gamma_2$. Figure 1.2 (a) shows the variations of the closed-loop poles for $0 \leq \gamma_1\gamma_2 \leq 3$. It is seen that when $\gamma_1\gamma_2 = 0$, the closed-loop poles are at -1 (two poles) and $-0.5 \pm j0.866$. As $\gamma_1\gamma_2$ increases, the pair of real poles split along the real axis: one towards $-\infty$ and the other towards ∞. The closed-loop pole, which moves towards ∞ as $\gamma_1\gamma_2$ increases, will cause closed-loop instability. Figure 1.2 (b) shows the variations of the closed-loop poles for $-3 < \gamma_1\gamma_2 \leq 0$. It is seen that the pair of closed-loop poles at -1 split to become a pair of complex poles, but they are confined to the

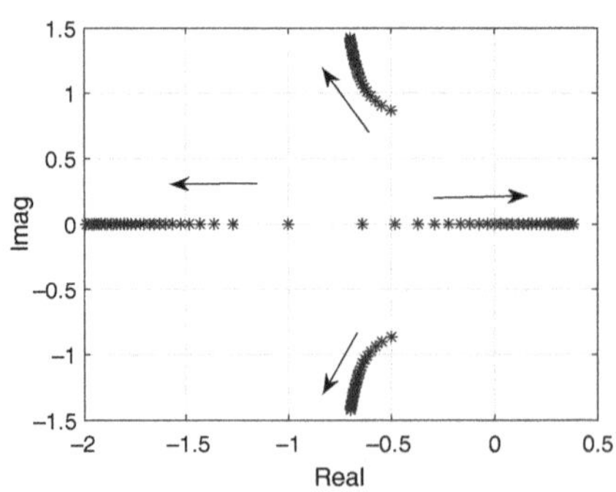

(a) Closed-loop poles for $0 \leq \gamma_1\gamma_2 < 3$

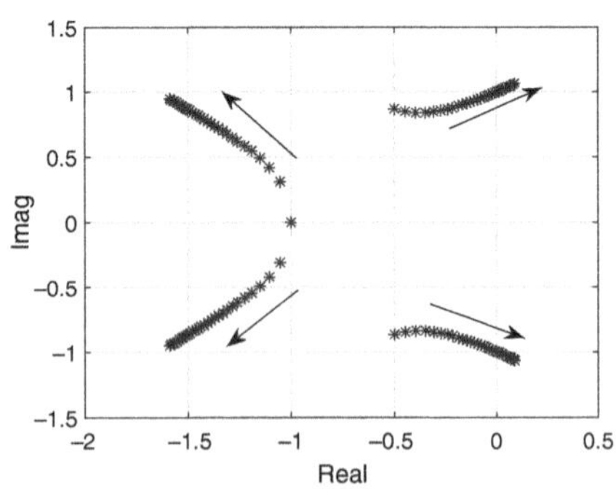

(b) Closed-loop poles for $-3 < \gamma_1\gamma_2 \leq 0$

Figure 1.2 The variations of closed-loop poles for $\gamma_1\gamma_2$.

left-half plane. In contrast, the pair of complex poles at $-0.5 \pm j0.866$ move towards right-half plane as $|\gamma_1\gamma_2|$ increases. The necessary condition (1.21) is eventually violated as $|\gamma_1\gamma_2|$ increases. By comparing Figure 1.2 (a) with Figure 1.2 (b), it seems that for the choice of $k_1 = k_2 = 1$, when γ_1 and γ_2 have an opposite sign, the closed-loop system can tolerate a larger change in the parameters γ_1 and γ_2.

In summary, when there are interactions in the system, single-loop PID controllers could not ensure closed-loop stability. Even if the closed-loop system is stable, the existence of interaction will cause uncertainty in the closed-loop dynamic response due to the changes of the closed-loop poles from what we have chosen in the original design.

(ii) Steady-state performance

Assume that the closed-loop system is stable, namely, the closed-loop poles obtained from the solutions of the characteristic equation (1.19) are strictly on the left-half of the complex plane. Under this assumption, to analyze closed-loop performance at steady-state, we will apply final value theorem. Assuming that the continuous time signal $y(t)$ has a Laplace transform $Y(s)$, final value theorem states that

$$\lim_{t\to\infty} y(t) = \lim_{s\to 0} sY(s)$$

if $sY(s)$ has all poles strictly on the left-half of the complex plane.

In the steady-state performance analysis, we will consider both reference signal and disturbance signal to be step signals.

A. Tracking step reference signals. We assume that both reference signals to the two outputs are unit step signals so that

$$R(s) = \begin{bmatrix} R_1(s) & R_2(s) \end{bmatrix}^T = \begin{bmatrix} \frac{1}{s} & \frac{1}{s} \end{bmatrix}^T$$

From (1.12) and (1.15), we obtain

$$sY(s) = sT(s)R(s) = \frac{1}{A_{cl}(s)} \begin{bmatrix} s(s+2)+k_2 & -k_2\gamma_1 \\ -k_1\gamma_2 & s(s+1)+k_1 \end{bmatrix} \begin{bmatrix} k_1 & k_2\gamma_1 \\ k_1\gamma_2 & k_2 \end{bmatrix} \begin{bmatrix} 1 \\ 1 \end{bmatrix}.$$

With the assumption that the closed-loop system is stable with all poles strictly on the left-half of the complex plane, from the application of final value theorem, the final values of the output signals become:

$$\lim_{t\to\infty} y(t) = \lim_{s\to 0} sY(s) = T(0) \begin{bmatrix} 1 \\ 1 \end{bmatrix} = \begin{bmatrix} 1 \\ 1 \end{bmatrix}, \tag{1.22}$$

where $T(0)$, the complementary sensitivity function at steady-state, is given by

$$T(0) = \frac{1}{k_1k_2(1-\gamma_1\gamma_2)} \begin{bmatrix} k_1k_2(1-\gamma_1\gamma_2) & 0 \\ 0 & k_1k_2(1-\gamma_1\gamma_2) \end{bmatrix} = \begin{bmatrix} 1 & 0 \\ 0 & 1 \end{bmatrix}. \tag{1.23}$$

This means that the closed-loop outputs will track their step reference signals without steady-state errors. It is important to emphasis that as long as the closed-loop system is stable, the single-loop integral controllers will produce the decoupled closed-loop output responses at steady-state because the complementary sensitivity matrix is an identity matrix at the steady-state, even though the plant itself has interactions between the variables.

B. Output disturbance rejection. We assume that the output disturbances are step disturbances with unknown amplitude, which are written as

$$D_o(s) = \begin{bmatrix} \frac{d_{m1}}{s} & \frac{d_{m2}}{s} \end{bmatrix}^T.$$

From (1.13) and (1.17), we have the output $Y(s)$ in response to the output disturbance as

$$\begin{aligned} Y(s) &= S(s)D_o(s) \\ &= \frac{1}{A_{cl}(s)} \begin{bmatrix} s(s+2)+k_2 & -k_2\gamma_1 \\ -k_1\gamma_2 & s(s+1)+k_1 \end{bmatrix} \begin{bmatrix} s(s+1) & 0 \\ 0 & s(s+2) \end{bmatrix} \begin{bmatrix} \frac{d_{m1}}{s} \\ \frac{d_{m2}}{s} \end{bmatrix}. \end{aligned}$$

With the assumption that the closed-loop system is stable, we apply final value theorem to determine the steady-state value of the output in response to the output disturbance, which gives

$$\lim_{t\to\infty} y(t) = \lim_{s\to 0} sS(s)D_o(s) = S(0) \begin{bmatrix} d_{m1} \\ d_{m2} \end{bmatrix} = \begin{bmatrix} 0 \\ 0 \end{bmatrix}.$$

This is because the sensitivity function $S(0)$ is a zero matrix at the steady-state. Therefore, the closed-loop response to the output disturbance is decoupled at the steady-state and the step output disturbances are rejected without steady-state errors.

C. Input disturbance rejection. We assume that the input disturbances are step signals with unknown amplitude, which has the following Laplace transform:

$$D_i(s) = \begin{bmatrix} \frac{d_1}{s} & \frac{d_2}{s} \end{bmatrix}^T.$$

From (1.14) and (1.18), we obtain the output response to the input disturbance as

$$\begin{aligned} Y(s) &= S_i(s)D_i(s) \\ &= \frac{1}{A_{cl}(s)} \begin{bmatrix} s(s+2)+k_2 & -k_2\gamma_1 \\ -k_1\gamma_2 & s(s+1)+k_1 \end{bmatrix} \begin{bmatrix} s & 0 \\ 0 & s \end{bmatrix} \begin{bmatrix} 1 & \gamma_1 \\ \gamma_2 & 1 \end{bmatrix} \begin{bmatrix} \frac{d_1}{s} \\ \frac{d_2}{s} \end{bmatrix}. \end{aligned} \tag{1.24}$$

With the assumption that the closed-loop system is stable, we apply final value theorem to determine the steady-state values of the output response, leading to

$$\lim_{t\to\infty} y(t) = \lim_{s\to 0} sS_i(s)D_i(s) = S_i(0) \begin{bmatrix} d_1 \\ d_2 \end{bmatrix} = \begin{bmatrix} 0 \\ 0 \end{bmatrix}, \tag{1.25}$$

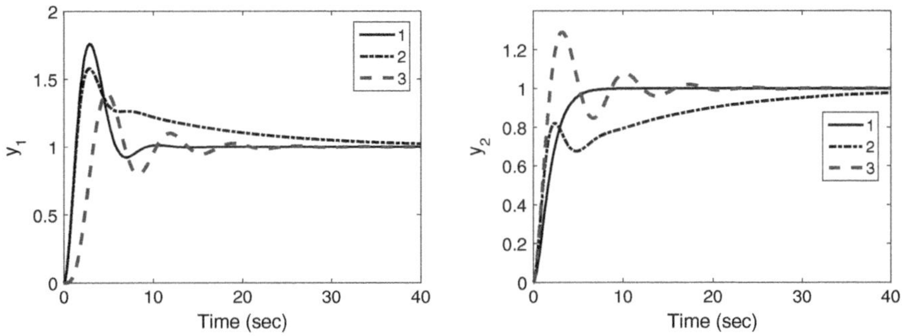

(a) Output y_1 (b) Output y_2

Figure 1.3 Closed-loop output responses. Key: line (1) $\gamma_1 = 1,\ \gamma_2 = 0$; line (2) $\gamma_1 = 1$, $\gamma_2 = 0.8$; line (3) $\gamma_1 = -1,\ \gamma_2 = 0.8$.

where the input disturbance sensitivity matrix is a zero matrix at the steady-state. Therefore, both input step disturbances are completely rejected at the steady-state. Because the input disturbance sensitivity matrix is a zero matrix, the closed-loop response to the input step disturbance is decoupled even though the plant has interactions.

Letting $k_1 = k_2 = 1$, we evaluate the closed-loop step response with $r_1(t) = r_2(t) = 1$. There are three cases under consideration: $\gamma_1 = 1;\ \gamma_2 = 0; \gamma_1 = 1$, $\gamma_2 = 0.8;\ \gamma_1 = -1,\ \gamma_2 = 0.8$. Figures 1.3(a)–(b) compare the closed-loop output responses for the three cases. It is seen from this figure that indeed when the closed-loop system is stable, both outputs successfully track the step reference signals without steady-state errors. However, the transient performance differs significantly when the γ_1 and γ_2 value changes. When $\gamma_1 = 1;\ \gamma_2 = 0$, the closed-loop output responses are ideal (see line (1)). When $\gamma_1\gamma_2 = 0.8$, there is a closed-loop pole close to the imaginary axis, and as a result, there is a long 'tail' for the closed-loop output response observed from both y_1 and y_2 (see line (2)). When $\gamma_1\gamma_2 = -0.8$, there are four complex poles, and as a result, the closed-loop output responses are oscillatory (see line (3)). Figures 1.4(a)–(b) compare the closed-loop control responses. It is seen from the control signal responses that the steady-state values of the control signals change with respect to the variation of the interactions in order to produce the closed-loop output responses without steady-state errors.

In closing this section, there are a few comments.

1. PID controllers are practical. A large amount of industrial control systems will utilize the decentralized control structure and produce reasonable dynamic performance.
2. When there are interactions in the system, the decentralized PID controllers could not ensure closed-loop stability and desired closed-loop dynamic

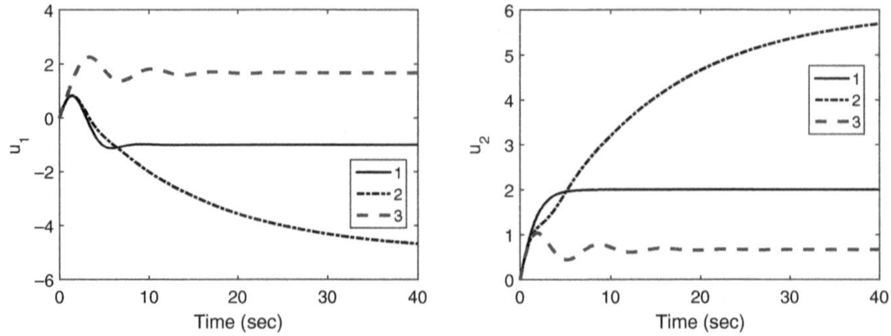

(a) Control u_1 (b) Control u_2

Figure 1.4 Closed-loop control response. Key: line (1) $\gamma_1\gamma_2 = 0$; line (2) $\gamma_1 = 1,\ \gamma_2 = 0.8$; line (3) $\gamma_1 = -1,\ \gamma_2 = 0.8$.

responses. Methods were proposed to improve robustness of the decentralized control system (see Hovd and Skogestad [1994], Chiu and Arkun [1991]), and to assess their performances (see Sourias and Manousiouthakis [1995], Liu et al. [2014], Goodwin et al. [2005]).

3. The techniques for improving the closed-loop performance of decentralized PID control systems include identifying the better pairing of inputs and outputs via the relative gain array (Bristol [1966]) and using feedforward compensation to reduce the effect of coupling (Wang et al. [2008], Hägglund et al. [2022]).

The state feedback control systems in this book belong to the class of control systems that have the natural capacity to handle multiple inputs and outputs with interactions.

1.3 Basics in State Feedback Control

1.3.1 State Feedback Control

The state feedback control systems are designed using state space models. The control signals are calculated using a feedback controller gain matrix together with the state variables. The following example is to show how a state space model is used to capture system dynamics with multi-input and multi-output.

Example 1.1

A three spring and double mass system Tongue [2002] is illustrated in Figure 1.5. In this figure, the two blocks are with mass M_1 and M_2, and there are two spring constants α_1 for the right and left springs, α_2 for the middle spring. The manipulated variables are the two applied forces u_1 and u_2 and the output variables are the

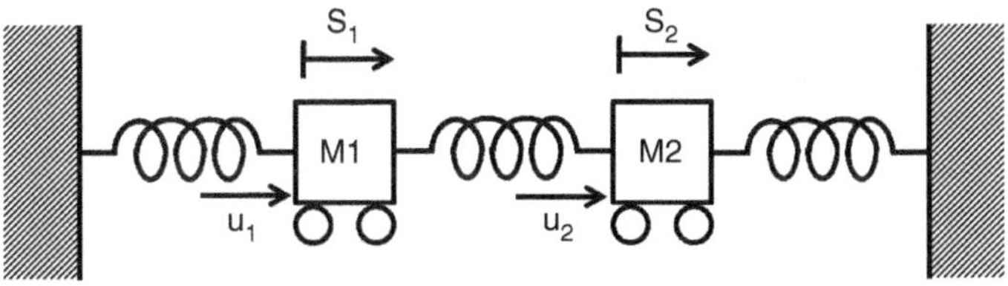

Figure 1.5 Three springs and double mass system (Example 1.1).

distances of the mass block movement y_1 corresponding to block one and y_2 corresponding to block two. The dynamics of this system are described by the following differential equations:

$$\frac{d^2y_1(t)}{dt^2} = -\frac{\alpha_1 + \alpha_2}{M_1} y_1(t) + \frac{\alpha_2}{M_1} y_2(t) + \frac{u_1(t)}{M_1} \tag{1.26}$$

$$\frac{d^2y_2(t)}{dt^2} = -\frac{\alpha_1 + \alpha_2}{M_2} y_2(t) + \frac{\alpha_2}{M_2} y_1(t) + \frac{u_2(t)}{M_2}. \tag{1.27}$$

Find the expression of the state space model for the three springs and double mass system.

Solution

To convert these two equations into a state space model, two additional variables are chosen to reduce the second derivatives to first derivatives.

We choose the first two state variables based on the first block: $x_1(t) = s_1(t)$, $x_2(t) = \dot{s}_1(t) = \dot{x}_1(t)$. Thus, $\dot{x}_2(t) = \ddot{x}_1(t)$. Similarly, $x_3(t) = s_2(t)$, and $x_4(t) = \dot{s}_2(t) = \dot{x}_3(t)$. Thus, $\dot{x}_4(t) = \ddot{x}_3(t)$. The output variables are chosen as $y_1(t) = s_1(t)$ and $y_2(t) = s_2(t)$. With the four state variables and output variables defined, the state space model is obtained as follows:

$$\begin{bmatrix} \dot{x}_1(t) \\ \dot{x}_2(t) \\ \dot{x}_3(t) \\ \dot{x}_4(t) \end{bmatrix} = \overbrace{\begin{bmatrix} 0 & 1 & 0 & 0 \\ -\frac{\alpha_1+\alpha_2}{M_1} & 0 & \frac{\alpha_2}{M_1} & 0 \\ 0 & 0 & 0 & 1 \\ \frac{\alpha_2}{M_2} & 0 & -\frac{\alpha_1+\alpha_2}{M_2} & 0 \end{bmatrix}}^{A} \begin{bmatrix} x_1(t) \\ x_2(t) \\ x_3(t) \\ x_4(t) \end{bmatrix} + \overbrace{\begin{bmatrix} 0 & 0 \\ \frac{1}{M_1} & 0 \\ 0 & 0 \\ 0 & \frac{1}{M_2} \end{bmatrix}}^{B} \begin{bmatrix} u_1(t) \\ u_2(t) \end{bmatrix}$$

$$\begin{bmatrix} y_1(t) \\ y_2(t) \end{bmatrix} = \overbrace{\begin{bmatrix} 1 & 0 & 0 & 0 \\ 0 & 0 & 1 & 0 \end{bmatrix}}^{C} \begin{bmatrix} x_1(t) \\ x_2(t) \\ x_3(t) \\ x_4(t) \end{bmatrix}.$$

The system has two input variables that are the two applied forces u_1 and u_2 and two output variables that are the distances of the mass block movement y_1 corresponding to block one and y_2 corresponding to block two. The system matrices are A, B, and C as illustrated.

Now, we turn our attention to how a state space model is used in the state feedback controller design.

Assuming that the system dynamics are described by the state space equation with appropriate dimensions for the matrices and vectors:

$$\begin{aligned} \dot{x}(t) &= Ax(t) + Bu(t) \\ y(t) &= Cx(t) \end{aligned} \tag{1.28}$$

a basic state feedback control configuration is illustrated in Figure 1.6, where K is the controller gain matrix. More specifically, in the state feedback control configuration, assuming that the number of state variables is n, there is only one input variable, and $x(t) = \begin{bmatrix} x_1(t) & x_2(t) & \dots & x_n(t) \end{bmatrix}^T$, the control signal $u(t)$ is calculated as a linear combination of the state variables, where

$$\begin{aligned} u(t) &= -(k_1 x_1(t) + k_2 x_2(t) + \dots + k_n x_n(t)) \\ &= -Kx(t), \end{aligned} \tag{1.29}$$

where K is a row vector defined as $K = \begin{bmatrix} k_1 & k_2 & \dots & k_n \end{bmatrix}$.

Clearly, if there are two input variables (for example, the three springs and double mass system), the control signal is a 2×1 vector and $u(t) = \begin{bmatrix} u_1(t) & u_2(t) \end{bmatrix}^T$. Then, the feedback controller gain matrix K will have two rows, that is

$$\begin{aligned} u_1(t) &= -(k_{11} x_1(t) + k_{12} x_2(t) + \dots + k_{1n} x_n(t)) \\ u_2(t) &= -(k_{21} x_1(t) + k_{22} x_2(t) + \dots + k_{2n} x_n(t)). \end{aligned}$$

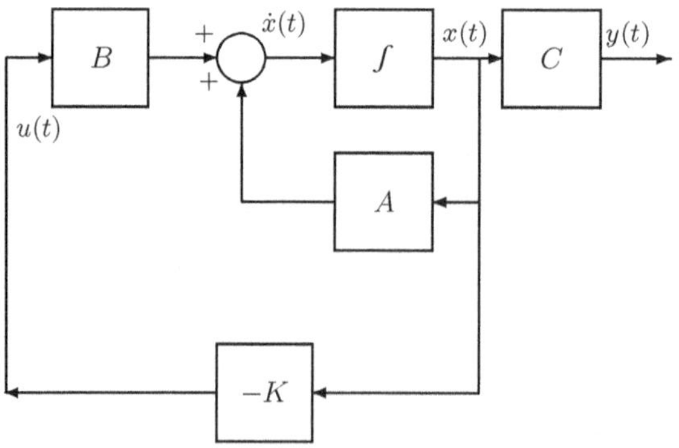

Figure 1.6 Block diagram of state feedback control system.

Or in a compact form,

$$u(t) = -Kx(t), \tag{1.30}$$

where the feedback control gain matrix K takes the form:

$$K = \begin{bmatrix} k_{11} & k_{12} \dots k_{1n} \\ k_{21} & k_{22} \dots k_{2n} \end{bmatrix}.$$

By comparing (1.29) with (1.30), it becomes clear that when using state feedback controller for a multi-input system, the formulation of the control signal remains the same regardless how many input signals the system has.

The closed-loop state feedback control system is obtained by substituting the control signal $u(t) = -Kx(t)$ into the original system model (1.28), leading to

$$\dot{x}(t) = (A - BK)x(t). \tag{1.31}$$

Given an initial state vector $x(0)$, the solution of the differential equation by integration leads to the closed-loop response of $x(t)$ at time $t \geq 0$, as

$$x(t) = e^{(A-BK)t}x(0). \tag{1.32}$$

In order to ensure that the closed-loop system is stable, the closed-loop system matrix $A - BK$ must have all its eigenvalues on the left-half of the complex plane, that is $real(\lambda_i(A - BK)) < 0$ for all i, where $\lambda_i(A - BK)$, $i = 1, \dots, n$, denote the eigenvalues of matrix $A - BK$ and are the solutions of the closed-loop characteristic equation:

$$\det(sI - (A - BK)) = 0.$$

There are three points to be noted from state feedback control.

1. The state feedback control is a negative feedback of the state variables.
2. There are no dynamics in the state feedback controller K.
3. There is no reference signal to the output, which means that the state feedback control system is designed for bringing the state vector to zero from an initial vector $x(0)$ when the closed-loop system is stable (see (1.32)).

When there is no reference signal, the state feedback control problem is called a regulator problem. The regulator problem is to find a state feedback control gain K so that the closed-loop response will decay to zero at a desired speed. For the three springs and double mass system, the problem may arise when there are some deviations from the desired positions y_1 and y_2 at time $t = 0$, which are the nonzero initial conditions. Bringing the initial deviations to the desired positions is a regulator problem, which is also the basic state feedback control problem.

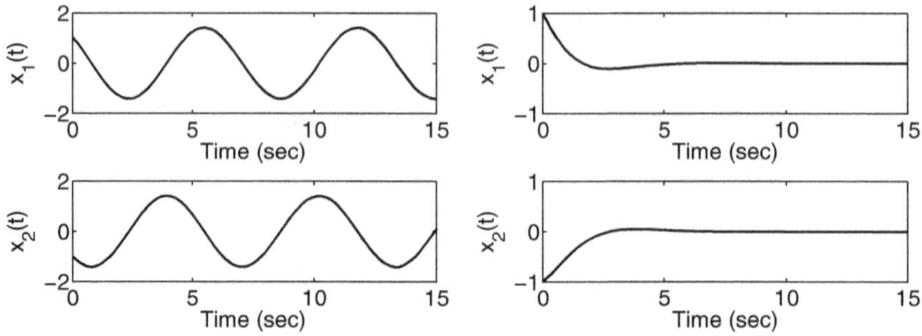

Figure 1.7 State variable responses (Example 1.2).

The following example illustrates the basic ideas about the state feedback control system.

Example 1.2

The state space model of an undamped oscillator is given by the following pair of differential equations, written in matrix form:

$$\overbrace{\begin{bmatrix} \dot{x}_1(t) \\ \dot{x}_2(t) \end{bmatrix}}^{\dot{x}(t)} = \overbrace{\begin{bmatrix} 0 & 1 \\ -\omega_0^2 & 0 \end{bmatrix}}^{A} \overbrace{\begin{bmatrix} x_1(t) \\ x_2(t) \end{bmatrix}}^{x(t)} + \overbrace{\begin{bmatrix} 0 \\ 1 \end{bmatrix}}^{B} u(t). \tag{1.33}$$

Design a state feedback controller so that the closed-loop system has a damping coefficient $\xi = 0.707$ and the natural frequency $w_n = \omega_0$. In this example, $\omega_0 = 1$ rad/s.

Solution

The open-loop system has a pair of eigenvalues on the imaginary axis with their locations at $\pm j\omega_0$. We can verify this by calculating the characteristic polynomial:

$$\det(sI - A) = \det \begin{bmatrix} s & -1 \\ \omega_0^2 & s \end{bmatrix} = s^2 + \omega_0^2 = 0.$$

Its solution gives the eigenvalues of the open-loop system as $s_{1,2} = \pm j\omega_0$.

With the initial condition of the state vector chosen as $x(0) = \begin{bmatrix} 1 & -1 \end{bmatrix}^T$, the open-loop state variable vector $x(t)$ is the solution of the differential equation (1.33), by assuming $u(t) = 0$, having the following exponential matrix form:

$$x(t) = e^{At}x(0). \tag{1.34}$$

Figure 1.7(a) shows the open-loop responses of the state variables. Indeed, without the intervention of feedback control, the undamped oscillator has sustained oscillations.

Now, with the state feedback control, the control signal u(t) is written as

$$u(t) = -\begin{bmatrix} k_1 & k_2 \end{bmatrix} \begin{bmatrix} x_1(t) \\ x_2(t) \end{bmatrix}.$$

Substituting this control signal into the state space model (1.33) gives the closed-loop system as

$$\overbrace{\begin{bmatrix} \dot{x}_1(t) \\ \dot{x}_2(t) \end{bmatrix}}^{\dot{x}(t)} = \overbrace{\begin{bmatrix} 0 & 1 \\ -\omega_0^2 - k_1 & -k_2 \end{bmatrix}}^{A-BK} \overbrace{\begin{bmatrix} x_1(t) \\ x_2(t) \end{bmatrix}}^{x(t)}. \tag{1.35}$$

It is seen from (1.35) that the state feedback control gives the freedom to position the closed-loop eigenvalues of the oscillator by changing the closed-loop system matrix $A - BK$. The next question is how to choose the controller coefficients k_1 and k_2.

With the requirement of closed-loop system having the damping coefficient $\xi = 0.707$ and the same natural frequency $w_n = \omega_0 = 1$, the desired closed-loop characteristic polynomial is chosen as

$$A_{cl}(s)^d = s^2 + 2\xi w_n s + w_n^2 = s^2 + 1.414s + 1. \tag{1.36}$$

The actual closed-loop characteristic polynomial is calculated using

$$A_{cl}(s) = \det(sI - (A - BK)) = \det \begin{bmatrix} s & -1 \\ \omega_0^2 + k_1 & s + k_2 \end{bmatrix} = s^2 + k_2 s + \omega_0^2 + k_1. \tag{1.37}$$

Thus, making the actual closed-loop characteristic polynomial equal to the desired one, $A_{cl}(s) = A_{cl}(s)^d$, the state feedback controller gain coefficients are obtained

$$k_1 = \omega_0^2 - \omega_0^2 = 0; \quad k_2 = 2\xi\omega_0 = 1.414.$$

With the same initial condition of the state vector, the closed-loop state vector $x(t)$ is the solution of the differential equation (1.35) with the following exponential matrix form:

$$x(t) = e^{(A-BK)t}x(0). \tag{1.38}$$

Figure 1.7(b) shows the closed-loop response of the state variables. In contrast to the open-loop response, the closed-loop system has achieved the damped response with the damping coefficient 0.707. It is interesting to note that when the natural frequency is unchanged, the control signal does not require the feedback action from the state variable $x_1(t)$. However, if the desired natural frequency is not equal to ω_0, then $k_1 \neq 0$ and the feedback action is required.

1.3.2 Controllability

The study of controllability is to answer the question under what conditions we can control a system to satisfy an arbitrary specification of desired closed-loop performance.

Definition of state being controllable. A state variable model of a dynamic system is said to be completely state controllable if, given any states $x(t_0)$, $x(t_f)$ and any initial time t_0, there exist a time $t_1 > t_0$ and an input $u(t)$ such that the system state $x(t)$ is transferred from the initial state $x(t_0)$ to the final state $x(t_1) = x(t_f)$. We will examine what the term 'controllability' means by examples.

A necessary and sufficient condition for a linear continuous time invariant system with n state variables to be completely controllable is that the controllability matrix L_c has rank of n, where L_c is defined as follows,

$$L_c = \begin{bmatrix} B & AB & A^2B & \dots & A^{n-2}B & A^{n-1}B. \end{bmatrix} \tag{1.39}$$

If the system has only one input, L_c is a square matrix with dimension $n \times n$ and the controllable condition $rank(L_c) = n$ is equivalent to $\det(L_c) \neq 0$. If the system has more than one input, say m inputs, then, the controllability matrix L_c has dimension of $n \times (m \times n)$. For the multi-input case, $rank(L_c) = n$ means that the number of independent columns in L_c is, at least, equal to n.

MATLAB function **ctrb.m** is very convenient for the calculation of L_c matrix and determining whether a system is controllable. L_c matrix is calculated using the following code:

```
Lc=ctrb(A,B);
```

followed by

```
q=rank(Lc);
```

If q equals to n, then the system is controllable.

The following examples illustrate what controllability means for control system design.

Example 1.3

Suppose that a two inputs and two states system is described by the differential equation:

$$\begin{bmatrix} \dot{x}_1(t) \\ \dot{x}_2(t) \end{bmatrix} = \begin{bmatrix} -\frac{\beta}{a_1} & \frac{\beta}{a_1} \\ \frac{\beta}{a_2} & -\frac{\beta}{a_2} \end{bmatrix} \begin{bmatrix} x_1(t) \\ x_2(t) \end{bmatrix} + \begin{bmatrix} \frac{1}{a_1} & 0 \\ 0 & \frac{1}{a_2} \end{bmatrix} \begin{bmatrix} u_1(t) \\ u_2(t) \end{bmatrix}, \tag{1.40}$$

where $a_1 \neq 0$, $a_2 \neq 0$. Determine if the system is controllable if $\beta = 0$.

Solution

We calculate the controllability matrix L_c as

$$L_c = \begin{bmatrix} B & AB \end{bmatrix} = \begin{bmatrix} \frac{1}{a_1} & 0 & -\frac{\beta}{a_1^2} & \frac{\beta}{a_1 a_2} \\ 0 & \frac{1}{a_2} & \frac{\beta}{a_1 a_2} & -\frac{\beta}{a_2^2} \end{bmatrix}. \quad (1.41)$$

Here, rank(L_c) means the number of independent columns. By visual inspection, it is seen that the first two columns are independent. Therefore, rank(L_c) = 2 regardless the value of β. The conclusion is that the system is controllable independent of β. In other words, β being zero does not affect the controllability of the system.

We continue from Example 1.3.

Example 1.4

We assume that we only have one input available for control purpose, say that the first input $u_1(t)$ is available, then the state space equation becomes:

$$\begin{bmatrix} \dot{x}_1(t) \\ \dot{x}_2(t) \end{bmatrix} = \begin{bmatrix} -\frac{\beta}{a_1} & \frac{\beta}{a_1} \\ \frac{\beta}{a_2} & -\frac{\beta}{a_2} \end{bmatrix} \begin{bmatrix} x_1(t) \\ x_2(t) \end{bmatrix} + \begin{bmatrix} \frac{1}{a_1} \\ 0 \end{bmatrix} u_1(t). \quad (1.42)$$

Examine if the system is still controllable if $\beta = 0$. Additionally, investigate the effect of $\beta = 0$ on the closed- loop control system.

Solution

The controllability matrix becomes,

$$L_c = \begin{bmatrix} \frac{1}{a_1} & -\frac{\beta}{a_1^2} \\ 0 & \frac{\beta}{a_1 a_2} \end{bmatrix} \quad (1.43)$$

where $\det(L_c) = \frac{\beta}{a_1^2 a_2}$. *It is seen that* $\det(L_c) = 0$ *if* $\beta = 0$. *Therefore, if there is only one control input, the system will loss controllability if* $\beta = 0$.

Let us try the state feedback control as a function of β. Here, the control signal $u(t)$ is expressed as

$$u(t) = -\begin{bmatrix} k_1 & k_2 \end{bmatrix} \begin{bmatrix} x_1(t) \\ x_2(t) \end{bmatrix}. \quad (1.44)$$

The closed-loop system becomes,

$$\dot{x}(t) = (A - BK)x(t), \quad (1.45)$$

where the system matrix is

$$A - BK = \begin{bmatrix} -\frac{\beta}{a_1} & \frac{\beta}{a_1} \\ \frac{\beta}{a_2} & -\frac{\beta}{a_2} \end{bmatrix} - \begin{bmatrix} \frac{1}{a_1} \\ 0 \end{bmatrix} \begin{bmatrix} k_1 & k_2 \end{bmatrix}$$

$$= \begin{bmatrix} -\frac{\beta+k_1}{a_1} & \frac{\beta}{a_1} - \frac{k_2}{a_1} \\ \frac{\beta}{a_2} & -\frac{\beta}{a_2} \end{bmatrix}. \tag{1.46}$$

Thus, the closed-loop characteristic polynomial is calculated using

$$\det(sI - (A - BK)) = \det \begin{bmatrix} s + \frac{\beta+k_1}{a_1} & -\frac{\beta}{a_1} + \frac{k_2}{a_1} \\ -\frac{\beta}{a_2} & s + \frac{\beta}{a_2} \end{bmatrix}$$

$$= \left(s + \frac{\beta + k_1}{a_1}\right)\left(s + \frac{\beta}{a_2}\right) + \frac{\beta}{a_2}\left(-\frac{\beta}{a_1} + \frac{k_2}{a_1}\right). \tag{1.47}$$

Now, if $\beta = 0$, then the second term of the characteristic equation becomes zero, leading to

$$\det(sI - (A - BK)) = s\left(s + \frac{k_1}{a_1}\right). \tag{1.48}$$

Assuming that the desired closed-loop polynomial is

$$A_{cl}(s)^d = s^2 + \alpha_1 s + \alpha_0,$$

for any arbitrary choice of $\alpha_1 > 0$ and $\alpha_0 > 0$, the actual characteristic polynomial (1.48) cannot be made equal to the desired characteristic polynomial determined by α_1 and α_0. In this case, the system becoming uncontrollable means that we cannot move the closed-loop pole at $s = 0$, no matter what choice we have for the controller coefficients k_1 and k_2.

In contrast, if $\beta \neq 0$, it can be verified that the closed- loop characteristic polynomial is equal to the desired characteristic polynomial if

$$k_1 = \left(\alpha_1 - \frac{\beta}{a_2}\right) a_1 - \beta$$

$$k_2 = \frac{a_1 a_2}{\beta}\left(\alpha_0 + \frac{\beta^2}{a_1 a_2} - \left(\alpha_1 - \frac{\beta}{a_2}\right)\frac{\beta}{a_2}\right).$$

This means that we can achieve the control performance specified by the desired closed-loop polynomial.

MATLAB function **ctrbf.m** decomposes a system that is uncontrollable into controllable and uncontrollable subspaces. It is useful to find out whether the uncontrollable modes are unstable. We call a system is stabilizable if its uncontrollable modes are stable regardless whether the controllable modes are stable or not. If the uncontrollable modes are unstable, then the system cannot be stabilized using state feedback control. We call this type of systems not stabilizable.

1.3.3 Food for Thought

1. If a state space model is controllable, would you expect that a desired closed-loop performance specification can be achieved through state feedback control?
2. From the computation of the controllability matrix L_c, would you expect that more input variables in general help improving controllability of the system?
3. If a state space model is not controllable, is it correct to say that there is one or more open-loop poles that could not be changed by the state feedback control?
4. If a state space model is not controllable, and you found out that the uncontrollable modes are stable, would you consider continuing using this model? What modifications would you propose for this model and the state feedback control law, if any?
5. If a state space model is not controllable, and the uncontrollable modes are unstable, would you continue using this model? Would you consider add an additional input signal as a modification to the system?

1.4 Pole-assignment Controller

Pole-assignment controller design is a very useful technique among state space control system design methods, which is simple to use because the locations of closed-loop poles are the desired closed-loop performance specification. We have illustrated this design method in Examples 1.2 and 1.4.

1.4.1 The Design Method

The state feedback closed-loop system is stable if the closed-loop system matrix $A - BK$ has all eigenvalues on the left-half of the complex plane. Additionally, the locations of the closed-loop eigenvalues directly affect the closed-loop response speed of the state vector $x(t)$. In the design of state feedback control system, the specifications of closed-loop system performance are often given in terms of the desired closed-loop eigenvalues.

For a single input system, the state feedback controller K is clearly a row vector and has a dimension equal to that of the state vector $x(t)$. Assuming that the state vector $x(t)$ is $n \times 1$, the task is to find its coefficients $k_1, k_2, \dots, k_n$ such that the closed-loop eigenvalues are at the desired locations. Since the closed-loop eigenvalues are the solutions of the closed-loop characteristic equation:

$$A_{cl}(s) = \det(sI - (A - BK)) = 0, \tag{1.49}$$

the feedback controller design is to match the desired closed-loop characteristic polynomial $A_{cl}(s)^d$ specified by the designer with the actual closed-loop characteristic polynomial $A_{cl}(s)$. This means to solve the polynomial equation:

$$\det(sI - (A - BK)) = A_{cl}(s)^d, \tag{1.50}$$

in order to find the controller coefficients $k_1, k_2, \ldots, k_n$. The design idea sounds intuitive and straightforward. What would be the difficult aspects associated with the state feedback controller design?

The difficulty is to find the determinant of the $n \times n$ matrix $sI - (A - BK)$ analytically. Furthermore, even with the analytical expression of the characteristic polynomial found, there is no guarantee that the controller coefficients will appear in a set of linear equations so that they can be solved easily.

The next question is whether there is a specific pair of A and B matrices so that the determinant of $sI - (A - BK)$ is calculated easily and the coefficients appear in a set of linear equations. The question is answered with the following example.

Example 1.5

Consider that a state space model is described by:

$$\begin{bmatrix} \dot{x}_1(t) \\ \dot{x}_2(t) \\ \dot{x}_3(t) \end{bmatrix} = \overbrace{\begin{bmatrix} 0 & 1 & 0 \\ 0 & 0 & 1 \\ -a_0 & -a_1 & -a_2 \end{bmatrix}}^{A} \begin{bmatrix} x_1(t) \\ x_2(t) \\ x_3(t) \end{bmatrix} + \overbrace{\begin{bmatrix} 0 \\ 0 \\ 1 \end{bmatrix}}^{B} u(t)$$
$$= Ax(t) + Bu(t). \tag{1.51}$$

Note the special structures of A and B matrices, which is called the controller canonical form.

The objective is to design a state feedback controller K using the controller canonical form. The desired closed-loop eigenvalues are specified by the solutions of the characteristic polynomial equation:

$$A_{cl}(s)^d = s^3 + \alpha_2 s^2 + \alpha_1 s + \alpha_0 = 0, \tag{1.52}$$

where the coefficients α_2, α_1 and α_0 are chosen such that the polynomial $A_{cl}(s)^d$ has all stable zeros that are at desired locations.

Solution

The feedback closed-loop system is formed by substituting $u(t) = -Kx(t)$ into the state space model (1.51) leading to

$$\dot{x}(t) = (A - BK)x(t), \tag{1.53}$$

where $K = \begin{bmatrix} k_1 & k_2 & k_3 \end{bmatrix}$. Thus, the closed-loop characteristic polynomial is calculated by evaluating the following determinant

$$\det(sI - (A - BK)) = \det \begin{bmatrix} s & -1 & 0 \\ 0 & s & -1 \\ a_0 + k_1 & a_1 + k_2 & s + a_2 + k_3 \end{bmatrix}. \tag{1.54}$$

Suppose that we choose the third row to calculate the determinant using Cramer's rule (see Kreyszig [2006]). Then,

$$\det(sI - (A - BK)) = (a_0 + k_1)C_{31} + (a_1 + k_2)C_{32} + (s + a_2 + k_3)C_{33}, \tag{1.55}$$

where $C_{3k} = (-1)^{3+k}M_{3k}$ $(k = 1, 2, 3)$ and M_{3k} is a determinant of order 2 that is the determinant of the submatrix obtained by omitting the third row and the corresponding column entry k. It is easy to see that

$$C_{31} = (-1)^{3+1} = 1$$
$$C_{32} = (-1)^{3+2}(-s) = s$$
$$C_{33} = (-1)^{3+3}s^2 = s^2.$$

With the controller canonical form, we obtain the closed-loop characteristic polynomial as a function of the controller coefficients:

$$\det(sI - (A - BK)) = s^3 + (a_2 + k_3)s^2 + (a_1 + k_2)s + a_0 + k_1. \tag{1.56}$$

Now, letting the actual closed-loop characteristic polynomial equal to the desired closed-loop characteristic polynomial ($A_{cl}(s) = A_{cl}(s)^d$) gives

$$s^3 + (a_2 + k_3)s^2 + (a_1 + k_2)s + a_0 + k_1 = s^3 + \alpha_2 s^2 + \alpha_1 s + \alpha_0. \tag{1.57}$$

By comparing the coefficients of (1.57), we obtain the state feedback controller coefficients as

$$k_1 = \alpha_0 - a_0$$
$$k_2 = \alpha_1 - a_1$$
$$k_3 = \alpha_2 - a_2. \tag{1.58}$$

In general, assuming that a state space model is described by the following differential equation:

$$\begin{bmatrix} \dot{x}_1(t) \\ \vdots \\ \dot{x}_{n-2}(t) \\ \dot{x}_{n-1}(t) \\ \dot{x}_n(t) \end{bmatrix} = \begin{bmatrix} 0 & 1 & \dots & 0 & 0 \\ \vdots & \ddots & \ddots & \vdots & \vdots \\ 0 & 0 & \dots & 1 & 0 \\ 0 & 0 & \dots & 0 & 1 \\ -a_0 & -a_1 & \dots & -a_{n-2} & -a_{n-1} \end{bmatrix} \begin{bmatrix} x_1(t) \\ \vdots \\ x_{n-2}(t) \\ x_{n-1}(t) \\ x_n(t) \end{bmatrix} + \begin{bmatrix} 0 \\ \vdots \\ 0 \\ 0 \\ 1 \end{bmatrix} u(t), \tag{1.59}$$

if we specify the desired closed-loop eigenvalues as the solutions of the characteristic polynomial:

$$A_{cl}(s)^d = s^n + \alpha_{n-1}s^{n-1} + \alpha_{n-2}s^{n-2} + \ldots + \alpha_1 s + \alpha_0 = 0,$$

then, the state feedback control law $u(t) = -Kx(t)$ *will position the eigenvalues of the closed-loop system matrix* $A - BK$ *at the desired positions, where*

$$K = \begin{bmatrix} k_1 & k_2 & k_3 & \ldots & k_{n-1} & k_n \end{bmatrix}$$

with

$$\begin{aligned} k_1 &= \alpha_0 - a_0 \\ k_2 &= \alpha_1 - a_1 \\ k_3 &= \alpha_2 - a_2 \\ &\vdots \\ k_{n-1} &= \alpha_{n-1} - a_{n-1} \\ k_n &= \alpha_n - a_n. \end{aligned}$$

1.4.2 Similarity Transformation for Controller Design

Assuming that the state space model is controllable for a single input system, the controllability matrix L_c is invertible. One of the general transformations is performed using a similarity transformation that will transform a state space model into a controller canonical form (see Kailath [1980]).

Suppose that we have the following state space model

$$\dot{x}(t) = Ax(t) + Bu(t). \tag{1.60}$$

We choose a new state vector $z(t)$ such that

$$x(t) = Tz(t),$$

where T is the similarity transformation matrix and T^{-1} is assumed to exist. The state space model (1.60) in terms of the state vector $z(t)$ is

$$T\dot{z}(t) = ATz(t) + Bu(t),$$

and multiplying T^{-1} from left gives:

$$\dot{z}(t) = T^{-1}ATz(t) + T^{-1}Bu(t). \tag{1.61}$$

If the pair of matrices $(T^{-1}AT,\ T^{-1}B)$ is in a controller canonical form, then the state feedback controller will be found using the state vector $z(t)$ in a straightforward manner:

$$u(t) = -\hat{K}z(t).$$

However, in terms of the original state vector $x(t)$, we will replace $z(t)$ by $T^{-1}x(t)$ as the transformation dictates, which is

$$u(t) = -\hat{K}z(t) = -\hat{K}T^{-1}x(t) = -Kx(t),$$

where the feedback controller $K = \hat{K}T^{-1}$.

How do we find such a transformation matrix T for the controller design? One of the approaches to construct the T matrix is based on the following steps (see Kailath [1980]), where we assume that the state space model has n state variables and has one input.

1. Form the controllability matrix $L_c = \begin{bmatrix} B & AB & \dots & A^{n-1}B \end{bmatrix}$.
2. Calculate the inverse of L_c to obtain L_c^{-1} where L_c^{-1} exists because the state space model is controllable.
3. Find the last row of L_c^{-1} and denote it γ. Construct the inverse of the similarity tranformation matrix, T^{-1} as

$$T^{-1} = \begin{bmatrix} \gamma \\ \gamma A \\ \vdots \\ \gamma A^{n-2} \\ \gamma A^{n-1} \end{bmatrix}. \tag{1.62}$$

4. The similarity transformation matrix T is obtained using $T = (T^{-1})^{-1}$ where T^{-1} is constructed using (1.62).

We will next verify the transformed pair $(T^{-1}AT,\ T^{-1}B)$ is a pair of matrices in the controller canonical form.

First note that the following relationship is true: $L_c^{-1}L_c = I$. This means that the last row of L_c^{-1}, γ, has the following relationship with the components in the controllability matrix as: $\begin{bmatrix} \gamma B & \gamma AB & \dots & \gamma A^{n-1}B \end{bmatrix}$. Because the last row of the identity matrix is all zero but the last element being unity, we have

$$\gamma B = 0 \tag{1.63}$$

$$\gamma AB = 0 \tag{1.64}$$

$$\vdots \tag{1.65}$$

$$\gamma A^{n-2}B = 0 \tag{1.66}$$

$$\gamma A^{n-1}B = 1 \tag{1.67}$$

Noting that the transformation matrix T^{-1} is constructed using (1.62), therefore, from (1.63)–(1.67), we conclude that

$$T^{-1}B = \begin{bmatrix} 0 & 0 & \dots & 0 & 1 \end{bmatrix}^T \tag{1.68}$$

Now, we will examine what is $T^{-1}AT$. Note from the construction of T^{-1} in (1.62) that

$$T^{-1}A = \begin{bmatrix} \gamma A \\ \gamma A^2 \\ \vdots \\ \gamma A^n \end{bmatrix} = \begin{bmatrix} \beta_1 \\ \beta_2 \\ \vdots \\ \beta_n \end{bmatrix}.$$

Letting $T = \begin{bmatrix} t_1 & t_2 & \dots & t_n \end{bmatrix}$ and noting that $T^{-1}T$ is an identity matrix, the matrix $T^{-1}AT$ with the first row to the $n-1$ row is determined by $\beta_k t_{k+1} = 1$ for $k = 1, 2, \dots, n-1$, but the rest of the elements being zero.

The last row of $T^{-1}AT$ becomes

$$\gamma A^n \begin{bmatrix} t_1 & t_2 & \dots & t_n \end{bmatrix} = \begin{bmatrix} \gamma A^n t_1 & \gamma A^n t_2 & \dots & \gamma A^n t_n \end{bmatrix}. \tag{1.69}$$

The Cayley-Hamilton theorem (Kreyszig [2006], Kailath [1980]) says that

$$A^n + a_{n-1}A^{n-1} + \dots + a_1 A + a_0 I = 0,$$

where the coefficients $a_0, a_1, a_2, \dots, a_n$ are defined by the characteristic polynomial of matrix A. Thus, we replace the matrix A^n in (1.69) by the following relationship:

$$A^n = -a_{n-1}A^{n-1} - \dots - a_1 A - a_0 I$$

Noting that T^{-1} is constructed using (1.62), effectively, γ, γA, γA^2, …, γA^{n-1} are the n rows in the T^{-1} matrix. Given that $T^{-1}T = I$, we have,

$$\begin{aligned} \gamma A^n t_1 &= -a_0 \\ \gamma A^n t_2 &= -a_1 \\ &\vdots \\ \gamma A^n t_n &= -a_n. \end{aligned}$$

Putting together the derivations, we obtain that

$$T^{-1}AT = \begin{bmatrix} 0 & 1 & \dots & 0 & 0 \\ \vdots & \ddots & \ddots & \vdots & \vdots \\ 0 & 0 & \dots & 1 & 0 \\ 0 & 0 & \dots & 0 & 1 \\ -a_0 & -a_1 & \dots & -a_{n-2} & -a_{n-1} \end{bmatrix}. \tag{1.70}$$

Equation (1.70) together with (1.68) shows that the pair of matrices $(T^{-1}AT,\ T^{-1}B)$ has the controller canonical form.

1.4.3 MATLAB Tutorial on Pole-assignment Controller

Although there are several MATLAB functions (for example, **place.m** and **acker.m**) that will design a state feedback controller using pole-assignment techniques, it is useful to learn how the actual design is performed for a single input system. In the following MATLAB tutorial it is shown that using the similarity transformation, it is quite straightforward to design a state feedback controller.

Tutorial 1.1 *Write a MATLAB function* **T2place.m** *for generating the state feedback controller gain matrix K.*

Step by Step

1. *Create a new file called T2place.m.*
2. *The following program sets the input and output variables to the function. Enter the following program into the file:*

```
function K=T2place(A,B,P);
%A,B are system matrices
%P is the desired characteristic polynomial
%(P=s^n+a_{n-1}s^{n-1}+\ldots+a_0)
%dimension of P is n+1
```

3. *Generate controllability matrix L_c using MATLAB function* **ctrb.m**, *and terminate the computation if the controllability matrix L_c is not full rank. Continue entering the following program into the file:*

```
 n=length(B);
Lc=ctrb(A,B);
m=rank(Lc);
if (m<n), disp('Uncontrollable'); return; end
```

4. *Assuming that the matrix L_c has full rank, we proceed to find its inverse using singular value decomposition, where the singular value decomposition is written as $L_c = V_1 D_1 U_1^T$, V_1 and U_1 are unitary matrices and $L_c^{-1} = U_1 D_1^{-1} V_1^T$. Singular value decomposition gives a better accuracy in computing the matrix inversion. Continue entering the following program into the file:*

```
[V1,D1,U1]=svd(Lc);
InLc=U1*inv(D1)*V1';
```

5. *Construct the row vector γ and use the MATLAB function* **ctrb.m** *to construct the inversion of the similarity transformation matrix, T^{-1}. Continue entering the following program into the file:*

```
gamma=InLc(n,:);
Tinv=ctrb(A',gamma')';
```

6. *Compute inverse of T^{-1} to obtain the similarity transformation matrix T. Continue entering the following program into the file:*

```
[V2,D2,U2]=svd(Tinv);
T=U2*inv(D2)*V2';
```

7. *Construct matrix $T^{-1}AT$. Entering the following program into the file:*

```
Astf=Tinv*A*T;
```

8. *Note that the desired closed-loop polynomial is expressed in terms of the descending order of s. It needs to be rearranged to ascending order of s. Enter the following program into the file:*

```
for kk=1:n;
     Pi(kk)=P(n+2-kk);
end
```

9. *The feedback controller $\hat{K}$ is computed and the feedback controller $K = \hat{K}T^{-1}$. Note that the coefficients in the last row of $T^{-1}AT$ have opposing signs to the coefficients of the characteristic polynomial. Enter the following program into the file:*

```
Khat=Pi+Astf(n,:);
K=Khat*Tinv;
```

To test the **T2place.m** *function, we use the following system matrices:*

$$A = \begin{bmatrix} -1 & 0 & 2 & 3 & 0 \\ 0 & 1 & 0 & 1 & 0 \\ 1 & -2 & -3 & 0 & 5 \\ 0 & 3 & 4 & -5 & 2 \\ 6 & 0 & 0 & 3 & 3 \end{bmatrix}; \quad B = \begin{bmatrix} 1 \\ 1 \\ 0 \\ 3 \\ 3 \end{bmatrix}.$$

The open-loop system has eigenvalues at $6.1630, -3.4304 \pm j3.6312, -5.5125, 1.2103$. *Say if we place all closed-loop eigenvalues at* -5, *then the state feedback controller K calculated using the MATLAB function* **T2place.m** *is*

$$K = \begin{bmatrix} 5.4049 & -8.2840 & 2.1339 & 1.7337 & 5.8927 \end{bmatrix}.$$

For this example, the desired closed-loop characteristic polynomial P is $(s+5)^5$, *which can be generated using the following MATLAB program:*

```
P1=conv([1 5],[1 5]);
P2=conv(P1,P1);
P=conv(P2,[1 5]);
```

Indeed, calculating the eigenvalues of the closed-loop system matrix $A - BK$ validates the results.

1.4.4 Food for Thought

1. Assume that a single input system has the following system matrices,

$$A = \begin{bmatrix} -2 & 0 \\ 0 & -1 \end{bmatrix}; \quad B = \begin{bmatrix} 1 \\ 2 \end{bmatrix}.$$

   Would you consider that it is an easy task to design a state feedback controller when using this model? If not, what might be the challenge?
2. What is the key requirement for the design method of a pole-assignment controller outlined in this section?
3. Would you be able to design a state feedback controller for the system described by the following transfer function

$$G(s) = \frac{s+1}{(s+3)(s+1)}?$$

   If not, what modification would you like to propose?
4. If the open-loop system is stable, what would happen to the state feedback controller if you choose the closed-loop eigenvalues equal to the open-loop eigenvalues? What conclusions would you be able to draw for the controller gain in relation to the locations of the open-loop eigenvalues and closed-loop eigenvalues?
5. If the open-loop system is stable, but the model quality is poor, would you choose the closed-loop eigenvalues closer to the open-loop eigenvalues to reduce the controller gain in order to improve closed-loop robust stability?

1.5 Linear Quadratic Regulator (LQR) Design

Different from the traditional pole-assignment controller design where the closed-loop performance is specified by the locations of the desired closed-loop poles, the Linear Quadratic Regulator (LQR) offers a new landscape for the desired closed-loop performance specification and controller solutions in which optimization underpins the design methodology.

1.5.1 Motivational Example

For a multi-input system, the pole-assignment controller design approach does not necessarily lead to a unique solution. In other words, for a set of desired closed-loop eigenvalues, there are many solutions of the state feedback controller. It is difficult for us to determine which ones are more suitable than others. The following example is used as illustration.

Example 1.6

A system is described by the state space model:

$$\dot{x}(t) = \begin{bmatrix} 2 & -11 & 12 & 31 \\ -3 & 13 & -11 & -33 \\ 4 & -25 & 14 & 51 \\ -3 & 16 & -11 & -36 \end{bmatrix} x(t) + \begin{bmatrix} -1 & -4 \\ 1 & 6 \\ -1 & -11 \\ 1 & 7 \end{bmatrix} u(t). \tag{1.71}$$

The control objective is to find the state feedback controller K to regulate the nonzero initial state x(0) to zero. The desired closed-loop poles are specified as $-5, -5, -3 \pm j3$. Find two state feedback controllers that will position the closed-loop eigenvalues to the desired positions and simulate the closed-loop responses to the non-zero initial state values.

Solution

In Bay [1999], a state feedback controller is found analytically as

$$K_1 = \begin{bmatrix} 8.6667 & -23.5 & 25.5 & 65.3333 \\ -4.6667 & -9.3333 & -2.6667 & 1.3333 \end{bmatrix}.$$

It can be verified that the closed-loop eigenvalues of the state feedback system are indeed $-5, -5, -3 \pm j3$.

Now, using MATLAB function **place.m**, *then the following state feedback controller is found:*

$$K_2 = \begin{bmatrix} -54.9826 & -252.0137 & 30.5728 & 240.6862 \\ 4.7615 & 29.1283 & -5.8864 & -32.0795 \end{bmatrix}.$$

It can also be verified that the closed-loop eigenvalues are at $-5, -5, -3 \pm j3$.

These two controllers have been designed to satisfy the performance specification, however they are completely different from each other. Naturally, we want to ask which one performs better in the closed-loop control. With the state initial conditions $x_1(0) = x_2(0) = x_3(0) = x_4(0) = 1$, and sampling interval $\Delta t = 0.001$ (sec), the closed-loop control results are presented in Figure 1.8. Figures 1.8(a)–(b) show the control signal responses. It is seen that the initial values from the controller K_1 are larger than those of K_2, but the control signals are with faster decay rate. The closed-loop response of the x_1 using K_1 is faster than that using K_2. The situation is reversed for x_2. The responses of x_3 and x_4 are similar. Clearly the simulation studies are not conclusive on which controller performs better and which one we should use.

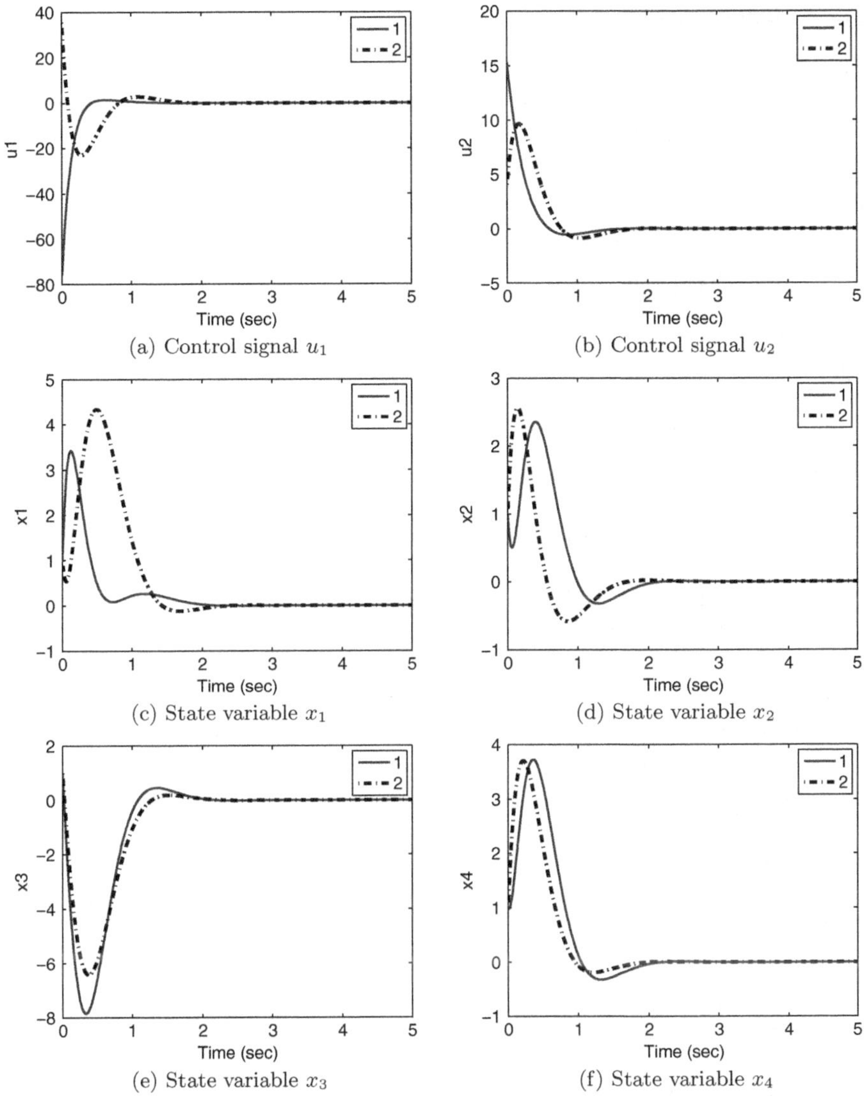

(a) Control signal u_1 (b) Control signal u_2

(c) State variable x_1 (d) State variable x_2

(e) State variable x_3 (f) State variable x_4

Figure 1.8 Closed-loop state response to state initial values (Example 1.6). Key: line (1) closed-loop control using K_1 and line (2) using K_2.

This example indicates that for multi-input systems, we need to seek alternative performance specification to find the state feedback controllers because of the non-uniqueness of the controller solutions when using pole-assignment controller design techniques.

1.5.2 Linear Quadratic Regulator Design

The general idea about linear quadratic regulator design is to use an optimal criterion in the selection of the state feedback control gain matrix. By defining and minimizing a quadratic objective function, the state feedback control law will be found uniquely and the closed-loop system eigenvalues will be determined. This is particularly useful in the situation where the system has more than one input.

Assume that the system is described by the following state space equation:

$$\dot{x}(t) = Ax(t) + Bu(t), \tag{1.72}$$

where $x(t)$ is an $n \times 1$ vector and $u(t)$ is an $m \times 1$ vector. With given initial condition $x(0)$, the objective is to find the state feedback control law:

$$u(t) = -Kx(t) \tag{1.73}$$

that will minimize the objective function:

$$J = \int_0^\infty \left(x(t)^T Qx(t) + u(t)^T Ru(t)\right) dt, \tag{1.74}$$

where Q and R are the weighting matrices with Q to be semi-positive definite ($Q \geq 0$) and R to be positive definite ($R > 0$). The roles of the weighting matrices in relation to the closed-loop performance will be discussed in the next section. The word of 'quadratic' is associated with the type of quadratic objective function and 'regulator' is associated with bringing the state initial values to zero.

How we find the controller gain matrix K to minimize the objective function has been well studied in the control engineering literatures for the past several decades (Anderson and Moore [1971], Kailath [1980], Chen [1998], Bay [1999], Goodwin et al. [2000]). An intuitive approach is outlined as follows.

Substituting the control law (1.73) into the objective function (1.74) leads to

$$\begin{aligned} J &= \int_0^\infty \left(x(t)^T Qx(t) + x(t)^T K^T RKx(t)\right) dt \\ &= \int_0^\infty x(t)^T \left(Q + K^T RK\right) x(t)dt. \end{aligned} \tag{1.75}$$

Now, the objective function becomes integral quadratic function of state vector $x(t)$.

For a given initial state vector $x(0)$, the solution of the closed-loop system is governed by the matrix exponential function:

$$x(t) = e^{(A-BK)t}x(0). \tag{1.76}$$

Thus, with the expression of $x(t)$ in closed-loop control, the objective function is further written as

$$J = x(0)^T \int_0^\infty e^{(A-BK)^T t}(Q + K^T RK)e^{(A-BK)t}dtx(0). \tag{1.77}$$

We define the matrix P as

$$P = \int_0^\infty e^{(A-BK)^T t} \left(Q + K^T RK\right) e^{(A-BK)t} dt. \tag{1.78}$$

With the definition of P matrix, the objective function with the closed-loop system simply becomes

$$J = x(0)^T P x(0). \tag{1.79}$$

Note that because of the nature of the exponential integral equation, we have the following equality:

$$\begin{aligned}
&\int_0^\infty (A - BK)^T e^{(A-BK)^T t} \left(Q + K^T RK\right) e^{(A-BK)t} dt \\
&+ \int_0^\infty e^{(A-BK)^T t} \left(Q + K^T RK\right) e^{(A-BK)t} (A - BK) dt \\
&= \int_0^\infty d[e^{(A-BK)^T t} (Q + K^T RK) e^{(A-BK)t}] \\
&= [e^{(A-BK)^T t} (Q + K^T RK) e^{(A-BK)t}]_0^\infty \\
&= -(Q + K^T RK).
\end{aligned} \tag{1.80}$$

where we have used the assumption that the integral equation converges.

Now, we rewrite (1.80) as the linear algebra equation:

$$(A - BK)^T P + P(A - BK) = -(Q + K^T RK). \tag{1.81}$$

Equation (1.81) is expressed in the form of a Lyapunov equation (Chen [1998]). The Lyapunov theorem states that all eigenvalues of $A - BK$ have negative real parts if and only if for any given positive definite symmetric matrix $(Q + K^T RK)$, the Lyapunov equation (1.81) has a unique symmetric solution P and P is positive definite. The proof of this theorem can be found in Chen [1998] and Bay [1999].

Because the matrix $A - BK$ is the closed-loop system matrix with state feedback controller K, if we can find the positive definite matrix P, then closed-loop state feedback control system is stable with all eigenvalues on the left-half of the complex plane. To solve the Lyapunov equation (1.81), we choose the state feedback controller gain matrix as

$$K = R^{-1} B^T P, \tag{1.82}$$

and noting that

$$K^T RK = PBR^{-1} RK = PBK,$$

the algebra equation (1.81) becomes the following algebraic Riccati equation:

$$A^T P + PA - PBR^{-1} B^T P + Q = 0. \tag{1.83}$$

With the optimal control signal $u(t)$ calculated as

$$u(t) = -R^{-1}B^T Px(t),$$

it can be verified that the minimum of the objective function is given by

$$J_{min} = x(0)^T Px(0). \tag{1.84}$$

A unique symmetric solution P exists and hence asymptotic stability of the closed-loop system is guaranteed if the pair (A, D) is observable where $Q = D^T D$ and if the pair (A, B) is controllable (Anderson and Moore [1971]).

1.5.3 Selection of *Q* and *R* Matrices

In the design of linear quadratic regulator, the closed-loop performance is specified by the weighting matrices Q and R. The following example is used to illustrate how to choose them in the specification of closed-loop performance.

Example 1.7
Consider the state space model of undamped oscillator:

$$\dot{x}(t) = \begin{bmatrix} 0 & 1 \\ -1 & 0 \end{bmatrix} x(t) + \begin{bmatrix} 0 \\ 1 \end{bmatrix} u(t). \tag{1.85}$$

The system has a pair of eigenvalues at $\pm j$. *Design a state feedback controller to bring the initial states* $x_1(0) = x_2(0) = 1$ *to zero by using state feedback control. Demonstrate the selection of* Q *matrix on the effect of closed-loop response.*

Solution
We choose $R = 1$, *and assume that* Q *is a diagonal matrix for a simple case. Then, the objective function becomes*

$$J = q_{11} \int_0^\infty x_1(t)^2 dt + q_{22} \int_0^\infty x_2(t)^2 dt + \int_0^\infty u(t)^2 dt.$$

The following MATLAB code is used to compute the state feedback controller gain K, *the Riccati solution* P *and the closed-loop eigenvalues* E:

```
[K,P,E]=lqr(A,B,Q,R);
```

where A *and* B *are the system matrices given in (1.85).* J_{min} *is calculated using*

$$J_{min} = [x_1(0) \;\; x_2(0)]\, P [x_1(0) \;\; x_2(0)]^T.$$

In the first case, we choose both diagonal elements in Q *to be 1, which is* $q_{11} = q_{22} = 1$. *Choice of this weighting matrix intuitively says that the integral squared errors of both variables are equally important. The MATLAB function* **lqr.m** *gives* *the state feedback gain matrix K_1 and the Riccati solution P_1:*

$$K_1 = \begin{bmatrix} 0.4142 & 1.3522 \end{bmatrix}; \; P_1 = \begin{bmatrix} 1.9123 & 0.4142 \\ 0.4142 & 1.3522 \end{bmatrix}.$$

The closed-loop eigenvalues are: $-0.6761 \pm j0.9783$. The minimum integral squared error is $J_{min} = 4.0929$.

In the second case, we decide that we need to have a smaller integral squared error for $x_1(t)$ in comparison to $x_2(t)$, thus we put more emphasis on the first state variable. For this purpose, we choose the first diagonal element to be 100 times larger than the second element: $q_{11} = 100$ and $q_{22} = 1$. The state feedback controller gain matrix and the Riccati solution are calculated as

$$K_2 = \begin{bmatrix} 9.0499 & 4.3703 \end{bmatrix}; \; P_2 = \begin{bmatrix} 43.9212 & 9.0499 \\ 9.0499 & 4.3703 \end{bmatrix}.$$

The closed-loop eigenvalues are calculated as $-2.1852 \pm 2.2967i$. The minimum integral squared error is $J_{min} = 66.3913$.

In the third case, we reverse the situation from the second case by putting more emphasize on the state variable x_2, thus, we select $q_{11} = 1$ and $q_{22} = 100$. The state feedback controller gain matrix and the Riccati solution are calculated as

$$K_3 = \begin{bmatrix} 0.4142 & 10.0413 \end{bmatrix}; \; P_3 = \begin{bmatrix} 14.2006 & 0.4142 \\ 0.4142 & 10.0413 \end{bmatrix},$$

and the closed-loop eigenvalues are $-0.1429, -9.8985$. The minimum integral squared error is $J_{min} = 25.074$.

Figures 1.9(a)–(b) compare the closed-loop state variable responses to the initial states. For the benchmark case, $Q(1,1) = Q(2,2) = 1$ (see line (1)). It is interesting to note that when $Q(1,1)$ increases to 100 while $Q(2,2) = 1$ (see line (2)), $x_1(t)$'s response becomes faster, however, there is a larger variation in $x_2(t)$. In reverse, when $Q(2,2)$ increases to 100 while $Q(1,1) = 1$, $x_1(t)$'s variation becomes smaller, however, there is a large 'tail' for $x_2(t)$ due to the eigenvalue at -0.1429. Figure 1.9(c) compares the three control signals used to produce the closed-loop state variable responses. In summary, we choose the Q matrix to reflect our desire on the integral squared errors. If we demand a smaller integral squared error, then we choose a larger weighting component in the corresponding diagonal element.

The off-diagonal elements in Q matrix are also important on the effect of closed-loop performance. One must try different combinations of elements in Q matrix to obtain desired closed-loop performance.

The next example is used to illustrate how to choose the weighting matrix R for obtaining the desired closed-loop control performance.

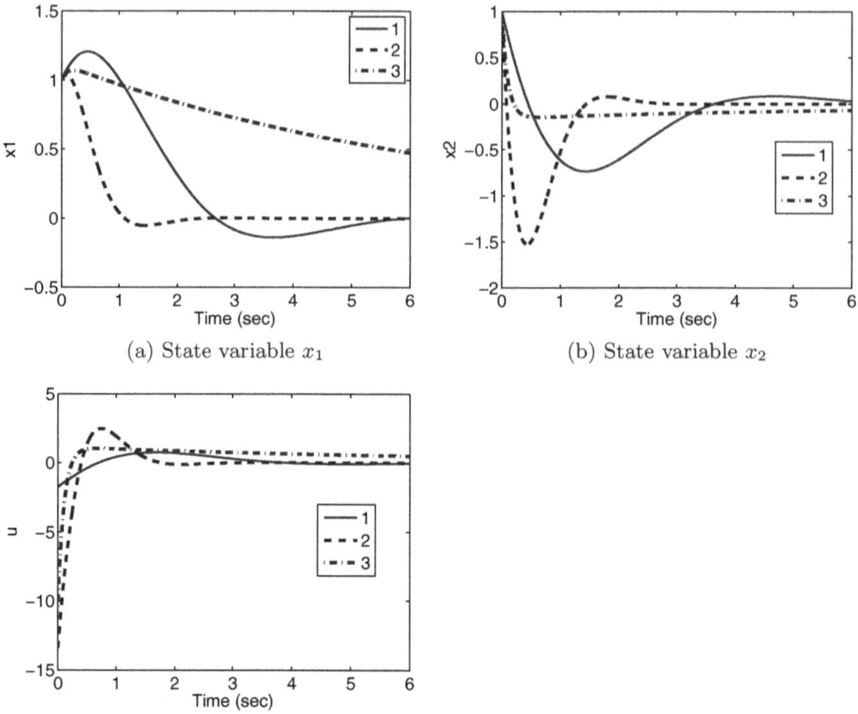

(a) State variable x_1

(b) State variable x_2

(c) Control signal u

Figure 1.9 Closed-loop state response to state initial conditions (Example 1.7). Key: line (1) closed-loop control using K_1; line (2) using K_2 and line (3) using K_3.

Example 1.8

Continue from Example 1.7. We choose Q as an identity matrix ($q_{11} = q_{22} = 1$). Illustrate the variation of R matrix on the closed-loop performance.

Solution

When the weighting R is reduced, the relative importance of the quantity $\int_0^\infty u(t)^2 dt$ is reduced in the objective function J. This means that this term can be allowed to be larger in the balance of the objective function. In reverse, when R is increased, the relative importance of the quantity $\int_0^\infty u(t)^2 dt$ is increased, meaning that this term is limited more in the balance of the objective function. In practice, we choose a smaller R if we wish to increase the closed-loop response speed and larger R to reduce the closed-loop response speed. As we have examined the case $R = 1$ in Example 1.7, we consider the three new choices below.

For the first case, $R = 0.1$ is selected. Using MATLAB **lqr.m** *function, we calculate the state feedback controller gain K_1 and the Riccati solution P_1:*

$$K_1 = \begin{bmatrix} 2.3166 & 3.8253 \end{bmatrix}; \; P_1 = \begin{bmatrix} 1.2687 & 0.2317 \\ 0.2317 & 0.3825 \end{bmatrix}.$$

For this selection, the closed-loop eigenvalues are $-2.4972, -1.3281$, and the minimum of the objective function $J_{min} = 2.1146$.

For the second case, $R = 0.01$ is selected. The state feedback controller K_2 and the Riccati solution P_2 are calculated as

$$K_2 = \begin{bmatrix} 9.0499 & 10.8674 \end{bmatrix}; \; P_2 = \begin{bmatrix} 1.0922 & 0.0905 \\ 0.0905 & 0.1087 \end{bmatrix}.$$

The closed-loop eigenvalues are $-9.8467, -1.0206$, and the minimum of the objective function is $J_{min} = 1.3818$.

For the third case, $R = 10$ is selected. The state feedback controller K_3 and the Riccati solution P_3 are calculated as

$$K_3 = \begin{bmatrix} 0.0488 & 0.4445 \end{bmatrix}; \; P_3 = \begin{bmatrix} 4.6624 & 0.4881 \\ 0.4881 & 4.4454 \end{bmatrix}.$$

The closed-loop eigenvalues are $-0.2223 \pm j0.9997$, and the minimum of the objective function is $J_{min} = 10.084$.

Comparing the four cases including the case from the previous example, we note that when the weighting coefficient R is reduced, the state feedback controller gain K increases. This is clearly observed from the variations of K_1 and K_2 as well as the one from the previous example. The second observation is that the closed-loop eigenvalues are changing with respect to the variation of R. When R is increased, the closed-loop system has a pair of complex eigenvalues with increased magnitude on their imaginary component (see the third case), meaning that the closed-loop eigenvalues are closer to the open-loop eigenvalues. Additionally, the minimum of the objective function J_{min} reduces as R reduces. Figures 1.10(a)–(b) show the state variable responses to the initial values $x_1(0) = x_2(0) = 1$. When $R = 10$, the closed-loop system has oscillatory response. From Figure 1.10(c), we can see that the reduction of R leads to a larger amplitude of the control signal.

As an exercise, one can try many combinations of Q and R matrices, for example, increasing and decreasing R for the second and third choice of Q in Example 1.7.

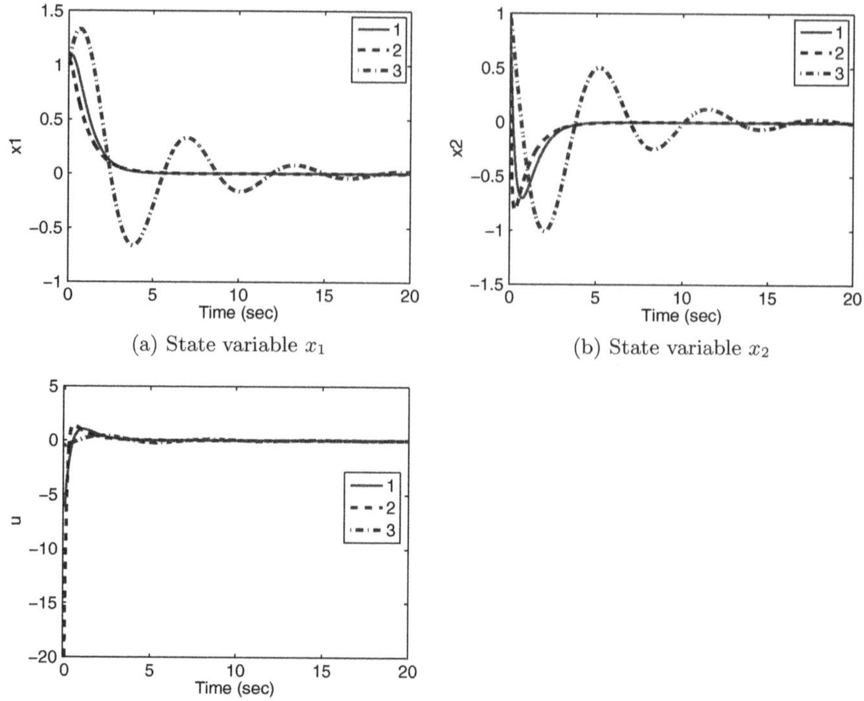

(a) State variable x_1 (b) State variable x_2 (c) Control signal u

Figure 1.10 Closed-loop state response to state initial conditions (Example 1.8). Key: line (1) closed-loop control using K_1, line (2) using K_2 and line (3) using K_3.

Example 1.9

We consider again to design a state feedback controller for the system used in Example 1.6, which has the following state space model:

$$\dot{x}(t) = \begin{bmatrix} 2 & -11 & 12 & 31 \\ -3 & 13 & -11 & -33 \\ 4 & -25 & 14 & 51 \\ -3 & 16 & -11 & -36 \end{bmatrix} x(t) + \begin{bmatrix} -1 & -4 \\ 1 & 6 \\ -1 & -11 \\ 1 & 7 \end{bmatrix} u(t). \tag{1.86}$$

Instead of using the desired closed-loop eigenvalue locations, we specify the desired performance in terms of the weighting matrices Q and R. We assume that the variations of all state variables are equally important, however, because of the actuators' characteristics, the first control signal should have a smaller variation in comparing with the second control signal.

Solution

We choose equal weights on the state variables so that the Q is an identity matrix. In order to illustrate how the weighting in the R matrix changes the control signals, we study two cases: (A) R is an identity matrix, and (B) the first diagonal element in R is increased to 10 in order to reduce the amplitude of the control signal for $u_1(t)$ whilst the second diagonal element is kept unchanged.

The following MATLAB program is used in Case A to compute the state feedback controller gain K and the closed-loop eigenvalues.

```
Q=eye(4,4);
R=eye(2,2);
[K,P,E]=lqr(A,B,Q,R);
```

Case A. *For both Q and R being diagonal matrices, the state feedback controller is calculated as*

$$K_1 = \begin{bmatrix} 0.4034 & -1.3628 & 1.6164 & 4.1022 \\ -0.6101 & 2.0416 & -1.9308 & -3.4373 \end{bmatrix},$$

and the closed-loop eigenvalues are $-15.2898, -1.5478 \pm j0.7115, -1.2019$.

Case B. *With Q being identity matrix and* $R(1,1) = 10$ *and* $R(2,2) = 1$, *we calculate the state feedback controller gain as*

$$K_2 = \begin{bmatrix} 0.0507 & -0.1888 & 0.2174 & 0.5497 \\ -0.5739 & 1.8895 & -1.7610 & -3.0129 \end{bmatrix},$$

and the closed-loop eigenvalues are $-15.1953, -1.3565 \pm j0.4374, -1.0978$.

When K_2 is compared with K_1, we see that the magnitudes of the components in the first row are reduced as the consequence of increasing the weight on the control signal $u_1(t)$. Additionally the closed-loop system by using K_2 has a slower dynamic response evident from the locations of the closed-loop eigenvalues. Figures 1.11(a)–(d) compare the closed-loop responses with the two controllers. They confirm that choosing a larger weight on the first control signal indeed reduces its amplitude, also produces slower closed-loop state responses. As an exercise, we can also try to put more emphasis on some of the state variables so that their variations are smaller. For instance, we can choose $q(1,1) = 10$, $q(2,2) = q(3,3) = q(4,4) = 1$.

1.5.4 LQR with Prescribed Degree of Stability

In Example 1.7, when $q(1,1) = 1$ and $q(2,2) = 100$, if we reduce the weighting R from 1 to 0.1, then the closed-loop eigenvalues will change from $-9.8985, -0.1429$

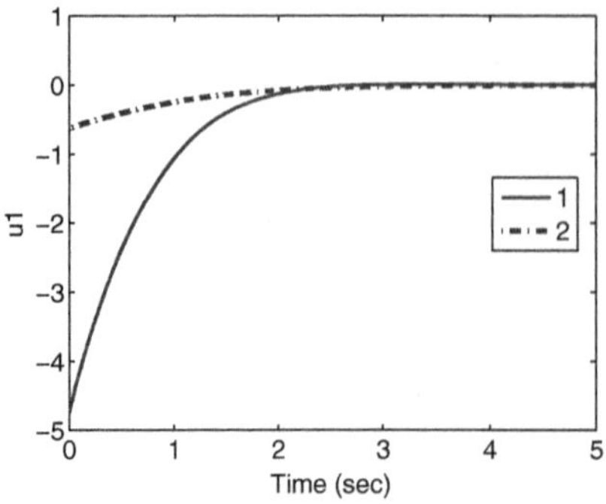

(a) Control signal u_1

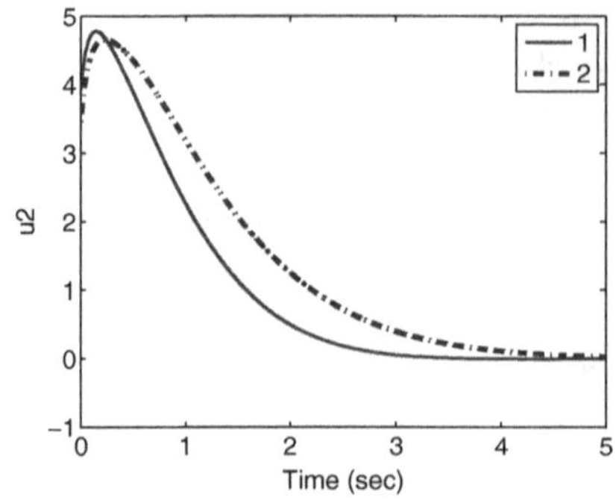

(b) Control signal u_2

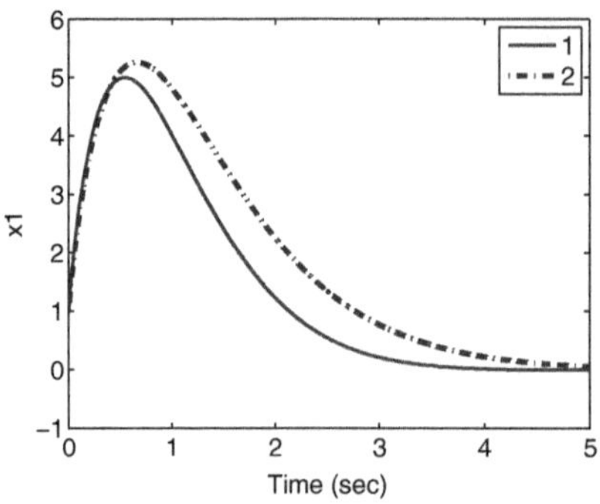

(c) State variable x_1

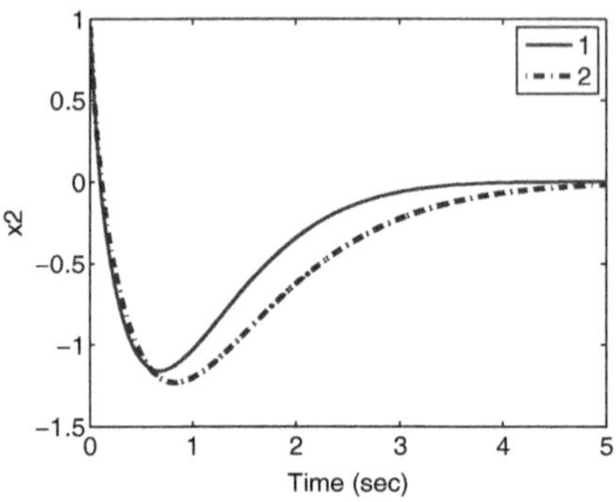

(d) State variable x_2

Figure 1.11 Closed-loop state response to state initial values (Example 1.9). Key: line (1) closed-loop control using K_1 and line (2) using K_2.

to −31.5910, −0.1050. Clearly, the effect on the closed-loop performance by simply reducing R is not adequate because the dominant eigenvalue moves closer to the imaginary axis from −0.1429 to −0.1050 instead of moving away from it. This is counter-intuitive since we would expect that as R reduces, the closed-loop response speed becomes faster. This simple exercise demonstrates that we need to fine tuning the elements in the Q matrix both diagonal and off-diagonal to achieve the desired closed-loop performance specifications. However, for this simple exercise, there are three parameters in the Q matrix to be adjusted (Q is a symmetric matrix). We intuitively understand how to adjust the diagonal elements to reflect our qualitative requirements as shown in Example 1.7. How to adjust the off-diagonal elements in Q is less obvious. Additionally, as the number of states increases, the number of off-diagonal elements in Q increases drastically. Thus, it is simply time consuming when we try to tune the Q matrix element-by-element.

The seminal work by Anderson and Moore [1971] has shown a simple, yet effective solution to put a constraint on the location of the slowest closed-loop eigenvalue. The idea of having a choice on the locations of the closed-loop eigenvalues when using LQR design is attractive because the closed-loop response speed is directly linked to their locations. This design method is called LQR with prescribed degree of stability.

In Anderson and Moore [1971], the objective function with α ($\alpha > 0$) is chosen to be:

$$J_\alpha = \int_0^\infty e^{2\alpha t} \left[x(t)^T Q x(t) + u(t)^T R u(t)\right] dt, \tag{1.87}$$

and it is minimized subject to

$$\dot{x}(t) = Ax(t) + Bu(t). \tag{1.88}$$

Note that the objective function (1.87) contains a time-varying scalar weight $e^{2\alpha t}$ in addition to the usual weight matrices Q and R. As the parameter $\alpha > 0$ is selected and the scalar $e^{2\alpha t}$ exponentially increases with time t, we put more and more emphasis on the states and the control signal as t increases. Additionally, in order for the objective function J_α to become finite, all states $x(t)$ and $u(t)$ at least will have the exponential decay rate $e^{-\alpha t}$. Thus, if we solve the optimization problem with the exponential data weighting, then both $x(t)$ and $u(t)$ will be guaranteed to have the minimum decay rate of $e^{-\alpha t}$. For linear time invariant systems, this is translated into the locations of all closed-loop eigenvalues (poles) being constrained to the left of the $-\alpha$ line on the complex plane.

In order to minimize J_α subject to the state space equation (1.88), we define the transformed variables:

$$x_\alpha(t) = e^{\alpha t} x(t); \;\; u_\alpha(t) = e^{\alpha t} u(t).$$

By substituting these transformed variables into the objective function J_α, it is rewritten to the following form:

$$J_\alpha = \int_0^\infty \left[x_\alpha(t)^T Q x_\alpha(t) + u_\alpha(t)^T R u_\alpha(t)\right] dt. \tag{1.89}$$

Now, we also need to match the system equation using the transformed variables $x_\alpha(t)$ and $u_\alpha(t)$. To this end, note that

$$\begin{aligned}\dot{x}_\alpha(t) &= \alpha e^{\alpha t} x(t) + e^{\alpha t} \dot{x}(t)\\ &= \alpha e^{\alpha t} x(t) + e^{\alpha t}(Ax(t) + Bu(t))\\ &= (A + \alpha I) x_\alpha(t) + B u_\alpha(t).\end{aligned} \tag{1.90}$$

With the initial condition $x_\alpha(0) = x(0)$, minimizing the exponentially weighted objective function J_α subject to the system equation (1.90) becomes a standard

LQR problem. Thus, as before, the optimal control $u_\alpha(t)$ is obtained through the solution of the algebraic Riccati equation with the transformed system $(A + \alpha I, B)$:

$$P(A + \alpha I) + (A + \alpha I)^T P - PBR^{-1}B^T P + Q = 0, \tag{1.91}$$

leading to

$$u_\alpha(t) = -R^{-1}B^T Px_\alpha(t). \tag{1.92}$$

Now, returning to the original control signal yields

$$\begin{aligned} u(t) = u_\alpha(t)e^{-\alpha t} &= -R^{-1}B^T Px_\alpha(t)e^{-\alpha t} \\ &= -R^{-1}B^T Px(t), \end{aligned} \tag{1.93}$$

where we have used the relationship: $x_\alpha(t)e^{-\alpha t} = x(t)$.

To establish that the closed-loop system has a prescribed degree of stability α, we consider the following properties.

1. We note that if the pair (A, D) is observable where $Q = D^T D$, then $(A + \alpha I, D)$ is observable; if the pair (A, B) is controllable, then $(A + \alpha I, B)$ is controllable. Hence, the solution of the Riccati equation (1.91) leads to asymptotic stability of the closed-loop system for the pair $(A + \alpha I, B)$. This means that for the transformed variable $x_\alpha(t)$, we have

   $$||x_\alpha(t)|| \to 0$$

   as $t \to \infty$.
2. Because the state variable $x(t)$ is in the relation to the transformed variable $x_\alpha(t)$ through

   $$x(t) = e^{-\alpha t}x_\alpha(t),$$

   this guarantees that all variables in $x(t)$ decay at least as fast as the rate of $e^{-\alpha t}$. This establishes that the exponentially weighted objective function produces a closed-loop system with a prescribed degree of stability α.
3. The asymptotic stability of the closed-loop system for the pair $(A + \alpha I, B)$ ensures that the closed-loop eigenvalues, for all k

   $$real\{\lambda_k(A + \alpha I - BK)\} < 0,$$

   where $K = R^{-1}B^T P$. This means that the closed-loop eigenvalues for the pair (A, B) must be at least, for all k

   $$real\{\lambda_k(A - BK)\} < -\alpha.$$

   This establishes that the closed-loop eigenvalues are on the left of the $s = -\alpha$ line in the complex plane.

4. We compare the two Riccati equations:

$$A^TP + PA - PBR^{-1}B^TP + Q = 0 \tag{1.94}$$

$$P(A + \alpha I) + (A + \alpha I)^TP - PBR^{-1}B^TP + Q = 0. \tag{1.95}$$

   The first equation (1.94) is used to solve the optimal controller without the exponential weighting whilst the second equation (1.95) is used with the exponential weighting. If we let

$$Q_\alpha = Q + 2\alpha P$$

   then, (1.95) becomes

$$A^TP + PA - PBR^{-1}B^TP + Q_\alpha = 0,$$

   which is identical to (1.94) except that the weighting matrix Q_α is a function of the Riccati solution P from (1.95). In essence, a simple means is provided to find the weighting matrix Q_α that ensures all the closed-loop eigenvalues on the left of $-\alpha$ line, and the actual objective function minimized is equal to

$$J = \int_0^\infty \left[x(t)^TQ_\alpha x(t) + u(t)^TRu(t)\right] dt. \tag{1.96}$$

In summary, if we wish to constrain all closed-loop eigenvalues to the left of $-\alpha$ ($\alpha > 0$) line on the complex plane, the state feedback controller gain K has the identical formula, except that the Riccati matrix P is solved using (1.91), where $(A + \alpha I)$ is used to replace the original system matrix A. Based on this observation, we will use the same MATLAB function **lqr.m** for the LQR design with a prescribed degree of stability α. This is illustrated in the following example.

We consider the undamped oscillator used in Example 1.7. In the third case, when $q(1,1) = 1$ and $q(2,2) = 100$ and $R = 1$, the pair of closed-loop eigenvalues are at -9.8985, -0.1429. Later, we also found that for this choice of Q matrix, reducing R could not move the slower closed-loop eigenvalue further away from the imaginary axis.

Example 1.10

Find the state feedback controller K for the undamped oscillator described by the following equation:

$$\dot{x}(t) = \begin{bmatrix} 0 & 1 \\ -1 & 0 \end{bmatrix} x(t) + \begin{bmatrix} 0 \\ 1 \end{bmatrix} u(t), \tag{1.97}$$

where the dominant closed-loop eigenvalue has real part to be less than -0.5.

Solution

The system matrices A and B are given by (1.97). We find the state feedback controller using MATLAB **lqr.m** *function with $\alpha = 0.5$ shown in the following program.*

```
alpha=0.5;
Q=[1 0;0 100];
R=1;
[K,P,E]=lqr(A+alpha*eye(2,2),B,Q,R);
eig(A-B*K)
```

Note that the closed-loop poles are calculated by using the controller gain K applying to the pair A and B matrices because the internal calculation of closed-loop poles from the **lqr.m** *function only leads to those in the transformed system. The* **lqr.m** *solution with $\alpha = 0.5$ gives the state feedback controller gain $K = \begin{bmatrix} 9.7202 & 11.4403 \end{bmatrix}$, and the closed-loop eigenvalues are $-1.0297, -10.4106$. All the closed-loop eigenvalues are moved to the left of -0.5 line.*

This is equivalent to the original optimization without the exponential weighting by choosing the $Q = Q_\alpha$ where Q_α is:

$$Q_\alpha = \begin{bmatrix} 113.9217 & 9.7202 \\ 9.7202 & 111.4403 \end{bmatrix}.$$

Here, Q_α is calculated with $Q_\alpha = Q + 2\alpha P$, and P is the Riccati solution from the **lqr.m** *function. It is noted that the Q_α has almost equal weights on the diagonal elements.*

Figures 1.12(a)–(c) show the closed-loop responses to the initial state $x_1(0) = x_2(0) = 1$ in comparison to the case without the prescribed degree of stability. Clearly, the slow decay problem in the $x_1(t)$ is overcome and the 'tail' in $x_2(t)$ due to the dynamic from the slow pole is also gone. However, the behavior of $x_2(t)$ has changed with a large undershoot, which is perhaps because the diagonal weight on x_2 has become almost the same as that on x_1. Figure 1.12 (c) shows that the control signal amplitude has increased.

Example 1.11

Continue from Example 1.9. The state space model is given in (1.86). In that example, with $Q = I$ and $R(1,1) = 10$ and $R(2,2) = 1$, the closed-loop eigenvalues are $-15.1953, -1.3565 \pm j0.4374, -1.0978$. The objectives for this example are to find the state feedback controller K such that all the closed-loop poles are positioned to the left of -1.5 line on the complex plane and compare their closed-loop performance.

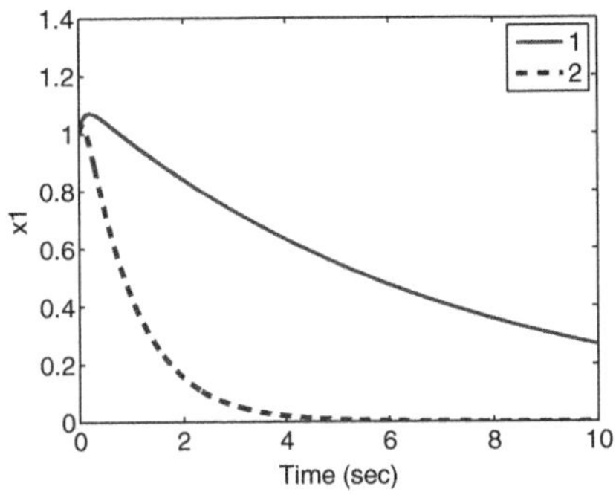

(a) State variable x_1

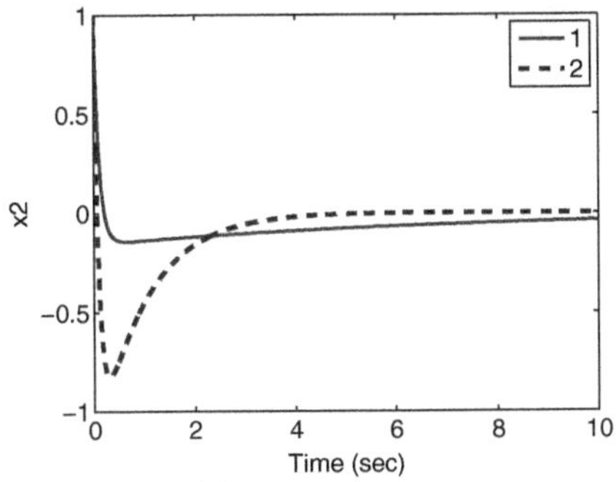

(b) State variable x_2

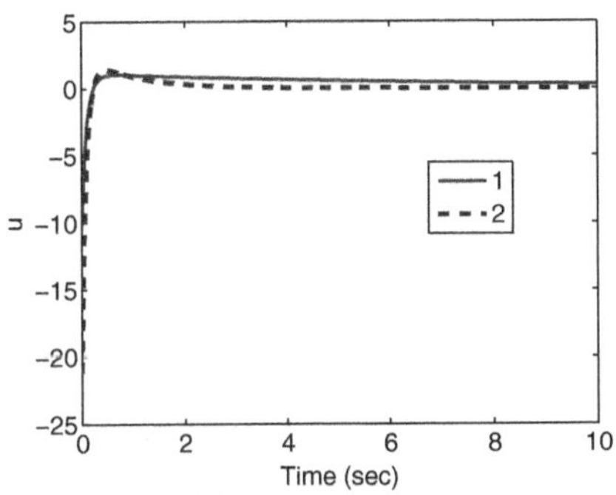

(c) Control signal u

Figure 1.12 Closed-loop state response to state initial values (Example 1.10). Key: line (1) $\alpha = 0$; line (2) $\alpha = 0.5$.

Solution

With $\alpha = 1.5$, $Q = I$ and $R(1,1) = 10$ and $R(2,2) = 1$, we find the optimal state feedback controller using the prescribed degree of stability as

$$K = \begin{bmatrix} 1.2054 & -4.1030 & 5.0689 & 12.2116 \\ -2.2744 & -10.6553 & 2.4547 & 13.6692 \end{bmatrix}$$

The closed-loop eigenvalues for the state feedback control system are -16.4406, $-2.1890 \pm j0.6721$, -1.8641, which satisfy the design requirement. Figures 1.13(a)–(d) show the first two closed-loop state variables and the control signals. It is seen from these figures that the control signals behave quite differently for the two cases. Note that the original choice of weighting matrix R with a larger coefficient on the first control signal had resulted in a smaller control signal $u_1(t)$. However,

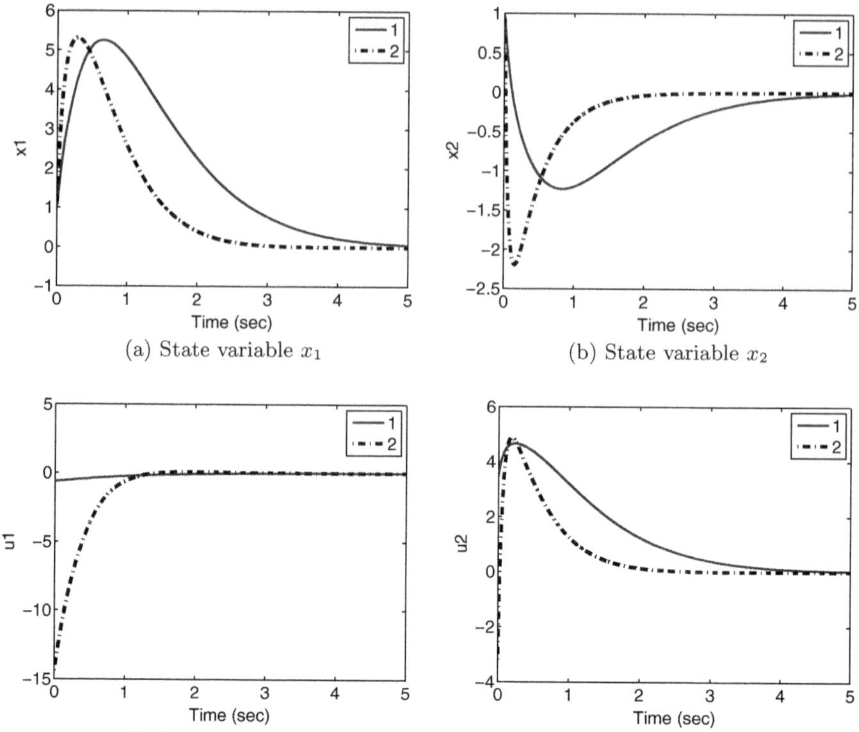

(a) State variable x_1 (b) State variable x_2 (c) Control signal u_1 (d) Control signal u_2

Figure 1.13 Closed-loop state response to state initial values. Key: line (1) closed-loop control using K_2 (from Example 1.9) and line (2) using K from this example.

using the design with the prescribed degree of stability, this is no longer the case (see Figures 1.13(c)–(d)). The dynamic response speeds for the state variables have increased.

1.5.5 Food for Thought

1. Do we have a unique solution for the state feedback controller K for a single input system by specifying desired closed-loop poles?
2. Can we always find a controller K to correctly position the desired closed-loop poles for a multi-input system?
3. Why are the Q and R matrices symmetric? What if they were not?
4. From the Examples 1.7–1.9, what are your observations for the choices of Q and R matrices on the effect of closed-loop performance?
5. When using the prescribed degree of stability in the LQR design, we have selected $\alpha \geq 0$. Can we choose α to be negative? What does it mean if α is negative?

6. When using the LQR with prescribed degree of stability, does the solution tell us how far away the closed-loop eigenvalues are from the $-\alpha$-line?
7. Do we need to change the B matrix when using the LQR with prescribed degree of stability?
8. Assume that we have chosen a pair of Q and R weighting matrices and found the Riccati solution P matrix and the closed-loop eigenvalues. Would it be an identical solution to the method with prescribed degree of stability if we adjust the Q matrix using

$$Q_{new} = Q + 2\alpha P?$$

1.6 Observer Design

1.6.1 Motivational Example for Observer

In state feedback controller design, the control law uses the feedback relationship:

$$u(t) = -Kx(t), \tag{1.98}$$

where we have assumed that the state vector

$$x(t) = \begin{bmatrix} x_1(t) & x_2(t) & \dots & x_n(t) \end{bmatrix}^T$$

is measured for feedback control. In reality, not all state variables are measured (or available). Some of them are even impossible to be measured. Furthermore, from saving cost point of view, it is beneficial to reduce the number of sensors used to measure the state variables.

Our objective is to construct the state vector $x(t)$ using available measurement. One might attempt to use the state space model in the way how we usually carry out the system simulation. For instance, since we know the system matrices (A, B) and the input signal $u(t)$, then the estimated state variable $\hat{x}(t)$ is expressed as

$$\frac{d\hat{x}(t)}{dt} = A\hat{x}(t) + Bu(t). \tag{1.99}$$

This approach would indeed work if the system is stable. The following example shows the scenarios of how this would work.

Example 1.12

There are three systems given here. All of them have the same input signal and the same B matrix.

Case A:

$$A = \begin{bmatrix} -3 & 1 \\ 2 & -3 \end{bmatrix}; \quad B = \begin{bmatrix} 1 \\ 2 \end{bmatrix}$$

Case B:

$$A = \begin{bmatrix} -3 & 1 \\ 2 & -1 \end{bmatrix}; \quad B = \begin{bmatrix} 1 \\ 2 \end{bmatrix}$$

Case C:

$$A = \begin{bmatrix} -3 & 1 \\ 2 & 0 \end{bmatrix}; \quad B = \begin{bmatrix} 1 \\ 2 \end{bmatrix}$$

The input signal $u(t)$ has an amplitude of ± 1. All three systems have the same initial state conditions as $x_1(0) = 1$ and $x_2(0) = 1$. The initial conditions for the estimated state variables are $\hat{x}_1(0) = 3$ and $\hat{x}_2(0) = 3$. Calculate the estimated states $\hat{x}(t)$ using (1.99) with the given initial conditions and the input signal, and compare the responses of the estimated states and the true states.

Solution

By using MATLAB **lsim.m** *function, we obtain the responses of the estimated states with the initial conditions and the input signal. Here, the sampling interval is selected as $\Delta t = 0.01$. Figure 1.14 shows the comparative results for the three cases. The numerical results are discussed as follows.*

1. *It is seen that in Case A, the estimated state variable $\hat{x}(t)$ converges to the true $x(t)$ in a relatively short time, in Case B, it takes a longer time for the estimated state variable $\hat{x}(t)$ to converge, however in Case C, the estimated state variable $\hat{x}(t)$ diverges from $x(t)$.*
2. *There is only one coefficient in the system matrix A different in the three cases. However, it changes the eigenvalues of A matrix. For Case A, the characteristic polynomial $\det(sI - A)$ is $(s + 3 - \sqrt{2})(s + 3 + \sqrt{2}) = (s + 1.586)(s + 4.14)$, for Case B, is $s^2 + 4s + 1 = (s + 3.732)(s + 0.268)$, and for Case C, is $s^2 + 3s - 2 = (s + 3.562)(s - 0.562)$. It is seen that although both Case A and Case B have stable systems, the dominant eigenvalue in Case A is -1.586, however, in Case B is -0.268, which has a slower decay rate in the latter case. In Case C, the system is unstable because there is an eigenvalue located at 0.562 (> 0), thus, the response $x(t)$ will not decay, and it seems that $\hat{x}(t)$ went up in hurry and $x(t)$ could not 'catch up' with it.*

This example shows that directly using the mathematical models to evaluate the system responses may have a pitfall, particularly when the system has a slow dynamic response where the estimated state variable takes a long time to converge to their true variable. This approach will fail if the system is unstable or has eigenvalues on the imaginary axis.

We will further examine the cases analytically by considering

$$\dot{x}(t) = Ax(t) + Bu(t) \tag{1.100}$$

$$\dot{\hat{x}}(t) = A\hat{x}(t) + Bu(t), \tag{1.101}$$

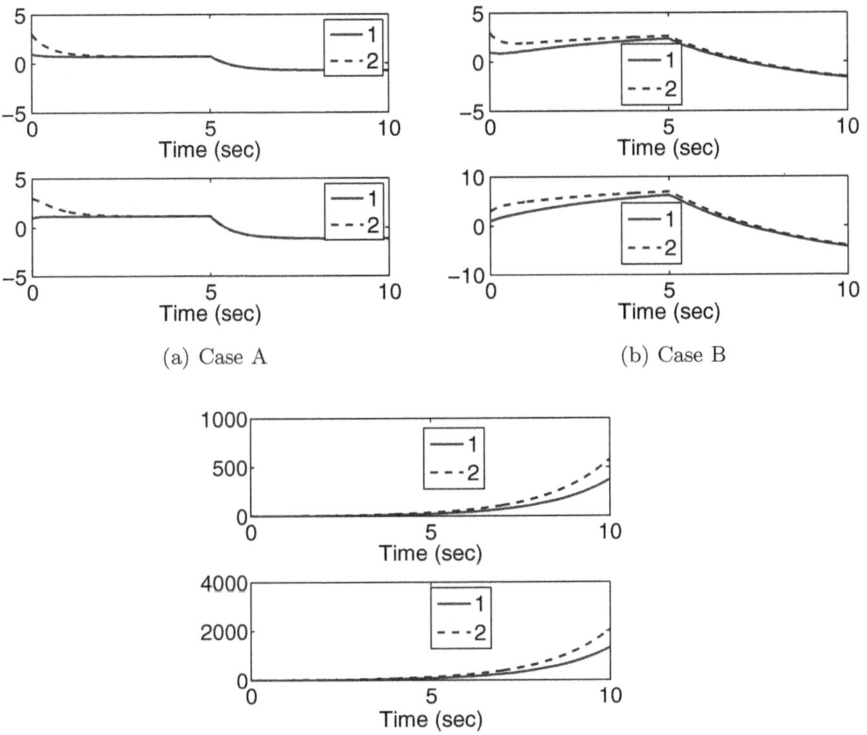

(a) Case A

(b) Case B

(c) Case C

Figure 1.14 Comparison between the true and estimated state variables using state space model (Example 1.12). Top figure $x_1(t)$, bottom figure $x_2(t)$. Key: line (1) $x(t)$; line (2) $\hat{x}(t)$.

and the error state is defined as

$$\tilde{x}(t) = x(t) - \hat{x}(t).$$

Subtracting (1.101) from (1.100) gives

$$\dot{x}(t) - \dot{\hat{x}}(t) = A(x(t) - \hat{x}(t))$$

which is

$$\dot{\tilde{x}}(t) = A\tilde{x}(t). \tag{1.102}$$

The solution of the differential equation is,

$$\tilde{x}(t) = e^{At}\tilde{x}(0), \tag{1.103}$$

for a given initial state error vector $\tilde{x}(0)$. This means that the error state variable $\tilde{x}(t) \neq 0$ if $\tilde{x}(0) \neq 0$. It basically says that if we take a wrong guess at the initial state variable $x(0)$, then there is an error between the true state variable $x(t)$ and the estimated state variable $\hat{x}(t)$. How the error $\tilde{x}(t)$ behaves is dependent on the eigenvalues of system matrix A as demonstrated by Example 1.12.

1.6.2 Observer Design

Example 1.12 provides us with 'food' for thought. It is apparent that if the error state vector $\tilde{x}(t)$ is to converge to zero, the error system described by (1.102) is required to be stable, and the convergence rate is dependent on the locations of the eigenvalues of the error system. Furthermore, the system dynamic model used by its own for the state estimation (see (1.101)) cannot achieve an arbitrary desired convergence rate specified by the user. Therefore, an additional mechanism is needed to give us a leverage in order to alter the system matrix of the error system. A control engineer's approach is to add a 'correction' term to the original dynamic model for the estimation of the state variable $x(t)$, leading to

$$\dot{\hat{x}}(t) = A\hat{x}(t) + Bu(t) + K_{ob}(y(t) - C\hat{x}(t)), \tag{1.104}$$

where the output $y(t) = Cx(t)$ is available through measurement and K_{ob} is a gain matrix chosen from the design. Equation (1.104) is commonly called the 'observer' equation, also K_{ob} $(n \times m)$ is called observer gain matrix where n is the number of states and m is the number of outputs. It is seen that when K_{ob} is zero, the observer equation falls back to the original approach (see (1.101)). However, the additional term $K_{ob}(y(t) - C\hat{x}(t))$ provides the key information that reflects how well the estimated state $\hat{x}(t)$ is doing via the error signal $y(t) - C\hat{x}(t)$. This is one of the examples of feedback mechanism at work.

To see what role of K_{ob} plays in the estimation of state variable $x(t)$, we examine the state error $\tilde{x}(t) = x(t) - \hat{x}(t)$. Subtracting (1.104) from (1.100) gives

$$\begin{aligned}\dot{\tilde{x}}(t) &= Ax(t) + Bu(t) - A\hat{x}(t) - Bu(t) - K_{ob}(y(t) - C\hat{x}(t)\\ &= (A - K_{ob}C)\tilde{x}(t).\end{aligned} \tag{1.105}$$

From (1.105) it is seen that first, the existence of K_{ob} provides a mechanism for altering the behavior of the error state $\tilde{x}(t)$ and hence, with the appropriate selection of K_{ob}, the system matrix of the error system $A - K_{ob}C$ could be made stable with desired eigenvalues, which then directly influences the convergence rate of the estimated state.

The observer error system (1.105) also gives a clue as to how the observer should be designed and how its performance should be quantified. Similar to the state feedback controller design, an obvious way is to place the eigenvalues of the observer error system matrix $A - K_{ob}C$ to desired locations. The following example illustrates observer design.

Example 1.13

A measured sinusoidal signal with known frequency ω_0, but unknown amplitude A_0 and phase θ_0 is corrupted by a zero mean bandlimited white noise. Design an observer *that will estimate the noise free sinusoidal signal from the measured signal $y(t)$, where the desired observer poles are positioned at $-\gamma\omega_0$ with $\gamma > 1$. Validate the observer design using simulation and illustrate the effect of γ on convergence rate and noise level of the estimated state in which we assume that $\omega_0 = 1$ and $\Delta t = 0.001$ (sec).*

Solution

Let $x_1(t) = A_0 \sin(\omega_0 t + \theta_0)$ and $x_2(t) = \dot{x}_1(t)$. Then,

$$\dot{x}_2(t) = \ddot{x}_1(t) = -A_0\omega_0^2 \sin(\omega_0 t + \theta_0) = -\omega_0^2 x_1(t).$$

This leads to the following state space equation that describes the sinusoidal signal:

$$\begin{bmatrix} \dot{x}_1(t) \\ \dot{x}_2(t) \end{bmatrix} = \begin{bmatrix} 0 & 1 \\ -\omega_0^2 & 0 \end{bmatrix} \begin{bmatrix} x_1(t) \\ x_2(t) \end{bmatrix};$$

$$y(t) = x_1(t) + \epsilon(t) = \begin{bmatrix} 1 & 0 \end{bmatrix} \begin{bmatrix} x_1(t) \\ x_2(t) \end{bmatrix} + \epsilon(t),$$

where $\epsilon(t)$ describes zero mean bandlimited white noise in continuous-time.

The estimated state variables $\hat{x}_1(t)$ and $\hat{x}_2(t)$ are obtained using the observer equation:

$$\begin{bmatrix} \dot{\hat{x}}_1(t) \\ \dot{\hat{x}}_2(t) \end{bmatrix} = \begin{bmatrix} 0 & 1 \\ -\omega_0^2 & 0 \end{bmatrix} \begin{bmatrix} \hat{x}_1(t) \\ \hat{x}_2(t) \end{bmatrix} + \begin{bmatrix} h_1 \\ h_2 \end{bmatrix} (y(t) - \hat{x}_1(t)) \tag{1.106}$$

with a chosen initial condition $[\hat{x}_1(0)\ \ \hat{x}_2(0)]^T$.

In order to design the observer, we consider the error system matrix,

$$\begin{aligned} A - K_{ob}C &= \begin{bmatrix} 0 & 1 \\ -\omega_0^2 & 0 \end{bmatrix} - \begin{bmatrix} h_1 \\ h_2 \end{bmatrix} \begin{bmatrix} 1 & 0 \end{bmatrix} \\ &= -\begin{bmatrix} h_1 & -1 \\ \omega_0^2 + h_2 & 0 \end{bmatrix}, \end{aligned} \tag{1.107}$$

and its characteristic polynomial,

$$A_{cl}(s) = \det(sI - (A - K_{ob}C)) = s^2 + h_1 s + \omega_0^2 + h_2. \tag{1.108}$$

With the observer poles chosen as $-\gamma\omega_0$ ($\gamma > 0$), the desired characteristic polynomial is

$$A_{cl}(s)^d = (s + \gamma\omega_0)^2 = s^2 + 2\gamma\omega_0 s + \gamma^2\omega_0^2. \tag{1.109}$$

By comparing the characteristic polynomial of the observer error system (1.108) with the desired characteristic polynomial, we obtain the observer gains as

$$h_1 = 2\gamma\omega_0; \quad h_2 = \gamma^2\omega_0^2 - \omega_0^2.$$

We consider two choices of γ. In the first case, $\gamma = 3$ is used in the calculation of observer gain, which gives $h_1 = 6$ and $h_2 = 8$. In the second case, $\gamma = 10$ is used leading to $h_1 = 20$ and $h_2 = 99$.

The observer is implemented by letting $\frac{d\hat{x}(t)}{dt} \approx \frac{\hat{x}(t_{k+1})-\hat{x}(t_k)}{\Delta t}$ *at the sampling time* t_k, *which leads to*

$$\hat{x}(t_{k+1}) = \hat{x}(t_k) + (A\hat{x}(t_k) + K_{ob}(y(t_k) - C\hat{x}(t_k))\Delta t, \tag{1.110}$$

where Δt *is the sampling interval and* $K_{ob} = \begin{bmatrix} h_1 & h_2 \end{bmatrix}^T$. *The equation (1.110) is solved iteratively with the input signal to the observer,* $y(t_k)$ *and* $\hat{x}(0) = \begin{bmatrix} 0 & 0 \end{bmatrix}^T$.

Figure 1.15(a) shows the measured sinusoidal signal $y(t)$ *used in the simulation study, which is clearly seen with a high noise level. To determine the quality of the estimated signal, we compare the estimated signal with the actual noise free signal used to generate the simulation data, although in real world application this noise free signal is unknown to us. Figure 1.15(b) illustrates the response of the estimated state and the noise free signal where* $\gamma = 10$. *It is seen that with the larger* γ, *the observer has a higher gain and the estimated state converged very fast*

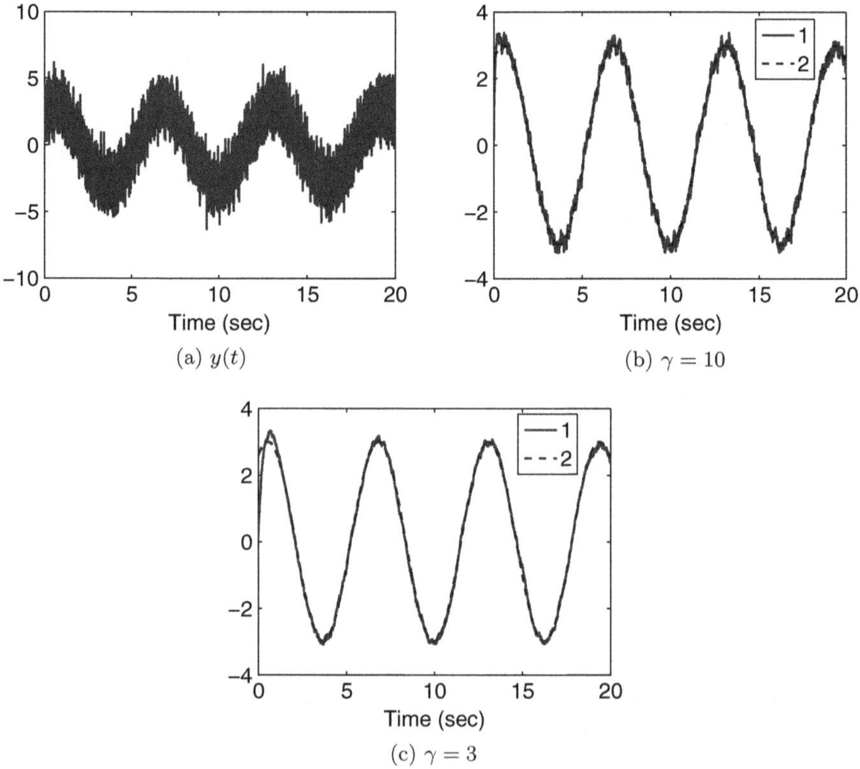

(a) $y(t)$ (b) $\gamma = 10$ (c) $\gamma = 3$

Figure 1.15 Estimation of sinusoidal signal from noisy measurement (Example 1.13). Key: line (1) $\hat{y}(t)$; line (2) noise free output.

to the noise free signal, however, there is a certain amount of noise still contained in the estimated signal. In contrast, when $\gamma = 3$ is used in the calculation of observer gain, the estimated state converged slowly to the noise free signal, but the estimated signal contained less noise in comparison with the higher γ case. In short, the simulation results point out that the selection of the closed-loop poles affects the convergence rate of the estimated state as well as its variation when the measurement has noise.

1.6.3 Observability

Observability is defined as follows. A state variable model of a dynamic system is said to be completely observable if, for any time t_0, there exists a time $t_1 > t_0$ such that a knowledge of the output $y(t)$ and input $u(t)$ in the time interval $t_0 \le t \le t_1$ is sufficient to determine the initial state $x(t_0)$ and as a consequence, $x(t)$ for all t between t_0 and t_1.

A necessary and sufficient condition for a linear time-invariant system to be completely observable if the observability matrix

$$L_o = \begin{bmatrix} C \\ CA \\ CA^2 \\ \vdots \\ CA^{n-1} \end{bmatrix}$$

has rank n, where n is the number of state variables.

The necessary and sufficient condition means that if it is a single output system, then the rank of L_o being equal to n means that $\det(L_o) \neq 0$. If it has more than one output, then the number of independent rows of L_o matrix is equal to n.

The following examples illustrate ideas about observability.

Example 1.14

Consider the dynamic system with the state space model:

$$\begin{bmatrix} \dot{x}_1(t) \\ \dot{x}_2(t) \end{bmatrix} = \begin{bmatrix} -\frac{\beta}{a_1} & \frac{\beta}{a_1} \\ \frac{\beta}{a_2} & -\frac{\beta}{a_2} \end{bmatrix} \begin{bmatrix} x_1(t) \\ x_2(t) \end{bmatrix} + \begin{bmatrix} \frac{1}{a_1} & 0 \\ 0 & \frac{1}{a_2} \end{bmatrix} \begin{bmatrix} u_1(t) \\ u_2(t) \end{bmatrix}$$

$$\begin{bmatrix} y_1(t) \\ y_2(t) \end{bmatrix} = \begin{bmatrix} \lambda & 0 \\ 0 & 1 \end{bmatrix} \begin{bmatrix} x_1(t) \\ x_2(t) \end{bmatrix},$$

where $a_1 \neq 0$, $a_2 \neq 0$. Investigate the observability of the system in terms of parameters β and λ.

Solution

The observability matrix is calculated as

$$L_o = \begin{bmatrix} \lambda & 0 \\ 0 & 1 \\ -\frac{\lambda\beta}{a_1} & \frac{\lambda\beta}{a_1} \\ \frac{\beta}{a_2} & -\frac{\beta}{a_2} \end{bmatrix}. \tag{1.111}$$

There are three cases as stated below.

Case A. *If $\lambda \neq 0$, $\beta = 0$, then the first two rows of the L_o matrix are independent, therefore, the system is observable.*

Case B. *If $\lambda = 0$, $\beta \neq 0$, then the number 2 and 4 rows are independent, therefore, the system is observable.*

Case C. *If $\lambda = 0$ and $\beta = 0$, then the number of independent row is 1. Therefore, the system is not observable.*

Example 1.15

Continue from Example 1.14, where we assume that $a_1 = a_2 = 1$, and $\lambda = 0$. Design an observer for this system and examine what happens when $\beta = 0$, which corresponds to Case C in Example 1.14. We assume that the desired closed-loop polynomial is specified as $s^2 + \alpha_1 s + \alpha_0$ with $\alpha_1 > 0$ and $\alpha_0 > 0$.

Solution

The state space model becomes

$$\begin{bmatrix} \dot{x}_1(t) \\ \dot{x}_2(t) \end{bmatrix} = \begin{bmatrix} -\beta & \beta \\ \beta & -\beta \end{bmatrix} \begin{bmatrix} x_1(t) \\ x_2(t) \end{bmatrix} + \begin{bmatrix} 1 & 0 \\ 0 & 1 \end{bmatrix} \begin{bmatrix} u_1(t) \\ u_2(t) \end{bmatrix}$$

$$y(t) = \begin{bmatrix} 0 & 1 \end{bmatrix} \begin{bmatrix} x_1(t) \\ x_2(t) \end{bmatrix}.$$

The eigenvalues of the system matrix A are determined by the solutions of the characteristic equation:

$$\det \begin{bmatrix} s+\beta & -\beta \\ -\beta & s+\beta \end{bmatrix} = s(s+2\beta) = 0, \tag{1.112}$$

which are located at $s_1 = 0$ and $s_2 = -2\beta$.

Choosing the observer gain vector $K_{ob} = \begin{bmatrix} h_1 & h_2 \end{bmatrix}^T$, the closed-loop characteristic polynomial for the observer is calculated as

$$\begin{aligned} A_{cl}(s) &= \det(sI - (A - K_{ob}C)) = (s+\beta)(s+\beta+h_2) + \beta(-\beta+h_1) \\ &= s^2 + (2\beta + h_2)s + \beta h_1 + \beta h_2, \end{aligned} \tag{1.113}$$

which is equal to the desired closed-loop characteristic polynomial

$$A_{cl}(s)^d = s^2 + \alpha_1 s + \alpha_0.$$

By comparing the closed-loop characteristic polynomials, we obtain:

$$h_2 = \alpha_1 - 2\beta;\ \ h_1 = \frac{1}{\beta}\alpha_0 - \alpha_1 + 2\beta$$

where $\beta \neq 0$.

However, if β *is to equal zero, the eigenvalues of the system matrix A are located on the origin of the complex plane,* $s_1 = s_2 = 0$ *(see (1.112)). Additionally, from (1.113) the closed-loop characteristic polynomial becomes*

$$A_{cl}(s) = s^2 + h_2 s.$$

By comparing the characteristic polynomial with the desired characteristic polynomial, we notice that there is no chance that these two polynomials can be made equal. This means that when the observability is lost in the case when $\lambda = \beta = 0$*, the eigenvalue of matrix A at the origin of the complex plane cannot be moved by choosing the observer gain* K_{ob}.

1.6.4 Duality between Controller and Observer

Examples 1.13 and 1.15 illustrate the design steps of an observer using pole-assignment technique. They are similar to those used in the state feedback controller design. There is a dual relationship between a controller and an observer, which will be discussed below.

The computation of the characteristic polynomial for observer design uses (1.108) as shown in Examples 1.13 and 1.15, which is

$$A_{cl}(s) = \det(sI - (A - K_{ob}C)). \tag{1.114}$$

Since for a matrix M, $\det M = \det M^T$, (1.114) can be re-written as

$$A_{cl}(s) = \det\ (sI - A + K_{ob}C)^T = \det(sI - A^T + C^T K_{ob}^T). \tag{1.115}$$

Now, we recall the computation of the closed-loop characteristic polynomial for a state feedback controller K given by (see Section 1.4)

$$A_{cl}(s) = \det(sI - A + BK). \tag{1.116}$$

When comparing (1.115) with (1.116), it is seen that both determinants have the identical structure. Therefore, what we need to do for the observer design is, instead of computing K_{ob}, that we compute K_{ob}^T by using the matrices A^T and C^T. This means that the methods and programs we have worked on for the controller design can be used for the observer design without any changes except that here we compute the transpose of the observer gain K_{ob}^T and use the transposes of system matrices A^T, C^T.

1.6.5 Observer Implementation

Although the observer is designed in continuous-time, its implementation is performed using either a computer or a micro-controller. Thus, the discretization of observer is needed. The key idea in the implementation is to approximate the estimated derivative $\dot{\hat{x}}(t)$ using forward difference.

Consider the observer dynamic model as

$$\dot{\hat{x}}(t) = A\hat{x}(t) + Bu(t) + K_{ob}(y(t) - C\hat{x}(t)). \tag{1.117}$$

At the sampling time t_k, with a sampling interval Δt the derivative $\dot{\hat{x}}(t)$ using Euler approximation is written as

$$\frac{d\hat{x}(t)}{dt} \approx \frac{\hat{x}(t_{k+1}) - \hat{x}(t_k)}{\Delta t}.$$

From this approximation, (1.117) becomes

$$\frac{\hat{x}(t_{k+1}) - \hat{x}(t_k)}{\Delta t} \approx A\hat{x}(t_k) + Bu(t_k) + K_{ob}(y(t_k) - C\hat{x}(t_k)), \tag{1.118}$$

leading to

$$\hat{x}(t_{k+1}) = \hat{x}(t_k) + \left(A\hat{x}(t_k) + Bu(t_k) + K_{ob}(y(t_k) - C\hat{x}(t_k))\right)\Delta t. \tag{1.119}$$

Equation (1.119) is commonly used in an observer implementation where at time t_k we will use the control signal $u(t_k)$, output signal $y(t_k)$ and estimated state $\hat{x}(t_k)$ to predict the next step of estimated state variable $\hat{x}(t_{k+1})$. This way, the estimated state $\hat{x}(t)$ is calculated recursively starting with a guess of the initial value of $\hat{x}(0)$.

The accuracy of the approximation is dependent on the sampling interval Δt. A higher order approximation of the derivative $\dot{\hat{x}}(t)$ will lead to a higher accuracy in the implementation, which then can tolerate a larger sampling interval Δt.

Consider the following approximation of the derivative $\dot{\hat{x}}(t)$ (Burden and Faires [1989]):

$$\frac{d\hat{x}(t)}{dt} \approx \frac{3\hat{x}(t_{k+1}) - 4\hat{x}(t_k) + \hat{x}(t_{k-1})}{2\Delta t},$$

leading to the approximation of the observer as

$$\frac{3\hat{x}(t_{k+1}) - 4\hat{x}(t_k) + \hat{x}(t_{k-1})}{2\Delta t} \approx A\hat{x}(t_k) + Bu(t_k) + K_{ob}(y(t_k) - C\hat{x}(t_k)). \tag{1.120}$$

Re-organizing (1.120) gives the discretization of the observer equation as

$$\hat{x}(t_{k+1}) = \frac{1}{3}(4\hat{x}(t_k) - \hat{x}(t_{k-1})) + \frac{2\Delta t}{3}\left(A\hat{x}(t_k) + Bu(t_k) + K_{ob}(y(t_k) - C\hat{x}(t_k))\right). \tag{1.121}$$

1.6.6 Food for Thought

1. Can you think of two engineering applications that might be beneficial from the application of 'soft sensors' to access the information that would be difficult to obtain using actual sensors?
2. If we know the initial condition vector $x(0)$ and the model is perfect, would you be able to construct the state vector $x(t)$ without the observer?
3. If there was an unknown constant disturbance d in the system, for instance,

   $$\dot{x}(t) = 3x(t) + d$$

   would the observer

   $$\dot{\hat{x}}(t) = 3\hat{x}(t) + K_{ob}(x(t) - \hat{x}(t))$$

   correctly estimate the state vector $x(t)$? We assume that K_{ob} is selected to produce a stable observer system. If not, what would you propose to do so that $\hat{x}(t)$ is correct in this situation?
4. Assume that a system is described by the first order differential equation:

   $$\dot{x}(t) = ax(t); \; y(t) = x(t).$$

   However, the parameter a changed over time and it is $0.8a$ in reality. Would the observer we originally designed in the form

   $$\dot{\hat{x}}(t) = a\hat{x}(t) + K_{ob}(x(t) - \hat{x}(t))$$

   be able to estimate $x(t)$ for the current system? Here we assume that K_{ob} still produced a stable observer system for the current system.
5. In the observer design, if the measurement noise is high, would you choose a higher observer gain corresponding to a set of fast closed-loop poles for the error system?
6. If you want the estimated states to converge faster, should you use a set of closed-loop poles corresponding to faster dynamic response?
7. If a state space model is not observable, is it correct to say that one or more poles of the observer error system cannot be moved by the choice of observer gain?
8. If a state space model is not observable, and you found out that the observable modes are stable, would you continue using this model for observer design? What modification would you propose if you want to continue using this model?
9. If a state space model is not observable and the unobservable modes are unstable, would you continue using this model for observer design? Would you consider adding additional sensing instruments to the system?
10. Is it correct to say that in general more measurements could improve system's observability?

11. Based on the duality between observer and controller, is it correct that we can use the controller design programs for observer design as long as we replace the system matrix A by A^T and input matrix B by C^T?
12. If we deployed the duality in the design, is it correct to check the closed-loop eigenvalues using either the system matrix $A^T - C^T K_{ob}^T$ or $A - K_{ob}C$?
13. In the observer design using LQR with a prescribed degree of stability, is it correct to use the pair $(A^T + \alpha I)$ and C^T together with the Q and R matrices?
14. When using the LQR with a prescribed degree of stability in an observer design, would you say that you have ensured the observer error system has an exponential decay rate of $e^{-\alpha t}$?

1.7 State Estimate Feedback Control System

The state estimate feedback control system is to combine the state feedback control system with the observer system. The closed-loop stability of such a system is ensured by the stability of the state feedback control system and the observer error system, which is stated in the separation principle.

1.7.1 State Estimate Feedback Control

In a state estimate feedback control, the basic idea is to combine the state feedback control scheme with the observer. Namely, replacing the state vector $x(t)$ by the estimated state vector $\hat{x}(t)$ yields the state estimate feedback control system. The control law is governed by the following equations:

$$u(t) = -K\hat{x}(t) \tag{1.122}$$

$$\dot{\hat{x}}(t) = A\hat{x}(t) + Bu(t) + K_{ob}(y(t) - C\hat{x}(t)), \tag{1.123}$$

where K is the state feedback controller gain matrix and K_{ob} is the observer gain matrix.

By substituting the control signal (1.122) into (1.123), we obtain the estimated states in terms of the output $y(t)$:

$$\begin{aligned}\dot{\hat{x}}(t) &= A\hat{x}(t) - BK\hat{x}(t) + K_{ob}y(t) - K_{ob}C\hat{x}(t)\\ &= (A - BK - K_{ob}C)\hat{x}(t) + K_{ob}y(t).\end{aligned} \tag{1.124}$$

From this, we will find the transfer function of the state estimate feedback controller by taking Laplace transform of (1.124), leading to

$$\hat{X}(s) = (sI - A + BK + K_{ob}C)^{-1}K_{ob}Y(s), \tag{1.125}$$

and the Laplace transform of the control signal

$$\begin{aligned} U(s) &= -K\hat{X}(s) \\ &= -K(sI - A + BK + K_{ob}C)^{-1}K_{ob}Y(s). \end{aligned} \tag{1.126}$$

Thus, the transfer function of the state estimate feedback controller is expressed as

$$C(s) = K(sI - A + BK + K_{ob}C)^{-1}K_{ob}, \tag{1.127}$$

which contains the feedback controller K and the observer gain K_{ob}.

1.7.2 Separation Principle

The question comes to our mind immediately is what the effect will be when we replace the measured state $x(t)$ by the estimated state $\hat{x}(t)$. To this effect, we examine the closed-loop system that consists of the observer and the controller. Assume that the state space model is

$$\dot{x}(t) = Ax(t) + Bu(t). \tag{1.128}$$

Substituting the state estimate feedback control signal $u(t) = -K\hat{x}(t)$ into the above model forms the closed-loop system:

$$\dot{x}(t) = Ax(t) - BK\hat{x}(t). \tag{1.129}$$

Equation (1.129) by itself does not reveal the closed-loop relationship because $x(t) \neq \hat{x}(t)$. So we define the error signal

$$\tilde{x}(t) = x(t) - \hat{x}(t).$$

From this, the estimated state vector $\hat{x}(t)$ is expressed as

$$\hat{x}(t) = x(t) - \tilde{x}(t). \tag{1.130}$$

Now, considering again (1.129) together with (1.130) leads to

$$\dot{x}(t) = Ax(t) - BK(x(t) - \tilde{x}(t)) = (A - BK)x(t) + BK\tilde{x}(t). \tag{1.131}$$

Additionally, from observer design, we have the observer error system:

$$\dot{\tilde{x}}(t) = (A - K_{ob}C)\tilde{x}(t). \tag{1.132}$$

Putting together (1.131) and (1.132) gives us the closed-loop system for the combined controller and observer system:

$$\begin{bmatrix} \dot{x}(t) \\ \dot{\tilde{x}}(t) \end{bmatrix} = \begin{bmatrix} A - BK & BK \\ 0_{n \times n} & A - K_{ob}C \end{bmatrix} \begin{bmatrix} x(t) \\ \tilde{x}(t) \end{bmatrix}, \tag{1.133}$$

where $0_{n \times n}$ is a zero matrix with dimension of $n \times n$ and n is the dimension of the state vector $x(t)$.

With the closed-loop system defined by (1.133), we can examine the stability and dynamic performance of the state estimate feedback control system. To this end, we look at the characteristic equation of the closed-loop system by examining:

$$\det\left(sI_{2n\times 2n} - \begin{bmatrix} A-BK & BK \\ 0_{n\times n} & A-K_{ob}C \end{bmatrix}\right)$$
$$= \det(sI_{n\times n} - A + BK)\det(sI_{n\times n} - A + K_{ob}C) = 0, \tag{1.134}$$

where we have used the property that the determinant of a block upper triangular matrix is equal to the multiplication of the determinants of block diagonal matrices. Equation (1.134) implies that

$$\det(sI_{n\times n} - A + BK) = 0 \tag{1.135}$$

$$\det(sI_{n\times n} - A + K_{ob}C) = 0. \tag{1.136}$$

As we recall, (1.135) is the closed-loop characteristic equation for the state feedback control system whilst (1.136) is the one for the observer. Hence, the solutions of (1.135) give the closed-loop eigenvalues of the state feedback control system and those of (1.136) lead to the closed-loop eigenvalues of the observer system. We have the following observation.

Observation. The set of eigenvalues of the combined closed-loop system with observer and controller consists of the eigenvalues of the control system and the observer error system. This means that the design of the state feedback control and the observer can be carried out separately (or independently), yet when they are put together as shown in (1.122) and (1.123), the closed-loop eigenvalues remain unchanged.

This rather simple, but useful observation is known as the separation principle. It is one of the cornerstones in the development of state feedback control system. It underpins the state estimate feedback control system and simplifies its design by decoupling the problems of controller design and observer design.

In general, the dynamic response for the observer is desired to be faster than the response of the closed-loop control system. This is reflected in the design of observer and controller by choosing the set of desired observer eigenvalues to be further away from the imaginary axis than those of the closed-loop control system.

1.7.3 Food for Thought

1. In the design of state estimate feedback control, is it correct to say that in the control law, we simply replace the true state $x(t)$ by the estimated state $\hat{x}(t)$ using an observer?
2. Why do you think that the separation principle plays such an important role in the design of a state feedback control system?

3. Suppose that a state space model has two state variables, one input and output:

$$\dot{x}(t) = Ax(t) + Bu(t); \quad y(t) = Cx(t).$$

How many closed-loop eigenvalues do we have if the control law is

$$u(t) = -Kx(t)?$$

How many closed-loop eigenvalues do we have if the control law is

$$u(t) = -K\hat{x}(t)?$$

Here $\hat{x}(t)$ is estimated using an observer.

4. In general, would you like the observer error system to have a faster dynamic response speed than that of the control system? If so, why do you think that this might be a good practice?
5. If a dynamic system has a lot of measurement noise, would you increase the observer gain or reduce the observer gain? Would you also increase or reduce controller gain accordingly?
6. Suppose that a complex system is described by a transfer function and we only have the information about input and output variables. Would you think that this is a good candidate for state estimate feedback control?

1.8 Summary

In this chapter, we have discussed feedback control systems designed using state space models. The control systems use the state variables for feedback control. If the state variables are not measured, an observer is used to estimate the state variables based on the information provided by the measured output variable as well as the control signal.

The other important aspects of this chapter are summarized as follows.

1. The state feedback control signal is calculated in linear combinations of controller gain coefficients and the state variables, which does not involve any differentiation or integration in its calculation. Hence, it is simple from the implementation point of view. The feedback controller gain is selected to produce a stable closed-loop system with satisfactory dynamic response.
2. The concept of pole-assignment controller design is introduced in the state feedback control system, where the closed-loop control performance is specified by the locations of the desired closed-loop poles. The controller gain is found numerically using MATLAB functions such as **place.m** or the program **T2place.m** written in Tutorial 1.1.
3. State estimation is a very important idea, which has found wide applications in the engineering field. When the state variables are not measured, the state

variables are estimated using an observer. The observer design is to ensure the stability of the observer error system.
4. The state feedback control system and the observer error system have a dual relationship. This means that we can use the controller design and analysis programs for the observer design and analysis by utilizing this dual relationship.
5. Examining the controllability and observability is to answer the questions whether we can achieve the desired closed-loop performance specified arbitrarily for the controller design and the observer design. If the system is not controllable, it means that there is one or more poles in the system that cannot be moved by the feedback control. Similarly, if the system is not observable, it means that there is one or more poles in the observer error system that cannot be changed by the observer gain.
6. Linear quadratic regulator offers a controller design method from the perspective of optimization. The design methodology is entirely different from the pole-assignment controller design. With the closed-loop performance specified by choosing the weighting matrices Q and R in the objective function, it provides intuitive interpretations and insight to the closed-loop performance. Once the weighting matrices are selected, the solutions of the feedback controller gain are obtained using MATLAB program **lqr.m**. Observer is designed with the same program by utilizing the duality between the observer and the controller.
7. The selection of the weighting matrices may not be straightforward for some applications. It proves to be very useful to apply the idea of prescribed degree of stability in the LQR design to achieve the desired closed-loop performance via the constraints on the closed-loop pole locations. The procedure involves a simple modification to the system matrix A and solves the original LQR problem as usual.
8. The state estimate feedback control system is a combination of the feedback control system and an observer system. We design the state feedback controller and the observer separately, and we put them together by replacing the unmeasured state variables with the estimated state variables. The closed-loop poles of the state estimate feedback control system consist of the control system poles and the observer error system poles.

The state estimate feedback control systems discussed in this chapter does not include integral action. It is not effective when tasks of reference tracking and disturbance rejection are required.

1.9 Further Reading

1. Text books in control engineering include, Bryson and Ho [1975], Kailath [1980], Chen [1998], Bay [1999], Franklin et al. [1991], Ogata [2002], Antsaklis

and Michel [1997], Zhou and Doyle [1998], Golnaraghi and Kuo [2010], Goodwin et al. [2000], Callier and Desoer [2012], Wolovich [2012], Golnaraghi and Kuo [2010].
2. The original papers about an observer were Luenberger [1964], Luenberger [1966], Luenberger [1971].
3. The topics of prescribed degree of stability for LQR design can be found in Anderson and Moore [1971], Anderson and Moore [1969], and the discrete-time version in Anderson and Moore [1979]. Its application in model predictive control system design can be found in Wang [2009].
4. The classical papers about pole-assignment controller design include Wonham [1967], Davison and Wonham [1968], Davison and Wang [1975], Kimura [1975], Furuta and Kim [1987].
5. Books about linear algebra and matrix computations include Watkins [2004], Demmel [1997], Trefethen and Bau III [1997], Lax [2007].
6. Books about PID control include Astrom and Hagglund [1995], Astrom and Hagglund [2006], Yu [2006], Johnson and Moradi [2005], Visioli [2006], Tan et al. [2012], Alfaro and Vilanova [2016], Wang [2020]. PID control of multi-variable systems is discussed in Wang et al. [2008].

Problems

Problem 1.1

1. Consider the system

$$\dot{x}(t) = \begin{bmatrix} -0.5 & 0 \\ 0 & -1 \end{bmatrix} x(t) + \begin{bmatrix} 0.5 \\ 1 \end{bmatrix} u(t)$$

   Is this system controllable? If it is yes, design a state feedback controller K such that the closed-loop poles are positioned at -1 and -2.
2. If the system is changed to

$$\dot{x}(t) = \begin{bmatrix} -1 & 0 \\ 0 & -1 \end{bmatrix} x(t) + \begin{bmatrix} 1 \\ 1 \end{bmatrix} u(t)$$

   is this system controllable? Can we design a state feedback controller K such that the closed-loop poles are at -2 and -3?

Problem 1.2 The state space equation for an electric circuit is given by

$$\begin{bmatrix} \dot{x}_1(t) \\ \dot{x}_2(t) \end{bmatrix} = \begin{bmatrix} -\frac{2}{RC} & \frac{1}{C} \\ -\frac{1}{L} & 0 \end{bmatrix} \begin{bmatrix} x_1(t) \\ x_2(t) \end{bmatrix} + \begin{bmatrix} \frac{1}{RC} \\ \frac{1}{L} \end{bmatrix} u(t)$$

$$y(t) = \begin{bmatrix} -1 & 0 \end{bmatrix} \begin{bmatrix} x_1(t) \\ x_2(t) \end{bmatrix}$$

1. Determine the eigenvalues (poles) of the system;
2. Determine the condition on the resistor R that would make the system have a pair of identical poles;
3. Is the system controllable and observable when the system has a pair of identical poles? Explain your answers.

Problem 1.3 Check controllability and observability of the following state space model:

$$\begin{bmatrix} \dot{x}_1 \\ \dot{x}_2 \\ \dot{x}_3 \\ \dot{x}_4 \end{bmatrix} = \begin{bmatrix} 0 & 1 & 0 & 0 \\ 3w^2 & 0 & 0 & 2w \\ 0 & 0 & 0 & 1 \\ 0 & -2w & 0 & 0 \end{bmatrix} \begin{bmatrix} x_1 \\ x_2 \\ x_3 \\ x_4 \end{bmatrix} + \begin{bmatrix} 0 \\ 1 \\ 0 \\ 0 \end{bmatrix} u(t) \tag{1.137}$$

$$y(t) = \begin{bmatrix} 0 & 0 & 1 & 0 \end{bmatrix} \begin{bmatrix} x_1 \\ x_2 \\ x_3 \\ x_4 \end{bmatrix} \tag{1.138}$$

where $w \neq 0$.

Problem 1.4 Assume that a dynamic system is described by the state space model

$$\begin{bmatrix} \dot{x}_1(t) \\ \dot{x}_2(t) \\ \dot{x}_3(t) \end{bmatrix} = \begin{bmatrix} 1 & 1 & 0 \\ 0 & 0 & 1 \\ 10 & 3 & -1 \end{bmatrix} \begin{bmatrix} x_1(t) \\ x_2(t) \\ x_3(t) \end{bmatrix} + \begin{bmatrix} 2 \\ 3 \\ 1 \end{bmatrix} u(t)$$

$$y(t) = \begin{bmatrix} 2 & 3 & 1 \end{bmatrix} \begin{bmatrix} x_1(t) \\ x_2(t) \\ x_3(t) \end{bmatrix} \tag{1.139}$$

where the state variables $x_1(t)$, $x_2(t)$, and $x_3(t)$ are not measured. Design a state estimate feedback control system to stabilize the system.

1. Here the closed-loop performance for the state estimate feedback control system is specified with the control system poles being located at $-1 \pm j1$ and -2 and the observer poles being -3, -3, and -3.
2. Simulate the closed-loop system performance with initial state variable values at $x_1(0) = x_2(0) = x_3(0) = 6$.

Problem 1.5 A mechanical system in a vehicle is described by the following differential equation:

$$\frac{dy^2(t)}{dt^2} = -1.6\frac{dy(t)}{dt} - 800y(t) + u(t)$$

where $y(t)$ is the position of the system and $u(t)$ is the control variable. The control objective is to make the acceleration $|\frac{dy^2(t)}{dt^2}|$ as small as possible. We can achieve this control objective by using state feedback control. We choose $x_1(t) = y(t)$ and $x_2(t) = \frac{dy(t)}{dt}$ and obtain the following state space equation:

$$\begin{bmatrix} \dot{x}_1(t) \\ \dot{x}_2(t) \end{bmatrix} = \begin{bmatrix} 0 & 1 \\ -1.6 & -800 \end{bmatrix} \begin{bmatrix} x_1(t) \\ x_2(t) \end{bmatrix} + \begin{bmatrix} 0 \\ 1 \end{bmatrix} u(t)$$

1. Find the open-loop poles for this mechanical system;
2. By selecting a pair of closed-loop poles at -3 and -4, find the state feedback controller K;
3. The measurement is the velocity of the mechanical system $\dot{y}(t)$, which contains a lot of noise. By selecting a pair of closed-loop poles at -6 and -7, find the observer gain K_{ob};
4. Write down the expression of the control signal $u(t)$ in terms of the estimated states and the observer equation for estimating the states.

Problem 1.6 A three spring and double mass system is illustrated in Figure 1.5 of Example 1.1. Its state space model is given as follows:

$$\begin{bmatrix} \dot{x}_1(t) \\ \dot{x}_2(t) \\ \dot{x}_3(t) \\ \dot{x}_4(t) \end{bmatrix} = \begin{bmatrix} 0 & 1 & 0 & 0 \\ -\frac{\alpha_1+\alpha_2}{M_1} & 0 & \frac{\alpha_2}{M_1} & 0 \\ 0 & 0 & 0 & 1 \\ \frac{\alpha_2}{M_2} & 0 & -\frac{\alpha_1+\alpha_2}{M_2} & 0 \end{bmatrix} \begin{bmatrix} x_1(t) \\ x_2(t) \\ x_3(t) \\ x_4(t) \end{bmatrix} + \begin{bmatrix} 0 & 0 \\ \frac{1}{M_1} & 0 \\ 0 & 0 \\ 0 & \frac{1}{M_2} \end{bmatrix} \begin{bmatrix} u_1(t) \\ u_2(t) \end{bmatrix} \tag{1.140}$$

$$\begin{bmatrix} y_1(t) \\ y_2(t) \end{bmatrix} = \begin{bmatrix} 1 & 0 & 0 & 0 \\ 0 & 0 & 1 & 0 \end{bmatrix} \begin{bmatrix} x_1(t) \\ x_2(t) \\ x_3(t) \\ x_4(t) \end{bmatrix} \tag{1.141}$$

The physical parameters are $M_1 = 2kg$, $M_2 = 4kg$, $\alpha_1 = 30N/m$ and $\alpha_2 = 90N/m$.

1. Determine the eigenvalues of the spring-mass system.
2. Determine the controllability and observability of this system.
3. Design a state estimate feedback control system to stabilize this spring mass plant using MATLAB **place.m** function where the closed-loop control system has poles at $-3, -3.1, -3.2, -3.3$ and the observer error system has poles -6, $-6.1, -6.2, -6.3$.
4. We assume that $x_1(0) = 0.5$ and $x_3(0) = 0.1$, $x_2(0) = x_4(0) = 0$, and $\hat{x}_1(0) = \hat{x}_2(0) = \hat{x}_3(0) = \hat{x}_4(0) = 0$. With sampling interval $\Delta t = 0.001$ (sec) and simulation time $T_{sim} = 6$ (sec), simulate the closed-loop response of the spring-mass

system and present the responses of the four state variables in response to the initial conditions of system.

5. Where are the closed-loop poles for this state estimate feedback control system?

Problem 1.7 A continuous time signal $y_0(t)$ is severely corrupted by noise, where $y_0(t)$ is described by the following continuous-time function

$$y_0(t) = a_m sin(\omega_0 t + \theta) + b_m e^{-\gamma t}$$

where the parameters $\omega_0 > 0$ and $\gamma > 0$ are known, but a_m, b_m and θ are unknown.

1. Write down the state space model that could lead to filtering the noise.
2. Design an observer to estimate the noise free signal $y_0(t)$, where all the closed-loop observer poles are chosen to be $-\alpha$ ($\alpha > 0$). (Hint: Consider the design using A^T and C^T, then follow the steps for controller canonical transformation).
3. Verify your design by using simulations and present the plots of the original noise data and comparison between the estimated $\hat{y}_0(t)$ and the original $y_0(t)$. (Hint: you need to construct the simulation with your selection of all known and unknown parameters, sampling interval Δt, α, noise).

2

Practical Multivariable Controllers in Continuous-time

2.1 Introduction

One of the primary objectives in feedback control is to compensate unexpected disruptive events of system operations. These events (or disturbances) are either expected or unexpected, and for most times, they are not measured. For instance, an electrical machine has a sudden change of electrical load. The state feedback control systems introduced in Chapter 1 lack the capability to reject the disturbances. Therefore, in their present forms, they have limited applications to practical problems.

The objective of this chapter is to introduce two types of practical multivariable controllers. These controllers are capable of tracking step reference signals and rejecting low frequency disturbances. In order to have these capabilities, the multivariable controllers will have integral control in addition to having an anti-windup implementation to deal with the events when the actuators reach their maximum or minimum values.

The chapter begins with the well-known classical approach to include integral action using state space model (see Section 2.2.1). This design is intuitive and easy to understand. But it lacks of an anti-windup mechanism when the control signal reaches its maximum or minimum limit, hence the integrator windup scenario occurring (see Section 2.2.2). The remainder of Section 2.2 is to modify the classical controller by considering the effect of disturbance in the controller design, leading to a practical multivariable controller that has integral action in addition to an anti-windup mechanism in implementation. Both MATLAB and Simulink tutorials are given in this section, and a detailed case study of drum boiler control is presented. Another classical approach to obtain integral action is through disturbance estimation with an observer, where the input disturbance is assumed to be a constant (see Section 2.3). With this disturbance-observer-based

State Feedback Control and Kalman Filtering with MATLAB/Simulink Tutorials, First Edition.
Liuping Wang and Robin Ping Guan.

Companion website: www.wiley.com/go/feedback19

approach, an anti-windup mechanism is naturally introduced. Both MATLAB and Simulink tutorials are given in this section for the design and implementation. This section is completed by presenting the disturbance-observer-based design for the case where all the state variables are measured. The next section of this chapter is to present a case study of drive train control of a wind turbine using both practical approaches (see Section 2.4).

2.2 Practical Controller I: Integral Action via Controller Design

The well-known classical approach to include integral action using state space model is to introduce the integrated output in the design model. However, this approach has the integrator windup problem when the control signal reaches its saturation limits.

2.2.1 The Original Control Law

We assume that a state space model is described by the following differential equations,

$$\dot{x}_p(t) = A_p x_p(t) + B_p u(t) \tag{2.1}$$

$$y(t) = C_p x_p(t), \tag{2.2}$$

where state vector $x_p(t)$ has the dimension $n \times 1$, both input $u(t)$ and output $y(t)$ have the same dimension $m \times 1$. A classical approach to generate integral action in state feedback control is to assume an additional state vector $z(t)$, where $\dot{z}(t) = y(t) - r(t)$, where $r(t)$ is the reference signal. Effectively, the state vector $z(t)$ is the integrated error where

$$z(t) = \int_0^t (y(\tau) - r(\tau))d\tau.$$

Then, the state space model (2.1) is augmented with the integrated state, yielding

$$\begin{bmatrix} \dot{x}_p(t) \\ \dot{z}(t) \end{bmatrix} = \overbrace{\begin{bmatrix} A_p & 0_{n\times m} \\ C_p & 0_{m\times m} \end{bmatrix}}^{A} \begin{bmatrix} x_p(t) \\ z(t) \end{bmatrix} + \overbrace{\begin{bmatrix} B_p \\ 0_{m\times m} \end{bmatrix}}^{B} u(t), \tag{2.3}$$

where $0_{p\times q}$ denotes a zero matrix with the dimension $p \times q$. The augmented A, B matrices are defined as the illustrations in (2.3).

This is the most commonly used approach in the literature to generate integral action in state feedback control. If the original system matrices A_p and B_p are controllable, then the augmented system matrices A and B are controllable, except for the case where the transfer function $C_p(sI - A_p)^{-1}B_p$ has a zero at the origin of the complex plane. When the transfer function $C_p(sI - A_p)^{-1}B_p$ has a zero on the origin of the complex plane, there is a pole-zero cancellation occurring in the augmented model, and the augmented model will loss its controllability due to the pole-zero cancellation. It may not be a simple task to find the zeros of a complex transfer function, and if this is the case, it is simpler to directly check the controllability of the augmented system.

Assuming that the augmented system is controllable, a state feedback controller K is designed using the model (2.3) to achieve desired closed-loop performance such as stable closed-loop system and desired transient response. Then, the state feedback control signal $u(t)$ is calculated using the feedback controller gain K, the state vector $x_p(t)$ and the integrated error $z(t)$ as

$$\begin{aligned} u(t) &= -\overbrace{\begin{bmatrix} K_p & K_y \end{bmatrix}}^{K} \begin{bmatrix} x_p(t) \\ z(t) \end{bmatrix} \\ &= -K_p x_p(t) - K_y \int_0^t (y(\tau) - r(\tau))d\tau, \end{aligned} \tag{2.4}$$

where the relationship $z(t) = \int_0^t (y(\tau) - r(\tau))d\tau$ is used.

It is clearly seen from the control law given by (2.4) that integral action has been introduced in the state feedback control.

If the state variables are not measured, then an observer is needed to generate the estimated state vector $\hat{x}_p(t)$. For observer design, we resort to the original state space model (2.1)-(2.2). Assume that an observer gain K_{ob} is chosen such that the observer error system matrix $A_p - K_{ob}C_p$ is stable with desired eigenvalues (or poles). Then, the state vector $x_p(t)$ is estimated using the observer equation,

$$\dot{\hat{x}}_p(t) = A_p\hat{x}_p(t) + B_p u(t) + K_{ob}(y(t) - C_p\hat{x}_p(t)). \tag{2.5}$$

By substituting the state vector $x_p(t)$ in the control law (2.4) with the estimated state vector, we obtain the state estimate feedback control law as

$$u(t) = -K_p\hat{x}_p(t) - K_y \int_0^t (y(\tau) - r(\tau))d\tau. \tag{2.6}$$

The control law (2.6) clearly shows that integral action is included in the computing of the control signal.

2.2.2 Integrator Windup Scenarios

The classical approach by using the control law (2.6) to introduce integral action in state feedback control is intuitive and simple for design and implementation. However, its biggest drawback is that when the control signal reaches either its maximum or minimum value, an integrator windup event will occur. As a result, the closed-loop control performance deteriorates when the control signal reaches its limits. This can be demonstrated using the following simple example.

Example 2.1

Assuming that a system has the following transfer function:

$$G(s) = \frac{(s-3)(s+5)}{s(s+1)(s+3)}$$

design a state estimate feedback control system for the purpose of step reference tracking and constant input disturbance rejection. All closed-loop poles for the controller are chosen to be −2 and all observer poles are chosen to be −10. Investigate the variations of closed-loop performance when the control signal is limited to ±5.

Solution

The transfer function is converted to the following state space model,

$$\dot{x}_p(t) = A_p x_p(t) + B_p u(t);\; y(t) = C_p x_p(t)$$

where the system matrices are

$$A_p = \begin{bmatrix} -4 & -3 & 0 \\ 1 & 0 & 0 \\ 0 & 1 & 0 \end{bmatrix};\; B_p = \begin{bmatrix} 1 \\ 0 \\ 0 \end{bmatrix};\; C_p = \begin{bmatrix} 1 & 2 & -15 \end{bmatrix}.$$

We will first produce the augmented system matrices A, B according to (2.3) for the controller design. By positioning the closed-loop control system poles at −2, the state feedback controller gain is calculated using the augmented model to give:

$$K = \begin{bmatrix} 4 & 22.0667 & 34.1333 & -1.0667 \end{bmatrix}$$

and with the closed-loop observer poles at −10, the observer gain is calculated based on A_p, C_p as:

$$K_{ob} = \begin{bmatrix} -20.0938 & -8.4896 & -4.2049 \end{bmatrix}^T.$$

The closed-loop control system performance is investigated using simulation studies. Choosing a reference step signal $r(t) = 1$ for $t \geq 0$ and an input unit step disturbance entering the system at $t = 100$ (sec), the closed-loop control system is simulated with and without the control signal amplitude limits imposed. Here the disturbance input matrix is selected as $B_d = \begin{bmatrix} 1 & 0 & 1 \end{bmatrix}^T$ for the simulation (see (2.7) for the description of input disturbance). Figures 2.1(a)-(b) show the closed-loop responses. It is clearly seen that the output follows the step reference signal without steady-state error. When the input disturbance enters the system (see Figure 2.1(a)),

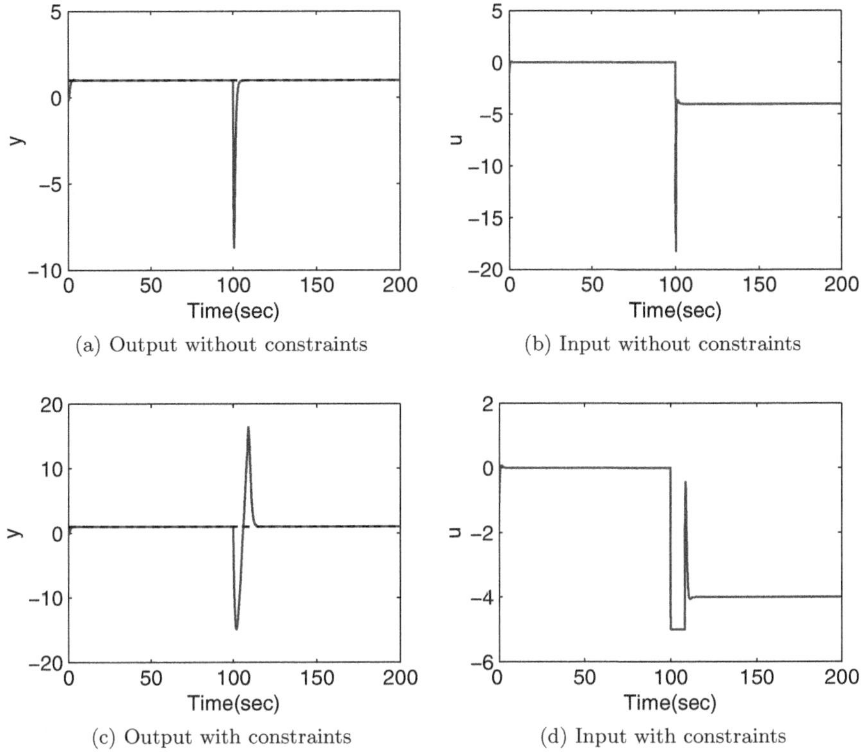

(a) Output without constraints (b) Input without constraints

(c) Output with constraints (d) Input with constraints

Figure 2.1 Comparison of closed-loop control system responses in presence of constraints (Example 2.1)

there is a large dip in the output with value at -8.5*, however the system recovers quickly to the reference signal within 4 (sec). The reason for this quick recovery is a large control signal generated by the controller for the disturbance rejection (see Figure 2.1(b)), which has a minimum of* -18 *exceeding the limit specified. In comparison, when the control amplitude is limited to* ± 5*, Figures 2.1(c)-(d) show that the closed-loop response performance deteriorates significantly for the disturbance rejection. It is seen from Figure 2.1(c) that the output initially drops to* -15 *due to the disturbance, then unexpectedly increases to 16 due to the integrator windup effect.*

2.2.3 Proposed Practical Multivariable Controller

To modify the original control law to overcome the problem with the integrator windup, we consider the effect of disturbance in the derivation of control law.

We assume that the system has the following state space model with an input constant disturbance vector $d(t)$:

$$\dot{x}_p(t) = A_p x_p(t) + B_p u(t) + B_d d(t) \tag{2.7}$$

$$y(t) = C_p x_p(t), \tag{2.8}$$

where $\dot{d}(t) = 0$. Taking differentiation of the original model (2.7)-(2.8) gives,

$$\ddot{x}_p(t) = A_p \dot{x}_p(t) + B_p \dot{u}(t) \tag{2.9}$$

$$\dot{y}(t) = C_p \dot{x}_p(t). \tag{2.10}$$

Note that the input to the state space model is $\dot{u}(t)$ and the output is $\dot{y}(t)$, however, the constant disturbance term, $B_d d(t)$, vanishes from this operation because $\dot{d}(t) = 0$. Then, the augmented model is to select the $y(t)$ as the additional state vector, leading to

$$\begin{bmatrix} \ddot{x}_p(t) \\ \dot{y}(t) \end{bmatrix} = \overbrace{\begin{bmatrix} A_p & 0_{n\times m} \\ C_p & 0_{m\times m} \end{bmatrix}}^{A} \begin{bmatrix} \dot{x}_p(t) \\ y(t) \end{bmatrix} + \overbrace{\begin{bmatrix} B_p \\ 0_{m\times m} \end{bmatrix}}^{B} \dot{u}(t). \tag{2.11}$$

We can also have the output equation as

$$y(t) = \overbrace{\begin{bmatrix} 0_{m\times n} & I_{m\times m} \end{bmatrix}}^{C} \begin{bmatrix} \dot{x}_p(t) \\ y(t) \end{bmatrix}. \tag{2.12}$$

It is clear that the system matrices in (2.11) are identical to those in (2.3), but the state vector and the input vector are different. The augmented state space model given by (2.11) and (2.12) was used in the continuous-time model predictive control (Wang [2009]). It was shown that the transfer function of the augmented model $G(s) = C(sI - A)^{-1}B$ has the following relationship with the plant transfer function $G_p(s) = C_p(sI - A_p)^{-1}B_p$:

$$G(s) = \frac{1}{s} G_p(s)$$

Thus, by assuming that the original system matrices A_p and B_p are controllable and the transfer function $C_p(sI - A_p)^{-1}B_p$ has no zero on the origin of the complex plane, the augmented system matrices A and B are controllable as the result. The state feedback controller K is designed using the augmented system matrices A, B to achieve desired closed-loop performance.

When using the state space model (2.11), the state feedback control law is expressed as

$$\dot{u}(t) = -K \begin{bmatrix} \dot{x}_p(t) \\ y(t) \end{bmatrix} = -\begin{bmatrix} K_p & K_y \end{bmatrix} \begin{bmatrix} \dot{x}_p(t) \\ y(t) \end{bmatrix}. \tag{2.13}$$

where the derivative of the state variable $\dot{x}_p(t)$ and the output $y(t)$ are used as the feedback variables. If the state vector $x_p(t)$ is measured, then the derivative of the state vector $\dot{x}_p(t)$ is available for the state feedback control. The closed-loop system is obtained by substituting (2.13) into (2.11), which is

$$\dot{x}(t) = (A - BK)x(t), \tag{2.14}$$

where $x(t) = \begin{bmatrix} \dot{x}_p(t)^T & y(t)^T \end{bmatrix}^T$. Therefore, for disturbance rejection, the controller K is designed to bring a non-zero initial state $x(0)$ to zero and the controller design is formulated as a regulator design. We can use either pole-assignment controller method or the LQR design method to find the controller K (see Chapter 1). The closed-loop stability and dynamic performance are determined by the closed-loop eigenvalues of the system matrix $A - BK$.

If the state variable $\dot{x}_p$ is unknown, an observer is constructed based on (2.9) and (2.10) to estimate it, which has the following form,

$$\ddot{\hat{x}}_p(t) = A_p\dot{\hat{x}}_p(t) + B_p\dot{u}(t) + K_{ob}(\dot{y}(t) - C_p\dot{\hat{x}}_p(t)). \tag{2.15}$$

The state estimate control law becomes

$$\dot{u}(t) = -\begin{bmatrix} K_p & K_y \end{bmatrix} \begin{bmatrix} \dot{\hat{x}}_p(t) \\ y(t) \end{bmatrix}, \tag{2.16}$$

where the measured state vector $\dot{x}_p(t)$ in (2.13) is replaced by the estimated state vector $\dot{\hat{x}}_p(t)$.

The next question is what the closed-loop system would be. To answer this question, we will examine the observer error system. Defining the state error $\dot{\tilde{x}}_p(t) = \dot{x}_p(t) - \dot{\hat{x}}_p(t)$, subtracting (2.15) from (2.9) gives the observer error system:

$$\ddot{\tilde{x}}_p(t) = (A_p - K_{ob}C_p)\dot{\tilde{x}}_p(t), \tag{2.17}$$

where we have replaced $\dot{y}(t)$ by $C_p\dot{x}_p(t)$. By substituting the state estimate control law (2.16) into the augmented model, we obtain:

$$\dot{x}(t) = (A - BK)x(t) + BK_p\dot{\tilde{x}}_p(t). \tag{2.18}$$

Combining (2.18) with (2.17) leads to the following closed-loop system for the state estimate feedback control:

$$\begin{bmatrix} \dot{x}(t) \\ \ddot{\tilde{x}}_p(t) \end{bmatrix} = \begin{bmatrix} A - BK & BK_p \\ 0_{n\times(n+m)} & A_p - K_{ob}C_p \end{bmatrix} \begin{bmatrix} x(t) \\ \dot{\tilde{x}}_p(t) \end{bmatrix}, \tag{2.19}$$

where $0_{n\times(n+m)}$ is a zero matrix with the dimension $n \times (n+m)$.

Note that the determinant of a block upper triangular matrix is equal to the multiplication of the determinants of block diagonal matrices. To determine the eigenvalues of the closed-loop system (2.19), we will examine:

$$\det(sI_{(n+m)\times(n+m)} - A + BK) = 0 \tag{2.20}$$

$$\det(sI_{n\times n} - A_p + K_{ob}C_p) = 0. \tag{2.21}$$

The solutions of polynomial equation (2.20) give the closed-loop eigenvalues of the state feedback control system and those of (2.21) lead to the closed-loop eigenvalues of the observer error system. In conclusion, the dynamic performance of the closed-loop system is dependent on the locations of the state feedback control system and the observer error system. The separation principle for regulator design in Section 1.7.2 of Chapter 1 holds for the disturbance rejection with integral action.

2.2.4 Anti-windup Implementation

When a reference signal $r(t)$ is introduced, the feedback control law (2.13) is modified to include it, leading to

$$\dot{u}(t) = -K_p \dot{x}_p(t) - K_y (y(t) - r(t)). \tag{2.22}$$

When the state vector $x_p(t)$ is not available, an observer is designed to estimate $\dot{x}_p(t)$. The control law with reference signal $r(t)$ becomes:

$$\dot{u}(t) = -K_p \dot{\hat{x}}_p(t) - K_y (y(t) - r(t)). \tag{2.23}$$

The state feedback controller with integral action can now be implemented with an anti-windup mechanism. Similar to the PID controller's anti-windup implementation in Wang [2020], the computation of the control signal will be based on computing the derivative (2.23) to allow the control signal saturation.

Now, at the sampling time t_k, the derivative of the control signal $\dot{u}(t)$ is approximated using the first order difference as,

$$\dot{u}(t_k) \approx \frac{u(t_k) - u(t_{k-1})}{\Delta t},$$

where Δt is the chosen sampling interval. From this, the control signal at sampling instant t_k is computed:

$$u(t_k) = u(t_{k-1}) + \Delta t \dot{u}(t_k), \tag{2.24}$$

where $u(t_k)$ and the $u(t_{k-1})$ are the current and past samples of the control signals. Accordingly, the observer equation is also discretized for digital implementation as

$$\begin{aligned} \dot{\hat{x}}_p(t_{k+1}) &= \dot{\hat{x}}_p(t_k) \\ &\quad + \left(A_p \dot{\hat{x}}_p(t_k) + B_p \dot{u}(t_k) + K_{ob}(\dot{y}(t_k) - C_p \dot{\hat{x}}_p(t_k))\right) \Delta t, \end{aligned} \tag{2.25}$$

where the input signals to the observer equation are the derivatives of the control signal and the output signal at the sampling time t_k. The derivative of the output signal is approximated using

$$\dot{y}(t_k) \approx \frac{y(t_k) - y(t_{k-1})}{\Delta t},$$

where $y(t_k)$ and the $y(t_{k-1})$ are the current and past samples of the output signals. With this approximation, the estimated derivative of the state variable at the sampling time t_{k+1} is calculated as,

$$\dot{\hat{x}}_p(t_{k+1}) = \dot{\hat{x}}_p(t_k) + \left(A_p\dot{\hat{x}}_p(t_k) + B_p\dot{u}(t_k) - K_{ob}C_p\dot{\hat{x}}_p(t_k)\right)\Delta t$$
$$+K_{ob}(y(t_k) - y(t_{k-1})). \tag{2.26}$$

It is interesting to note that although the computation is performed for the derivative of the estimated states, the input to the observer is not the derivative of the output. The fact that the difference of the output between the two sample data points is used in the implementation makes the proposed approach feasible in the presence of measurement noise because the measurement noise will not be amplified due to a small sampling interval Δt.

Furthermore, in the situation of severe noise, the remedy for its attenuation is to design an observer to estimate the augmented state vector $x(t)$ that contains both $\dot{x}_p(t)$ and $y(t)$ based on (2.11) and (2.12), which gives

$$\dot{\hat{x}}(t) = A\hat{x}(t) + B\dot{u}(t) + \overline{K}_{ob}(y(t) - C\hat{x}(t)), \tag{2.27}$$

where $\overline{K}_{ob}$ is the observer gain designed using the augmented A and C matrices. This observer implementation eliminates the difference operation of the output signal.

Recall the classical state feedback control system has the integrator windup problem when the control signal reaches saturation limits as demonstrated by Example 2.1. The problem was caused by the unstable implementation of the integrator using the integral equation (see (2.6)). The proposed approach avoids using the integral equation, instead using the computation of the derivatives. Key to the anti-windup mechanism is to stop the integral action as soon as the control signal reaches the saturation limits, and to inform the observer that the control signal has been saturated.

Suppose that the saturation limits for the control signal are u^{min} and u^{max}, at time t_0, the initial values for $\dot{\hat{x}}_p(t_0)$ and the measured output $y(t_0)$ and the reference signal $r(t_0)$, as well as the past information of the output and input $y(t_{-1})$ and $u(t_{-1})$ are available. The implementation of state estimate feedback control law with anti-windup mechanism is summarized as follows.

1. Calculate the derivative of the feedback control signal

$$\dot{u}(t_k) = -K_p\dot{\hat{x}}_p(t_k) - K_y(y(t_k) - r(t_k)). \tag{2.28}$$

2. Calculate the control signal

$$u(t_k) = u(t_{k-1}) + \dot{u}(t_k)\Delta t.$$

3. Impose the saturation limits on the control signal with the following computations.

(a) For the ith control signal, saturation limits are imposed:

$$u_i(t_k)^{act} = \begin{cases} u_i^{min}, & \text{if } u_i(t_k) < u_i^{min}; \\ u_i(t_k), & \text{if } u_i^{min} \le u_i(t_k) \le u_i^{max}; \\ u_i^{max}, & \text{if } u_i(t_k) > u_i^{max}, \end{cases}$$

where $i = 1, 2, \ldots, m$.

(b) Update the derivative of the control signal using the actual control signal $u(t_k)^{act}$ that contains the saturated elements:

$$\dot{u}(t_k)^{act} = \frac{u(t_k)^{act} - u(t_{k-1})^{act}}{\Delta t}.$$

4. Compute the estimated state vector for the sampling time t_{k+1} using the $\dot{u}(t_k)^{act}$ as

$$\begin{aligned}\dot{\hat{x}}_p(t_{k+1}) = \dot{\hat{x}}_p(t_k) + \left(A_p\dot{\hat{x}}_p(t_k) + B_p\dot{u}(t_k)^{act} - K_{ob}C_p\dot{\hat{x}}_p(t_k)\right)\Delta t \\ + K_{ob}(y(t_k) - y(t_{k-1})).\end{aligned}$$

5. As the sampling time progresses to t_{k+1}, repeat the computation from the first step.

It is emphasized that the integral action is stopped as soon as the saturation limits are reached for two samples consecutively because the derivative of the control signal automatically becomes zero. Additionally the observed states are calculated in reflection of the saturation limits by updating with the correct $\dot{u}(t_k)$.

Example 2.2

Continue from the Example 2.1, however, use the modified control law with the anti-windup mechanism proposed in this section. When the control signal violates its lower limit, show that the anti-windup mechanism implementation avoids the oscillation of the output signal and compare the differences of the results from the two approaches.

Solution

With $u^{min} = -5$, as shown in Example 2.1, the control signal reached the saturation limit when the disturbance entering the system at $t = 100$ (sec). The closed-loop state feedback control system is simulated using the anti-windup implementation under identical simulation conditions as in Example 2.1.

Figures 2.2(a)-(b) compare the output signal and the control signal responses to the step input disturbance entering the system at $t = 100$ (sec). We have the following observations.

1. *It is seen from Figure 2.2(a) that in both cases, the output dropped to -15 due to the disturbance and the control signal reaching its limit. However, with*

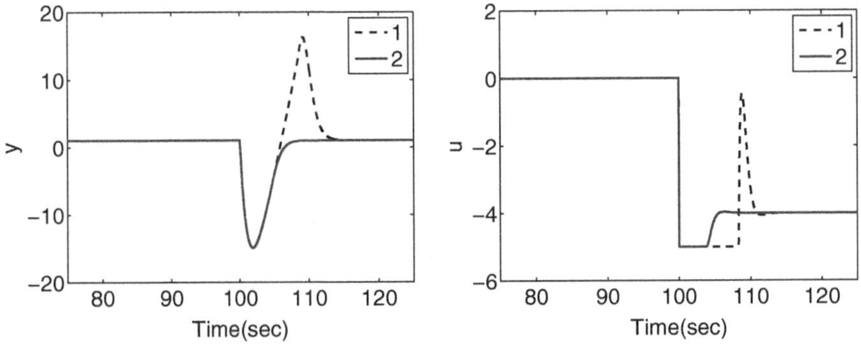

(a) Output signal

(b) Control signal

Figure 2.2 Comparison of closed-loop response (Example 2.2). Key: line (1) without anti-windup mechanism, line (2) with anti-windup mechanism

the implementation of anti-windup mechanism, within 7 seconds, the output recovers to the reference signal smoothly and without overshoot. In contrast, without the anti-windup mechanism, the output overshoot to 16 from the original control law before recovering to the reference signal. It is also seen that without the anti-windup mechanism, the whole recovery process from disturbance takes about 14 (sec) to complete, which is twice as much as the case with the anti-windup mechanism.

2. *From Figure 2.2(b) it is seen that the control signals are identical to the disturbance rejection for* $t < 103$ *(sec). After this time period, the control signal with anti-windup mechanism comes out from the saturation and increases smoothly to the desired steady-state. In comparison, without the anti-windup mechanism, the control signal remains to be saturated for additional 4 (sec), and comes out from the saturation by a sudden jump from* -5 *to* -0.62*. This behavior of the control signal results in the overshoot on the output signal.*

This example demonstrates that the implementation of anti-windup mechanism is important to ensure satisfactory closed-loop performance in the presence of control signal saturation.

2.2.5 MATLAB Tutorial on Design and Implementation

This section will present the design and simulation procedure for state estimate control systems with integral action and anti-windup mechanism. There are four tutorials associated.

The first tutorial presents how to generate the augmented system matrices $(A,\ B,\ C)$ based on (2.11) and (2.12).

Tutorial 2.1 *Write a MATLAB function* **aug4IC.m** *that will produce the augmented model for the control system design.*

Step by Step

1. *Create a new file called* **aug4IC.m**.
2. *The input variables to the function are the plant system matrices: A_p, B_p, C_p. The output variables from this function are the augmented system matrices A,B and C. Enter the following program into the file:*

```
function [A,B,C]=aug4IC(Ap,Bp,Cp)
```

3. *Check the numbers of inputs, outputs and state variables. Continue entering the following program into the file:*

```
[m1,n1]=size(Cp);
[n1,n_in]=size(Bp);
```

4. *Initialize the system matrices by using zero matrices. Continue entering the following program into the file:*

```
n=n1+m1;
A=zeros(n,n);
B=zeros(n,n_in);
C=zeros(m1,n);
```

5. *Create the block matrices A_p and C_p in A. Continue entering the following program into the file:*

```
A(1:n1,1:n1)=Ap;
A(n1+1:n1+m1,1:n1)=Cp;
```

6. *Create the block matrix B_p in B. Continue entering the following program into the file:*

```
B(1:n1,:)=Bp;
```

7. *Create the identity matrix in C matrix. Continue entering the following program into the file:*

```
C(:,n1+1:n)=eye(m1,m1);
```

8. *Test the function with the following system with two inputs and two outputs and two states:*

```
Ap=[1 -1; 2 1];
Bp=[1 2;2 1];
Cp=[1 0; 0 1];
```

9. *Check A matrix against the following answer:*

$$A = \begin{bmatrix} 1 & -1 & 0 & 0 \\ 2 & 1 & 0 & 0 \\ 1 & 0 & 0 & 0 \\ 0 & 1 & 0 & 0 \end{bmatrix}$$

The second tutorial presents the implementation of the controller in MATLAB simulation environment.

Tutorial 2.2 *Write a MATLAB function that will implement the state feedback control law with integral action and perform closed-loop simulation for reference tracking and disturbance rejection in the presence of control signal amplitude constraints. The starting point is that we have the continuous-time system model (A_p, B_p, C_p) and have designed the state feedback controller K and the observer K_{ob}. In the implementation, we do not need the augmented system matrices A, B and C, however they are required in the design of the state feedback controller K.*

Step by Step

1. *Create a new file called* **simu4IC.m**.
2. *The input variables to the function are the plant system matrices: A_p, B_p, C_p, the number of samples for the simulation (Nsim) with sampling interval Deltat, controller K, and observer gain K_{ob}, reference signal sp and disturbance signal dis. The dimension of sp is $m_1 \times Nsim$ and the dimension of dis is $n_{in} \times Nsim$ where m_1 is the number of outputs and n_{in} is the number of inputs. Umax and Umin are the vectors containing the limits for the maximum and minimum values of each component in the control signal. The output variables from this function are output y_1 and control signal u_1. Enter the following program into the file:*

```
function [y1,u1]=simu4IC(Ap,Bp,Cp,Deltat,Nsim,K,Kob,
sp,dis,Umin,Umax)
```

3. *Check number of states, number of outputs and number of inputs. Continue entering the following program into the file:*

```
[m1,n]=size(Cp);
[n,n_in]=size(Bp);
```

4. *Set initial conditions for the controller implementation and simulation. All initial conditions are chosen to be zero for simplicity. Continue entering the following program into the file:*

```
uPast=zeros(n_in,1);
yPast=zeros(m1,1);
```

```
y=zeros(m1,1);
xhat=zeros(n,1);
x=zeros(n,1);
```

5. *Perform the simulation iteratively, which corresponds to a real-time control system. Continue entering the following program into the file:*

```
for kk=1:Nsim;
```

6. *At the sampling instant kk that corresponds to time t_k, with measured output y(kk) and the reference signal sp(kk) as well as the estimated state variable xhat, we calculate the derivative of the control signal. Continue entering the following program into the file:*

```
udot=-K*[xhat;y-sp(:,kk)];
```

7. *Calculate the derivative of the output signal $\dot{y}$ at the sampling instant kk with the past output and the measured output. Continue entering the following program into the file:*

```
yd=(y-yPast)/Deltat;
```

8. *Calculate the control signal. Continue entering the following program into the file:*

```
u=uPast+udot*Deltat;
```

9. *Impose constraints on the control signal and calculate the derivative of the control signal in preparation for the observer update. This is the most important step for the implementation of anti-windup mechanism. Continue entering the following program into the file:*

```
for i=1:n_in
if (u(i,1)>Umax(i,1)) u(i,1)=Umax(i,1);
udot(i,1)=(u(i,1)-uPast(i,1))/Deltat; end
if (u(i,1)<Umin(i,1)) u(i,1)=Umin(i,1);
udot(i,1)=(u(i,1)-uPast(i,1))/Deltat; end
end
```

10. *Update the estimated state variable with derivative of the control signal that carries the information of saturation and the derivative of the output signal. Continue entering the following program into the file:*

```
xhat=xhat+(Ap*xhat+Kob*(yd-Cp*xhat)+Bp*udot)*Deltat;
```

11. *Prepare the control signal calculation for the next sampling period by updating uPast and yPast. Continue entering the following program into the file:*

```
yPast=y;
uPast=u;
```

12. *The above steps 6-11 capture the implementation of the control system. We save the control signal and output signal from this function. Continue entering the following program into the file:*

```
y1(:,kk)=y;
u1(:,kk)=u;
```

13. *Since we are in a simulation environment, the control signal calculated is sent to the plant model for closed-loop simulation. In the closed-loop simulation, we add the input disturbance to the system. If the input disturbance matrix is known to be B_d, then we can add this information to the closed-loop simulation. This completes one-cycle of computation in the closed-loop simulation. Continue entering the following program into the file:*

```
x=x+(Ap*x+Bp*u+Bp*dis(:,kk))*Deltat;
y=Cp*x;
end
```

14. *Test the function by following Tutorial 2.3.*

The third tutorial validates the design and simulation procedures using a simple example.

Tutorial 2.3 *The system under test was presented in Example 2.1, having the following transfer function:*

$$G(s) = \frac{(s-3)(s+5)}{s(s+1)(s+3)}.$$

All closed-loop poles for the controller are chosen to be -2 and all observer poles are chosen to be -10.

Step by Step

1. *Create a new file called Ex4StateEstim.m.*
2. *Input the transfer function of the system. Enter the following program into the file:*

```
num=conv([1 -3],[1 5]);
den1=conv([1 1],[1 3]);
den=conv([1 0],den1);
```

3. *Convert the transfer function into state space model. Continue entering the following program into the file:*

```
[Ap,Bp,Cp,Dp]=tf2ss(num,den);
```

4. *Generate the augmented state space model by following Tutorial 2.1. Continue entering the following program into the file:*

```
[A,B,C]=aug4IC(Ap,Bp,Cp);
```

5. *Design the controller using the pole-assignment design method using* **T2place.m** *written in Tutorial 1.1. Continue entering the following program into the file:*

```
T1=conv([1 2],[1 2]);
T=conv(T1,[1 2]);
T=conv(T,[1 2]);
K=T2place(A,B,T);
```

6. *Design the observer using the pole-assignment design method. Continue entering the following program into the file:*

```
P1=conv([1 10],[1 10]);
P=conv(P1,[1 10]);
Kob=T2place(Ap',Cp',P)';
```

7. *Prepare the closed-loop simulation with the reference signal and disturbance signal. Continue entering the following program into the file:*

```
Deltat=0.01;
Nsim=20000;
t=0:Deltat:(Nsim-1)*Deltat;
sp=ones(1,Nsim);
dis=[zeros(1,Nsim/2) ones(1,Nsim/2)];
Umax=5;
Umin=-5;
```

8. *Perform the closed-loop simulation using MATLAB function* **simu4IC.m**. *Continue entering the following program into the file:*

```
[y1,u1]=simu4IC(Ap,Bp,Cp,Deltat,Nsim,K,Kob,
sp,dis,Umin,Umax);
```

9. *Present the simulation results. Continue entering the following program into the file:*

```
figure(1)
plot(t,y1)
```

```
xlabel('Time(sec)')
ylabel('y')
figure(2)
plot(t,u1)
ylabel('u')
xlabel('Time(sec)')
```

10. *Validate this closed-loop controller design and simulation by comparing the results shown in Example 2.2. To obtain the identical results, the input disturbance matrix $B_d = \begin{bmatrix} 1 & 0 & 1 \end{bmatrix}^T$ is needed for which we can make a small modification to the MATLAB program* **simu4IC.m**. *We can change the maximum and minimum values of the control signal to observe variations of the closed-loop response.*

The next tutorial is to illustrate how to implement the state feedback control algorithm with integral action in real-time by producing a Simulink function, where anti-windup mechanism is incorporated for control signal constraints.

Tutorial 2.4 *The core of this activity is to produce a MATLAB embedded function that can be used in a Simulink simulation as well as in xPC Target implementation. This embedded function is based on the control algorithm discussed in this section. The entire embedded MATLAB function completes one cycle of computation for the control signal. For every sampling period, it will repeat the same computation procedure.*

Step by Step

1. *Create a new Simulink file called sfcI.slx*
2. *In Simulink's directory of User-Defined Functions, find the icon of the embedded MATLAB function and copy it to the sfcI model.*
3. *Click on the icon of the embedded function, and define the input and output variables and the controller parameters to the model so that the embedded function has the following form:*

```
function u=sfcI(r,y,Ap,Bp,Cp,Deltat,K,Kob,Umin,Umax)
```

where u is the calculated control signal at the sampling time t_k, the first two elements (r and y) among the input variables are the reference signal and the output signal at sampling time t_k, A_p,B_p and C_p are the system matrices from the plant model, Deltat is the sampling interval, K is the controller gain and K_{ob} is the observer gain, and Umin and Umax are the upper and lower limits imposed on the control signal.

4. *We need to edit the input and output data ports in order to let the embedded function know which input ports are the real-time variables and which are the parameters. This editing task is performed using Model Explorer.*
   - *click on 'r', on Scope, select 'input', assign port '1' and size '-1', complexity 'Inherited', type 'Inherit: Same as Simulink'. Repeat the same editing procedure for the output signal 'y'.*
   - *The rest of the inputs to the embedded function are the parameters required in the computation. Click on 'Ap', on Scope, select 'Parameter' and click 'Tunable' and click 'Apply' to save the changes. Repeat the same editing procedure for the rest of the parameters.*
   - *To edit the output port from the embedded function, click on 'u', on Scope, select 'Output', Port '1', Size '-1', Sampling Model 'Sample based', Type 'Inherit: Same as Simulink', and click on 'Apply' to save the changes.*
5. *In the following, the program will declare those variables that are stored in the embedded function during each iteration for their dimensions and initial values. 'uPast' is the past control signal ($u(t_{k-1})$), 'yPast' is the past output signal ($y(t_{k-1})$). For multi-input and multi-output systems, their dimensions need to be defined precisely. As an example, we consider a single input and single output system. Continue entering the following program into the file:*

```
persistent uPast
if isempty(uPast)
    uPast=0;
end
persistent yPast
if isempty(yPast)
   yPast=0;
end
```

6. *The estimated state variables will also be stored in the embedded function. We need to precisely define the dimension of $\hat{x}(t_k)$ for the embedded function. As an example, we choose the number of estimated state variables to be 3. Continue entering the following program into the file:*

```
persistent xhat
if isempty(xhat)
    xhat=zeros(3,1);
end
```

7. *After defining the variables to be stored, the calculation begins for the control signals at the sampling time t_k. The first step is to compute the derivatives of the control signal and the output signal. Continue entering the following program into the file:*

```
udot=-K*[xhat;y-r];
yd=(y-yPast)/Deltat;
```

8. *Construct the control signal using the past control signal value. Continue entering the following program into the file:*

```
u=uPast+udot*Deltat;
```

9. *If the control signal exceeds its maximum or minimum values, impose constraints on the amplitude of the control signal and update the derivative of the control signal with the saturation information. Enter the following program into the file:*

```
if (u>Umax) u=Umax; udot=(u-uPast)/Deltat; end
if (u<Umin) u=Umin; udot=(u-uPast)/Deltat; end
```

10. *Calculate the estimated state variables for the next sampling period* t_{k+1}. *Continue entering the following program into the file:*

```
xhat=xhat+(Ap*xhat+Kob*(yd-Cp*xhat)+Bp*udot)*Deltat;
```

11. *Update uPast and yPast in preparation for the next sampling period and complete the embedded function. Continue entering the following program into the file:*

```
uPast=u;
yPast=y;
```

12. *Validate this embedded function by building a Simulink simulation program based on Example 2.2 and compare the simulation results. In the Simulink program, we may need to add a discrete-time delay to connect the plant output to y in this program.*

2.2.6 Application to Drum Boiler Control

Drum boiler is one of the processes in an electrical power generation plant, which presents an important control problem for nuclear and conventional power generations including using gas and coal. Boiler control problem has been studied in the control engineering community since the early 1960s (see Nicholson [1964]). Figure 2.3 shows a simplified schematic diagram of a drum boiler system (Åström and Bell [2000]). The heat Q is applied to the risers that are the tubes connecting the lower drum to the upper (steam) drum. As the water heats and gravity forces the saturated steam to rise, the risers carry water and steam to the upper drum where an outlet is present. This process causes a circulation in the riser-drum-downcomer loop. There are many nonlinear models

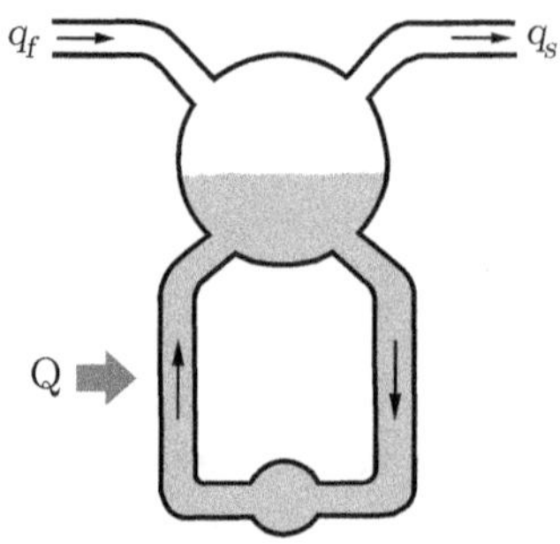

Figure 2.3 Schematic diagram of a drum boiler (Åström and Bell [2000]).

that have been developed for the drum boiler process (Åström and Bell [2000]). But the linear models for control system design are often obtained using system identification techniques with respect to specific operating conditions either based on experimental data or simulated data from a nonlinear model that has high fidelity. In Tan et al. [2002], an H^∞ controller was designed for a drum boiler. In this case study we will use the same transfer function model given in Tan et al. [2002] and design a state feedback controller with constrained control signal implementation.

For a drum boiler control system, it has three inputs and three outputs. The three input signals are feedwater flow rate (u_1 (kg/s)), fuel flow rate (u_2 (kg/s)) and attemperator spray flow rate (u_3 (kg/s)). Here attemperators are used to control steam temperature from the boiler. The three outputs are drum water level (y_1 (m)), drum pressure (y_2 (MPa)) and steam temperature (y_3 (^oC)).

The model used here is obtained from simulation data (see Tan et al. [2002]) based on a nonlinear simulator. The operating conditions for obtaining the transfer function models are

$$\begin{bmatrix} u_{10} \\ u_{20} \\ u_{30} \end{bmatrix} = \begin{bmatrix} 40.68 \\ 2.102 \\ 0 \end{bmatrix} kg/s; \begin{bmatrix} y_{10} \\ y_{20} \\ y_{30} \end{bmatrix} = \begin{bmatrix} 1.0(m) \\ 6.45(MPa) \\ 466.7(^oC) \end{bmatrix}. \tag{2.29}$$

With respect to the operating conditions, the following transfer function model is obtained:

$$G(s) = \begin{bmatrix} G_{11}(s) & G_{12}(s) & G_{13}(s) \\ G_{21}(s) & G_{22}(s) & G_{23}(s) \\ G_{31}(s) & G_{32}(s) & G_{33}(s) \end{bmatrix} \tag{2.30}$$

where

$$G_{11}(s) = \frac{10^{-3}(-0.16s^2 + 0.052s + 0.0014)}{s^2 + 0.0168s}, \; G_{12}(s) = \frac{10^{-3}(3.1s - 0.032)}{s^2 + 0.0215s},$$

$$G_{13}(s) = 0, \; G_{21}(s) = \frac{-10^{-3} \times 0.0395}{s + 0.018}, \; G_{22}(s) = \frac{10^{-3} \times 2.51}{s + 0.0157},$$

$$G_{23} = \frac{10^{-3}(0.588s^2 + 0.2015s + 0.0009)}{s^2 + 0.0352s + 0.000142},$$

$$G_{31}(s) = \frac{-0.00118s + 0.000139}{s^2 + 0.01852s + 0.000091},$$

$$G_{32}(s) = \frac{0.448s + 0.0011}{s^2 + 0.0127s + 0.000095}, \quad G_{33}(s) = \frac{0.582s - 0.0243}{s^2 + 0.1076s + 0.00104}.$$

The transfer function $G_{11}(s)$ that governs the relationship between the feedwater flow rate and the drum level is a proper transfer function with a second order polynomial for both numerator and denominator. This means that for any input flow rate change there is an immediate change on the drum water level, which is not realistic for the physical system. In order to reflect the response of the physical system, we add a small time constant to this relationship that could be regarded as actuator dynamics, which gives

$$G_{11}(s) = \frac{10^{-3}(-0.16s^2 + 0.052s + 0.0014)}{(s^2 + 0.0168s)(0.1s + 1)}.$$

The same procedure is applied to $G_{23}(s)$ that governs the relationship between attemperator spray flow rate and drum pressure, leading to

$$G_{23}(s) = \frac{10^{-3}(0.588s^2 + 0.2015s + 0.0009)}{(s^2 + 0.0352s + 0.000142)(0.1s + 1)}.$$

The overall control objective is to maintain the steady-state operations of the system for plant safety and achieving desired efficiency. More specifically, the drum water level y_1 must be maintained at a desired constant level to prevent drum components from overheating or flooding with respect to fluctuations in feedwater flow rate; the drum pressure y_2 must be maintained at a constant desired level for the boiler operation with respect to variations of steam demand; the steam temperature must be maintained at a desired level to prevent the heaters from overheating and wet steam from entering turbines subject to variations in fuel flow rate. In order to achieve the control objectives, the feedback control system must have integral action to maintain the constant steady-state operating conditions for all three outputs and to have the capacity to change their operating conditions when required. The fluctuations in feedwater flow rate, the variations of steam demand and the variations of fuel flow rate are process disturbances in low frequency, which will also be overcome by the integral action in the feedback controller.

The control signals have the following operational constraints:

$$0 \le u_1 \le 120; \; 0 \le u_2 \le 7; \; 0 \le u_3 \le 10. \tag{2.31}$$

For the control system design, we perform the following computations.

1. To convert the transfer function model into a state space model, we will use the MATLAB function **tf.m** to produce a three input and three output transfer

functions and obtain a minimal state space realization using MATLAB function **ss.m**. The state space model (A_p, B_p, C_p, D_p) that will be used for the control system design and simulation is produced with the MATLAB function **ssdata.m**. Because of the added dynamics to describe the actuators, D_p matrix is zero, which is important for the control system simulation to avoid an algebraic loop.

2. To design the state feedback controller with integral action, we use the MATLAB function **aug4IC.m** produced in Tutorial 2.1 to produce the augmented state space model A, B, C. We choose $Q = C^T C$ and $R = 0.1I$ to produce the state feedback controller gain K using MATLAB function **lqr.m**.
3. To design the observer, we use the state space model (A_p^T, C_p^T) where Q_{ob} is identity matrix and $R_{ob} = 0.1I$ and compute the observer gain K_{ob}^T with MATLAB **lqr.m**.

The next step in the case study is to simulate the closed-loop responses for disturbance rejection and reference tracking. For simulation studies, we will use the MATLAB program **simu4IC.m** produced in Tutorial 2.2.

Maintaining steady-state operation in presence of disturbance.
The steady-state condition in the simulation study is dictated by the operating conditions (2.29) that were used to obtain the transfer function model. Namely, the drum water level y_1 is maintained at 1 (m), drum pressure y_2 is maintained at 6.45 (MPa) and steam temperature y_3 is maintained at 466.7 ^oC. To reflect the fluctuations in feedwater flow rate, we add step disturbances to the control signal u_1 with amplitude variations between $\pm 5\%$ of the steady-state of the control signal u_{10} (see top figure in Figure 2.4). A sinusoidal disturbance is added to the control signal u_2 to reflect the variations in fuel flow rate (see middle figure in Figure 2.4),

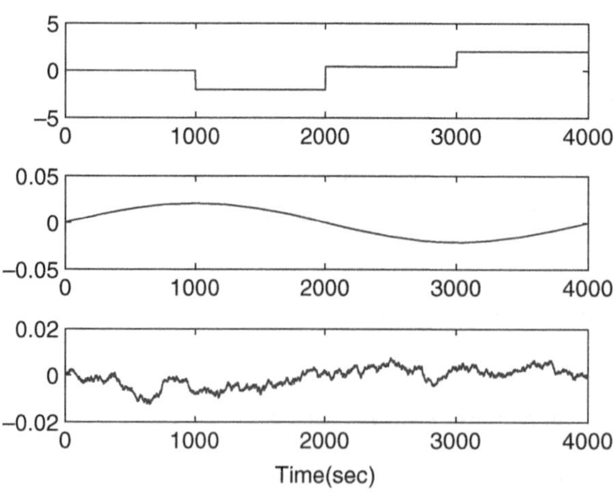

Figure 2.4 Input and output disturbances. Top figure: input disturbance on the feedwater flow rate; middle figure: input disturbance on fuel flow rate; bottom figure: output disturbance on drum pressure.

which has the magnitude of 10% of the steady-state control signal u_{20}. An output disturbance with low frequency variations is added to the output y_2 to reflect the variations of steam demand (see bottom figure in Figure 2.4). In the simulation study, the following control signal constraints are imposed:

$$-40.68 \leq u_1 \leq 79.32; -2.102 \leq u_2 \leq 4.898; \; 0 \leq u_3 \leq 10.$$

The constraints are calculated by subtracting the plant operational constraints given in (2.31) from the operational conditions of the control signals. Using the MATLAB program **simu4IC.m** produced in Tutorial 2.2 and adding the disturbance signals and zero reference signals, in the presence of operational constraints, the closed-loop simulation results for disturbance rejection are obtained. Upon completion of the simulation, the steady-state values of the outputs and control signals are added to reflect the plant operation in closed-loop control for disturbance rejection. The comments for the simulation results follow.

1. The drum level is shown together with ±0.8% bound in the top plot of Figure 2.5, the drum pressure is shown in the middle plot with ±0.2% bound and the steam temperature is shown in the bottom plot with ±0.2% bound. It is seen from these plots that all three outputs are tightly maintained at the steady-state operations with respect to both input and output disturbances.
2. The closed-loop control signals are shown in Figure 2.6, where top figure shows the water flow rate, middle figure the fuel flow rate and bottom figure attemperator spray flow rate. It is seen from these figures that all upper bounds on the control signals are satisfied, however, the attemperator spray flow rate is saturated at the lower limit from time to time.

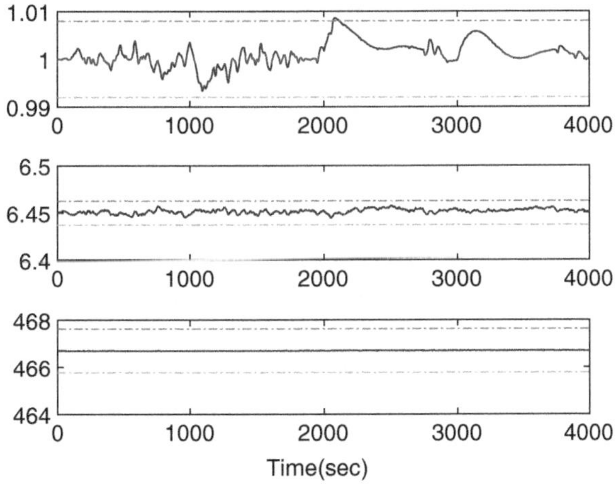

Figure 2.5 Output signal in presence of disturbance. Top figure: drum level (m) with ±0.8% bound; middle figure: drum pressure (MPa) with ±0.2% bound; bottom figure: steam temperature (^oC) with ±0.2% bound.

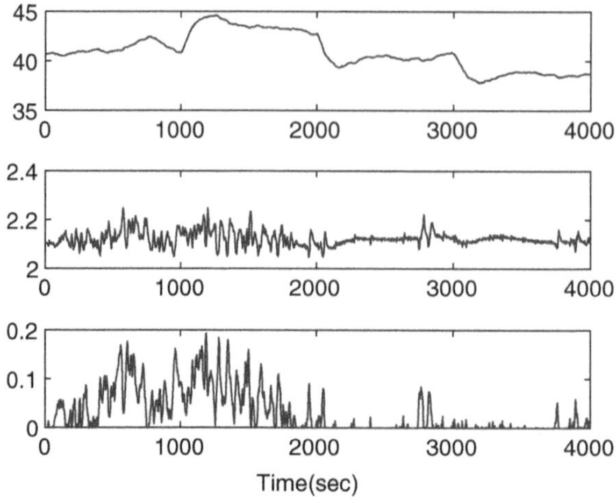

Figure 2.6 Control signal for disturbance rejection. Top figure: feedwater flow rate (kg/s); middle figure: fuel flow rate (kg/s); bottom figure: attemperator spray flow rate (kg/s).

Changing operational conditions in presence of disturbance

In Tan et al. [2002], changing operational conditions is specified by 10% increasing of drum level, 5% increasing of drum pressure and 10% decreasing of steam temperature. We will use the same step signals in the closed-loop control simulation while adding the same disturbance signals as before. The control signal constraints are also maintained in the closed-loop simulation. We begin the simulation with the 10% increase of the drum level while maintaining the same operating conditions for the drum pressure and steam temperature. At the half of the simulation time, we increase the drum pressure by 5% and decrease the steam temperature by 10%. There are a few comments for the simulation results.

1. Figure 2.7 shows the closed-loop output responses for the reference change in the presence of disturbances. It is seen that all three outputs successfully track the step reference signals without steady-state errors. In Åström and Bell [2000], it stated that one of the difficulties to control is the existence of complicated shrink and swell dynamics for the boiler drum system and such dynamics create non-minimum phase behavior. Indeed, this non-minimum phase behavior is seen in the closed-loop responses of drum level (top figure) and drum pressure (middle figure) when the reference changes for drum pressure and steam temperature are applied at the half of the simulation time.
2. Figure 2.8 shows the three closed-loop control signals for the reference tracking and disturbance rejection. It is seen that the upper limits of the control signals are all satisfied with significant margins, however, the fuel flow rate u_2 has reached its lower limit during the operating conditions change at the half of the simulation time and attemperator spray flow rate u_3 has has reached its lower limit during the first half of the operation.

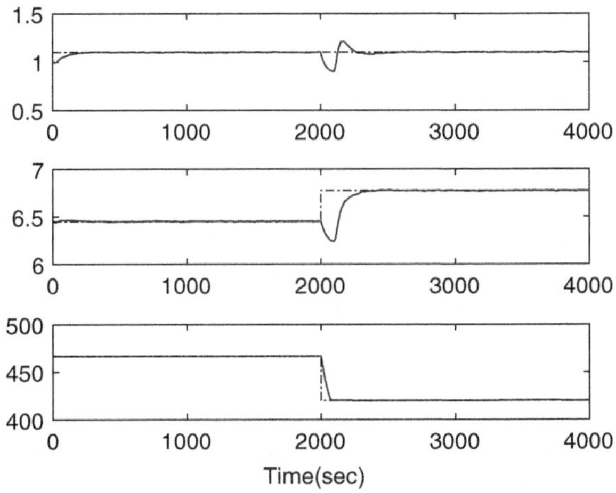

Figure 2.7 Output signal (solid line) and reference signal (dashed-line). Top figure: drum level (*m*); middle figure: drum pressure (*MPa*); bottom figure: steam temperature (^oC).

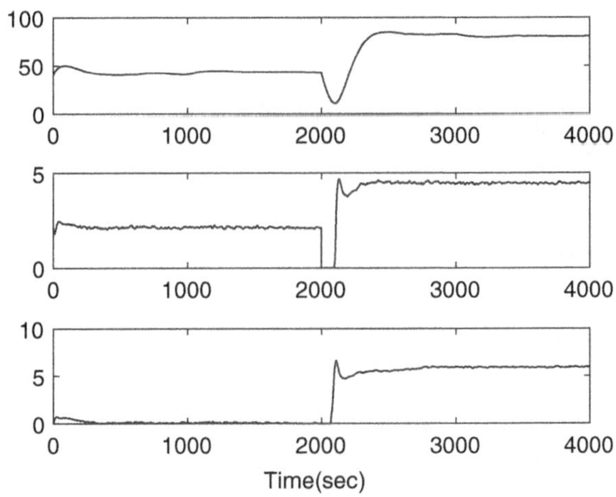

Figure 2.8 Control signal. Top figure: feedwater flow rate (*kg*/*s*); middle figure: fuel flow rate (*kg*/*s*); bottom figure: attemperator spray flow rate (*kg*/*s*).

In comparison to the original results obtained using H^∞ controller and PID controller presented in Tan et al. [2002], it is reasonable to conclude that the closed-loop control performance for the drum boiler is significantly improved because the output responses are not oscillatory, yet with faster response speed. Also, the design and implementation are not complicated in the state feedback framework. Finally the plant operational constraints are imposed in the implementation with anti-windup mechanisms.

2.2.7 Food for Thought

1. Is it correct to say that the integrator windup events only occur when the control signal amplitude reaches its limit?

2. Is it possible to avoid the integrator windup problem by carefully selecting the amplitude of the reference signal $r(t)$? How about a disturbance signal?
3. If you use LQR design for the controller, what would be the choice of Q matrix if the objective function is selected as

$$J = \int_0^\infty (y(t)^T y(t) + \dot{u}(t)^T R\dot{u}(t))dt$$

4. What is the closed-loop system for the modified approach? Is the closed-loop system identical to that from the original state feedback control system presented in Section 2.2? If not, what is the difference?
5. In the anti-windup implementation, is it correct to say that when the control saturation is reached, it is essential to recalculate the derivative $\dot{u}(t)$ and estimate the states with the correct $\dot{u}(t)$ that contains the saturation information?
6. Is it correct that when the saturation is maintained for two consecutive samples $\dot{u}(t) = 0$ and the observer becomes

$$\ddot{\hat{x}}_p(t) = A_p\dot{\hat{x}}_p(t) + K_{ob}(\dot{y}(t) - C_p\dot{\hat{x}}_p(t)) \ ?$$

2.3 Practical Controller II: Integral Action via Observer Design

Another classical approach to obtain integral action is through disturbance estimation where the disturbance is assumed to be a constant.

2.3.1 Integral Control via Disturbance Estimation

In this approach, we assume the existence of a constant input disturbance $d(t)$. With this assumption, a state space model is described by the following differential equations,

$$\dot{x}_p(t) = A_p x_p(t) + B_p(u(t) + d(t)) \tag{2.32}$$

$$y(t) = C_p x_p(t), \tag{2.33}$$

where state vector $x_p(t)$ has the dimension $n \times 1$, both input $u(t)$ and output $y(t)$ have the same dimension $m \times 1$. Additionally, the state vector $x_p(t)$ and the disturbance $d(t)$ are not measured. In order to estimate them, the state space model is augmented with the constant input disturbance as an additional state vector, leading to

$$\begin{bmatrix} \dot{x}_p(t) \\ \dot{d}(t) \end{bmatrix} = \overbrace{\begin{bmatrix} A_p & B_p \\ 0_{m\times n} & 0_{m\times m} \end{bmatrix}}^{A} \begin{bmatrix} x_p(t) \\ d(t) \end{bmatrix} + \overbrace{\begin{bmatrix} B_p \\ 0_{m\times m} \end{bmatrix}}^{B} u(t) \tag{2.34}$$

$$y(t) = \overbrace{\begin{bmatrix} C_p & 0_{m\times m} \end{bmatrix}}^{C} \begin{bmatrix} x_p(t) \\ d(t) \end{bmatrix} \tag{2.35}$$

where n and m denote the number of states and number of inputs and outputs and the augmented system matrices $A,\ B,\ C$ are illustrated in the equations. It can be shown that this augmented model is not controllable, however, it is observable if the original pair of system matrices $A_p,\ C_p$ is observable and if the plant transfer function $C_p(sI - A_p)^{-1}B_p$ does not have a zero on the origin of the complex plane.

By assuming that the augmented system is observable, an observer can be designed to estimate both state vector $x_p(t)$ and the disturbance $d(t)$ based on (2.34) and (2.35). With the observer, the estimated $\hat{x}_p(t)$ and $\hat{d}(t)$ are expressed in the following form:

$$\begin{bmatrix} \dot{\hat{x}}_p(t) \\ \dot{\hat{d}}(t) \end{bmatrix} = \overbrace{\begin{bmatrix} A_p & B_p \\ 0_{m\times n} & 0_{m\times m} \end{bmatrix}}^{A} \begin{bmatrix} \hat{x}_p(t) \\ \hat{d}(t) \end{bmatrix} + \overbrace{\begin{bmatrix} B_p \\ 0_{m\times m} \end{bmatrix}}^{B} u(t) + K_{ob}\left(y(t) - \overbrace{\begin{bmatrix} C_p & 0_{m\times m} \end{bmatrix}}^{C} \begin{bmatrix} \hat{x}_p(t) \\ \hat{d}(t) \end{bmatrix} \right), \tag{2.36}$$

where the observer gain K_{ob} is selected based on the A and C matrices such that the observer error system matrix $(A - K_{ob}C)$ is stable with satisfactory dynamic performance.

The feedback controller K_p is designed using the pair $A_p,\ B_p$ matrices such that the closed-loop system matrix $A_p - B_pK_p$ is stable and has all eigenvalues at desired locations. The state estimate feedback control law uses both estimated state vector $\hat{x}_p(t)$ and the input disturbance $\hat{d}(t)$. More specifically, the control signal $u(t)$ is computed as

$$u(t) = -K_p\hat{x}_p(t) - \hat{d}(t). \tag{2.37}$$

We can use either pole-assignment controller method or the LQR design method to find the observer gain K_{ob} and the controller K_p (see Chapter 1).

The next important question is where the closed-loop eigenvalues are. To answer this question, first we substitute the control signal (2.37) into the plant model (2.32), leading to

$$\dot{x}_p(t) = (A_p - B_pK_p)x_p(t) + B_pK_p\tilde{x}_p(t) + B_p\tilde{d}(t), \tag{2.38}$$

where $\tilde{x}_p(t) = x_p(t) - \hat{x}_p(t)$ and $\tilde{d}(t) = d(t) - \hat{d}(t)$. Second, subtracting (2.36) from (2.34) gives the observer error system:

$$\dot{\tilde{x}}(t) = (A - K_{ob}C)\tilde{x}(t), \tag{2.39}$$

where $\tilde{x}(t) = \left[\tilde{x}_p(t)^T \;\; \tilde{d}(t)^T\right]^T$. Third combining (2.38) with (2.39) gives the closed-loop system for the state estimate feedback control:

$$\begin{bmatrix} \dot{x}_p(t) \\ \dot{\tilde{x}}(t) \end{bmatrix} = \begin{bmatrix} A_p - B_pK_p & \overline{B_pK_p} \\ 0_{(n+m)\times n} & A - K_{ob}C \end{bmatrix} \begin{bmatrix} x_p(t) \\ \tilde{x}(t) \end{bmatrix}, \tag{2.40}$$

where $\overline{B_pK_p} = \begin{bmatrix} B_pK_p & B_p \end{bmatrix}$. Note that the system matrix in (2.40) has a block upper triangular structure. To determine the eigenvalues of the closed-loop system (2.40), the following two characteristic equations are examined:

$$\det(sI_{n\times n} - A_p + B_pK_p) = 0 \tag{2.41}$$

$$\det(sI_{(n+m)\times(n+m)} - A + K_{ob}C) = 0. \tag{2.42}$$

The solutions of polynomial equation (2.41) give the closed-loop eigenvalues of the state feedback control system and those of (2.42) lead to the closed-loop eigenvalues of the observer error system. The separation principle for regulator design in Section 1.7.2 of Chapter 1 holds for the disturbance estimation in state feedback control.

The feedback control law using (2.36) and (2.37) is for disturbance rejection. However, because the control law itself in (2.37) does not involve the output signal $y(t)$, in order to include the reference signal $r(t)$, we will modify the observer equation by subtracting the reference signal $r(t)$ from the output $y(t)$. Thus, if $r(t) \neq 0$, the state observer is written as

$$\begin{bmatrix} \dot{\hat{x}}_p(t) \\ \dot{\hat{d}}(t) \end{bmatrix} = \overbrace{\begin{bmatrix} A_p & B_p \\ 0_{m\times n} & 0_{m\times m} \end{bmatrix}}^{A} \begin{bmatrix} \hat{x}_p(t) \\ \hat{d}(t) \end{bmatrix} + \overbrace{\begin{bmatrix} B_p \\ 0_{m\times m} \end{bmatrix}}^{B} u(t)$$
$$+ K_{ob}\left(y(t) - r(t) - \overbrace{\begin{bmatrix} C_p & 0_{m\times m} \end{bmatrix}}^{C} \begin{bmatrix} \hat{x}_p(t) \\ \hat{d}(t) \end{bmatrix} \right). \tag{2.43}$$

How does this state estimate feedback control lead to a control system having integral action? Let us express the observer gain K_{ob} as K_{ob}^p and K_{ob}^d where K_{ob}^p and K_{ob}^d correspond to the observer gain for the state vector and the disturbance vector, respectively. Then, from (2.43), it is found that

$$\dot{\hat{d}}(t) = K_{ob}^d(y(t) - r(t) - C_p\hat{x}_p(t))$$

which gives

$$\hat{d}(t) = K_{ob}^{d} \int_{0}^{t} (y(\tau) - r(\tau) - C_p \hat{x}_p(\tau))d\tau. \tag{2.44}$$

Substituting (2.44) into the control law (2.37) reveals that

$$u(t) = -K_p \hat{x}_p(t) + K_{ob}^{d} C_p \int_{0}^{t} \hat{x}_p(\tau)d\tau + K_{ob}^{d} \int_{0}^{t} (r(\tau) - y(\tau))d\tau. \tag{2.45}$$

The integral action is clearly seen to have taken place on the error signal between the reference and the output as well as the estimated state variables.

2.3.2 Anti-windup Mechanism

One of the advantages of using the state estimate feedback control law given by (2.37) is that it has a built-in anti-windup mechanism. This is because the integral action comes from the estimation of the constant input disturbance $\hat{d}(t)$, and as a result, the controller structure does not explicitly contain an integrator. Additionally, the estimated constant disturbance $\hat{d}(t)$ could also be regarded as the steady-state values of the control signal, and $u(t)$ could be the actual control signal for implementation without consideration of its steady-state value.

Suppose that the control signal $u(t_k)$ is constrained between u^{min} and u^{max}, where u^{min} and u^{max} contain the saturation limits for all input signals. We also assume that the initial conditions of $\hat{x}_p(t_0)$ and $\hat{d}(t_0)$ are given as the starting point. The implementation of the state estimate feedback control system in the presence of control signal constraints is summarized as below.

1. Calculate the feedback control signal at the sampling time t_k:

$$u(t_k) = -K_p \hat{x}_p(t_k) - \hat{d}(t_k). \tag{2.46}$$

2. Impose the saturation limits on the control signal with the following computations.
   For the ith control signal

$$u_i(t_k)^{act} = \begin{cases} u_i^{min}, & \text{if } u_i(t_k) < u_i^{min}; \\ u_i(t_k), & \text{if } u_i^{min} \le u_i(t_k) \le u_i^{max}; \\ u_i^{max}, & \text{if } u_i(t_k) > u_i^{max}, \end{cases}$$

   where $i = 1, 2, \ldots, m$.
3. Compute the estimated state vector $\hat{x}(t_{k+1})$ with the actual control signal $u(t_k)^{act}$:

$$\hat{x}(t_{k+1}) = \hat{x}(t_k) + \left(A\hat{x}(t_k) + Bu(t_k)^{act} + K_{ob}(y(t_k) - r(t_k) - C\hat{x}(t_k))\right) \Delta t.$$

4. As the sampling time progresses to t_{k+1}, repeat the computation from the first step.

Note that the last m components in the estimated states $\hat{x}(t_k)$ correspond to the estimated $\hat{d}(t_k)$, where m is the number of inputs.

2.3.3 MATLAB Tutorial on Design and Implementation

This section will present three MATLAB/Simulink tutorials associated with the state feedback control with disturbance estimation.

Tutorial 2.5 *In this tutorial, we will write a MATLAB function* **aug4DOB.m** *that will give the augmented model for the disturbance-observer-based control system design.*

Step by Step

1. *Create a new file called* **aug4DOB.m**.
2. *The input variables to the function are the plant system matrices:* A_p, B_p, C_p. *The output variables from this function are the augmented system matrices A,B and C. Enter the following program into the file:*

```
function [A,B,C]=aug4DOB(Ap,Bp,Cp)
```

3. *Check the numbers of inputs, outputs and state variables. Continue entering the following program into the file:*

```
[m1,n1]=size(Cp);
[n1,n_in]=size(Bp);
```

4. *Initialize the system matrices by using zero matrices. Continue entering the following program into the file:*

```
n=n1+n_in;
A=zeros(n,n);
B=zeros(n,n_in);
C=zeros(m1,n);
```

5. *Create the block matrices* A_p *and* B_p *in A. Continue entering the following program into the file:*

```
A(1:n1,1:n1)=Ap;
A(1:n1,n1+1:n1+n_in)=Bp;
```

6. *Create the block matrix* B_p *in B. Continue entering the following program into the file:*

```
B(1:n1,:)=Bp;
```

7. *Create the block matrix C_p in C matrix. Continue entering the following program into the file:*

```
C(:,1:n1)=Cp;
```

8. *Test the function with the following system with two inputs and two outputs and two states:*

```
Ap=[1 -1; 2 1];
Bp=[1 2;2 1];
Cp=[1 0; 0 1];
```

9. *Check A matrix against the following answer:*

$$A = \begin{bmatrix} 1 & -1 & 1 & 2 \\ 2 & 1 & 2 & 1 \\ 0 & 0 & 0 & 0 \\ 0 & 0 & 0 & 0 \end{bmatrix}$$

The following MATLAB tutorial illustrates closed-loop simulation using the disturbance-observer-based controller.

Tutorial 2.6 *Write a MATLAB function that will implement the disturbance-observer-based feedback control law that has integral action and perform closed-loop simulation. Anti-windup mechanism is incorporated in the implementation. We assume that we have the continuous-time system model (A_p, B_p, C_p) and have designed the state feedback controller K and the observer K_{ob}.*

Step by Step

1. *Create a new file called* **simu4DOB.m**.
2. *The input variables to the function are the plant system matrices A_p, B_p, C_p, the augmented system matrices A, B, and C, the number of samples for the simulation (Nsim) with sampling interval Deltat, controller K, and observer gain K_{ob}, reference signal sp and disturbance signal dis. The dimension of sp is $m_1 \times Nsim$ and the dimension of dis is $n_{in} \times Nsim$ where m_1 is the number of outputs and n_{in} is the number of inputs. Umax and Umin are the vectors containing the limits for the maximum and minimum values of each component in the control signal. The output variables from this function are output y_1 and control signal u_1. Enter the following program into the file:*

```
function [y1,u1]=simu4DOB(Ap,Bp,Cp,A,B,C,Deltat,Nsim,
K,Kob,sp,dis,Umin,Umax)
```

3. *Check number of states, number of outputs and number of inputs. Continue entering the following program into the file:*

```
[m1,n1]=size(Cp);
[n,n_in]=size(B);
```

4. *Set initial conditions for the controller implementation and simulation. All initial conditions are chosen to be zero for simplicity. Continue entering the following program into the file:*

```
y=zeros(m1,1);
xhat=zeros(n,1);
x=zeros(n1,1);
```

5. *Perform the simulation iteratively, which corresponds to a real-time control system. Continue entering the following program into the file:*

```
for kk=1:Nsim;
```

6. *At the sampling instant kk that corresponds to time t_k, calculate the control signal using the estimated state and the disturbance. Continue entering the following program into the file:*

```
u=-K*xhat(1:n1,1)-xhat(n1+1:n,1);
```

7. *Impose constraints on the control signal. Continue entering the following program into the file:*

```
for i=1:n_in
if (u(i,1)>Umax(i,1)) u(i,1)=Umax(i,1); end
if (u(i,1)<Umin(i,1)) u(i,1)=Umin(i,1); end
end
```

8. *With the measured output y and the reference signal sp(kk) as well as the estimated state variable xhat, we update the estimated state variable using the control signal that carries the information of saturation. Continue entering the following program into the file:*

```
xhat=xhat+(A*xhat+Kob*(y-sp(:,kk)-C*xhat)+B*u)*Deltat;
```

9. *The above steps 6-8 capture the implementation of the disturbance-observer-based control system. We save the control signal and output signal from this function. Continue entering the following program into the file:*

```
y1(:,kk)=y;
u1(:,kk)=u;
```

10. *Since we are in a simulation environment, the control signal calculated is sent to the plant model for closed-loop simulation. In the closed-loop simulation, we add the input disturbance to the system. If the input disturbance matrix is known to be B_d, then we can add this information to the closed-loop simulation. This completes one-cycle of computation in the closed-loop simulation. Continue entering the following program into the file:*

```
x=x+(Ap*x+Bp*u+Bp*dis(:,kk))*Deltat;
y=Cp*x;
end
```

The following tutorial illustrates the closed-loop simulation using Simulink.

Tutorial 2.7 *This tutorial is to illustrate how to implement the disturbance-observer-based control algorithm in real-time simulation, where anti-windup mechanism is incorporated for control signal constraints. The core of this activity is to produce a MATLAB embedded function that can be used in a Simulink simulation as well as in xPC Target implementation. This embedded function is based on the control algorithm discussed in this section. The entire embedded MATLAB function completes one cycle of computation for the control signal. For every sampling period, it will repeat the same computation procedure.*

Step by Step

1. *Create a new Simulink file called sfcD.slx*
2. *In Simulink's directory of User-Defined Functions, find the icon of the embedded MATLAB function and copy it to the sfcD model.*
3. *Click on the icon of the embedded function, and define the input and output variables and the controller parameters to the model so that the embedded function has the following form:*

```
function u=sfcD(r,y,A,B,C,deltat,K,Kob,Umin,Umax)
```

   where u is the calculated control signal at the sampling time t_k, the first two elements (r and y) among the input variables are the reference signal and the output signal at sampling time t_k, A,B and C are the augmented system matrices that are used in the state estimation, deltat is the sampling interval, K is the controller gain and K_{ob} is the observer gain, and Umin and Umax are the upper and lower limits imposed on the control signal.
4. *We need to edit the input and output data ports in order to let the embedded function know which input ports are the real-time variables and which are the parameters. This editing task is performed using Model Explorer.*

- *click on 'r', on Scope, select 'input', assign port '1' and size '-1', complexity 'Inherited', type 'Inherit: Same as Simulink'. Repeat the same editing procedure for the output signal 'y'.*
- *The rest of the inputs to the embedded function are the parameters required in the computation. Click on 'A', on Scope, select 'Parameter' and click 'Tunable' and click 'Apply' to save the changes. Repeat the same editing procedure for the rest of the parameters.*
- *To edit the output port from the embedded function, click on 'u', on Scope, select 'Output', Port '1', Size '-1', Sampling Model 'Sample based', Type 'Inherit: Same as Simulink', and click on 'Apply' to save the changes.*

5. *In the following, the program will declare the variables stored in the embedded function during each iteration for their dimensions and initial values. For this control algorithm, the estimated state variables will be stored in the embedded function. We need to precisely define the dimension of $\hat{x}(t_k)$ for the embedded function. As an example, we consider a single input and single output system with three state variables, which gives the dimension of $\hat{x}$ to be 4×1. Continue entering the following program into the file:*

```
persistent xhat
if isempty(xhat)
xhat=zeros(4,1);
end
```

6. *After defining the variables to be stored, the calculation begins for the control signals at the sampling time t_k. The first three variables in $\hat{x}$ correspond to the estimated state variables and the final variable corresponds to the disturbance. Continue entering the following program into the file:*

```
u=-K*xhat(1:3,1)-xhat(4,1);
```

7. *If the control signal exceeds its maximum or minimum values, impose constraints on the amplitude of the control signal. Enter the following program into the file:*

```
if u<Umin;u=Umin;end
if u>Umax;u=Umax;end
```

8. *Calculate the estimated state variables for the next sampling period t_{k+1} and complete the embedded function. Continue entering the following program into the file:*

```
xhat=xhat+(A*xhat+Kob*(y-r-C*xhat)+B*u)*deltat;
```

9. *Validate this embedded function by building a Simulink simulation program based on Example 2.3 and compare the simulation results. In the Simulink*

program, we may need to add a discrete-time delay to connect the plant output to y in this program.

To demonstrate this design indeed has embedded the anti-windup mechanism, we examine the following the example.

Example 2.3
We consider a system with the following transfer function

$$G(s) = \frac{(s-3)(s+5)}{s(s+1)(s+3)}.$$

Design a disturbance-observer-based control system. All the observer poles are selected as −10 and the controller poles are selected as −2. Compare the closed-loop control responses with and without the control limit $u(t) \geq -5$.

Solution
By positioning the closed-loop control poles at −2, the feedback controller gain is found to be

$$K_p = \begin{bmatrix} 2 & 9 & 8 \end{bmatrix}$$

The observer gain with the positioning of all the closed-loop poles at −10 is calculated as

$$K_{ob} = \begin{bmatrix} -104.9896 & -50.5382 & -16.1377 & -666.6667 \end{bmatrix}^T.$$

The closed-loop system is simulated using a constant reference signal and a unit step input disturbance entering the system $t = 100$ (sec). Figures 2.9 (a)-(b) show the closed-loop control simulation with and without the constraint $u(t) > -5$ imposed. The simulation results are discussed as follows.

1. *When the control system encounters this step disturbance, a large control signal was generated to overcome it. Here, the control signal amplitude reaches −36.54, then goes back to a positive value within 0.5 (sec) before settling down to its steady-state of −4. Without the limit on the control signal amplitude, the closed-loop output takes about 2.8 (sec) to come back to the original steady-state. Therefore, without limit on the control signal, the closed-loop system rejects the constant disturbance very quickly. However, the superior control performance demands a control implementation mechanism to allow a large control amplitude (−36.54) in addition to a fast response time 0.5 (sec).*
2. *In comparison, when the control implementation mechanism is restricted to have its amplitude to be greater or equal to −5, it is seen from Figures 2.9(a)-(b) that the large control amplitude is limited to −5, and the control signal holds on to this position for 4 (sec) before gradually returns to its steady state of −4. Because of the*

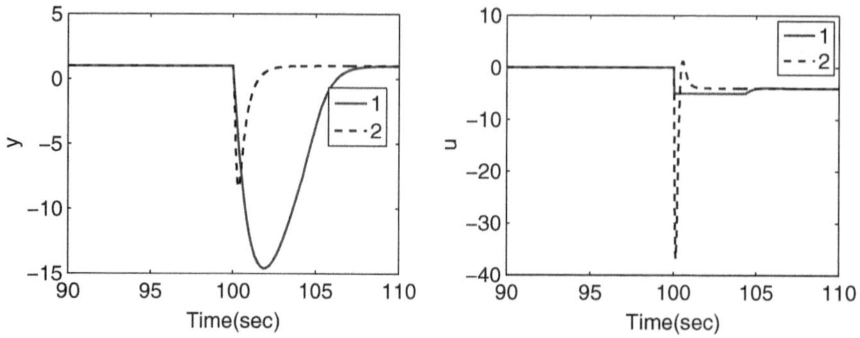

(a) Output (b) Control

Figure 2.9 Constrained state estimate feedback control for unit step response in the presence of input disturbance (Example 2.3). Key: line (1) closed-loop response in the presence of constraint; line (2) closed-loop response without constraint

limited control action, the closed-loop output takes about 7.2 (sec) to recover back to the original steady-state.

3. *It is emphasized that the approach using estimate of a constant input disturbance in observer to generate integral action has automatically embedded anti-windup mechanism. The action of this anti-windup mechanism is evident from the smooth response of the control signal and the non-oscillatory behavior of the output signal.*

2.3.4 Application to Sugar Mill Control

A sugar mill model was presented in Goodwin et al. [2000] for a case study of control system design. In the study, a single stage of a milling train was described by the following continuous-time transfer function model:

$$\begin{bmatrix} y_1(s) \\ y_2(s) \end{bmatrix} = \begin{bmatrix} g_{11}(s) & g_{12}(s) \\ g_{21}(s) & g_{22}(s) \end{bmatrix} \begin{bmatrix} u_1(s) \\ u_2(s) \end{bmatrix}, \tag{2.47}$$

where the output y_1 is the mill torque, and y_2 is the buffer chute height; the input u_1 is the flap position and input u_2 is the turbine speed reference signal. The four transfer functions are given as

$$g_{11}(s) = \frac{-5}{25s+1};\ g_{12}(s) = \frac{s^2 - 0.005s - 0.005}{s(s+1)(0.1s+1)};$$
$$g_{21}(s) = \frac{1}{25s+1};g_{22}(s) = \frac{-0.0023}{s}.$$

This process has significant multivariable interactions and nonminimum-phase behaviour. Also, the plant contains integrators. We choose to use MATLAB function **place.m** for both controller and observer design. For the controller design, we select the desired closed-loop poles as −0.5, −1, −1.5, −2, which gives the controller gain:

$$K_p = \begin{bmatrix} 11.0847 & 6.4527 & 2.7540 & 6.2585 \\ -1.8650 & -0.3979 & 0.0670 & -0.0410 \end{bmatrix}$$

The closed-loop observer poles for estimating disturbance are selected as −1, −1.5, −2, −2.5, −3, −3.5 and the corresponding observer gain is

$$K_{ob} = \begin{bmatrix} -0.0016 & -0.0087 \\ -0.2983 & 3.8855 \\ -0.0612 & -0.8220 \\ -0.0187 & 0.1964 \\ -0.0110 & 0.2876 \\ -0.1869 & 2.4296 \end{bmatrix} \times 10^3$$

We evaluate the closed-loop performance for disturbance rejection with sampling interval $\Delta t = 0.001$ seconds. The closed-loop simulations have been performed with input disturbances $d_1(t) = 0$ and $d_2(t) = 0.1$ for $0 \leq t \leq 14$(sec) and $d_1(t) = -0.1$ and $d_2(t) = 0$ for $14 < t \leq 26$(sec). Also, in the simulation studies, the control signals are constrained to satisfy $-10 \leq u_1(t) \leq 10$ and $-6 \leq u_2(t) \leq 6$. Figure 2.10 shows the closed-loop responses for the disturbance rejection. They have confirmed that the control signals are within the constraints and the step disturbances are rejected. One can follow the MATLAB Tutorials 2.5 and 2.6 to confirm the simulation results.

2.3.5 Design for Systems with Known States

In many applications, such as the control of electrical drives and power converters, the state variables are measured. It is unnecessary to estimate the state variables that are available through measurement. The approach with embedding integrator through state estimation is modified to accommodate these applications. We assume the full measurement of the state vector $x_p(t)$ and the system is described by the following state space model:

$$\dot{x}_p(t) = A_p x_p(t) + B_p(u(t) + d(t)) \tag{2.48}$$

$$y(t) = C_p x_p(t), \tag{2.49}$$

where $d(t)$ is the input disturbance with its dimension equal to that of the control signal $u(t)$. In order to produce integral action in the control system, the input disturbance vector is assumed to be constant so that its derivative satisfies:

$$\dot{d}(t) = 0_{m\times 1} \tag{2.50}$$

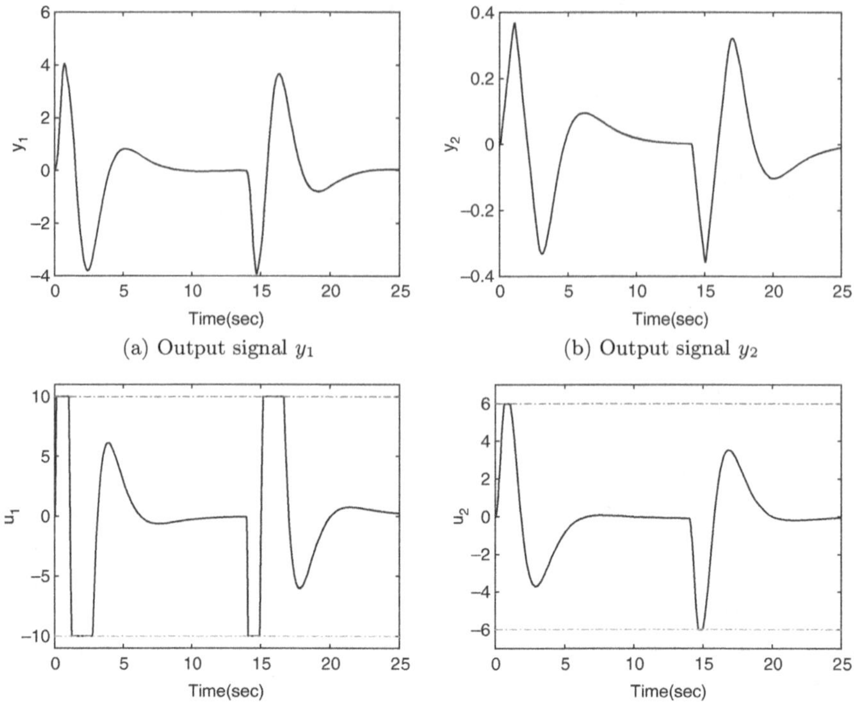

(a) Output signal y_1 (b) Output signal y_2

(c) Constrained control signal u_1 (d) Constrained control signal u_2

Figure 2.10 Disturbance rejection in sugar mill control.

$0_{m\times 1}$ is a zero vector. In the following, we will discuss how to estimate the disturbance vector $d(t)$, how to calculate the control signal $u(t)$, and how to implement the control system.

Estimation of disturbance

From (2.48), we obtain the input disturbance term:

$$B_p d(t) = \dot{x}_p(t) - A_p x_p(t) - B_p u(t). \tag{2.51}$$

Now, we will estimate the disturbance vector $d(t)$ using the state equation (2.50) and the output equation (2.51). To this end, we construct the estimate $\hat{d}(t)$ using the observer equation:

$$\begin{aligned}\dot{\hat{d}}(t) &= A_d\hat{d}(t) + K_{ob}(\dot{x}_p(t) - A_p x_p(t) - B_p u(t) - B_p\hat{d}(t))\\ &= K_{ob}(\dot{x}_p(t) - A_p x_p(t) - B_p u(t) - B_p\hat{d}(t)),\end{aligned} \tag{2.52}$$

where the system matrix A_d is a zero matrix with dimension $m \times m$, and m is the number of inputs. To find the observer gain K_{ob}, we assume that the pair of matrices (A_d, B_p) is observable. Then, K_{ob} is found to produce a stable observer error system with desired eigenvalues, which has the following form:

$$\dot{\tilde{d}}(t) = -K_{ob}B_p\tilde{d}(t), \tag{2.53}$$

where $\tilde{d}(t)$ is the error state defined as $\tilde{d}(t) = d(t) - \hat{d}(t)$. This error system is derived by subtracting (2.52) from (2.50), followed by the substitution of the output equation (2.51).

The observer equation (2.52) can not be implemented directly because it requires the derivative of the state vector $\dot{x}_p(t)$, which needs the information ahead of the current sampling time. Note that (2.52) can be re-written as

$$\dot{\hat{d}}(t) - K_{ob}\dot{x}_p(t) = -K_{ob}A_p x_p(t) - K_{ob}B_p u(t) - K_{ob}B_p\hat{d}(t) + K_{ob}B_p K_{ob}x_p(t) - K_{ob}B_p K_{ob}x_p(t). \tag{2.54}$$

Now, we choose the intermediate variable $\hat{z}(t)$ defined as

$$\hat{z}(t) = \hat{d}(t) - K_{ob}x_p(t).$$

The observer equation (2.54) is expressed in terms of the intermediate variable $\hat{z}(t)$:

$$\dot{\hat{z}}(t) = -K_{ob}B_p\hat{z}(t) - K_{ob}A_p x_p(t) - K_{ob}B_p u(t) - K_{ob}B_p K_{ob}x_p(t). \tag{2.55}$$

The righthand side of (2.55) is free of the derivative $x_p(t)$ and it can be implemented in discrete-time.

Assuming sampling interval Δt, and the first order approximation of the derivative $\dot{\hat{z}}(t)$, at sampling time t_k, based on current information given for $\hat{z}(t_k)$, $u(t_k)$ and $x_p(t_k)$, the one-step-ahead calculation of the estimated state $\hat{z}(t_{k+1})$ becomes

$$\hat{z}(t_{k+1}) = \hat{z}(t_k) - \Delta t K_{ob}(B_p\hat{z}(t_k) + A_p x_p(t_k) + B_p u(t_k) + B_p K_{ob}x_p(t_k)). \tag{2.56}$$

The estimation of $\hat{d}(t_k)$ is re-constructed as

$$\hat{d}(t_k) = \hat{z}(t_k) + K_{ob}x_p(t_k).$$

Calculation of the control signal

To design state feedback control with integral action, we consider the original state space equation (2.48). Letting the intermediate control signal $\tilde{u}(t) = u(t) + d(t)$, the state equation is re-written as

$$\dot{x}_p(t) = A_p x_p(t) + B_p\tilde{u}(t). \tag{2.57}$$

We design the state feedback controller K_p such that the closed- loop system

$$\dot{x}_p(t) = (A_p - B_p K_p)x_p(t)$$

is stable with satisfactory dynamic performance. Here the intermediate state feedback control signal is

$$\tilde{u}(t) = -K_p x_p(t).$$

Now, substituting the unknown disturbance $d(t)$ with the estimated $\hat{d}(t)$, the actual control signal is then calculated as

$$u(t) = \tilde{u}(t) - \hat{d}(t) = -K_p x_p(t) - \hat{d}(t). \tag{2.58}$$

Implementation of the control law

At the initial sampling time t_0, we assume that the initial vectors $\hat{z}(t_0)$ and $x_p(t_0)$ are given. At the sampling time t_k, given the current state measurement $x_p(t_k)$ and $\hat{z}(t_k)$, the implementation of the control law consists of the following steps:

1. Calculate the estimated disturbance $\hat{d}(t_k)$ by using,

$$\hat{d}(t_k) = \hat{z}(t_k) + K_{ob} x_p(t_k).$$

2. Calculate the state feedback control signal

$$u(t_k) = -K_p x_p(t_k) - \hat{d}(t_k).$$

3. Update the estimated intermediate variable $\hat{z}(t_{k+1})$:

$$\hat{z}(t_{k+1}) = \hat{z}(t_k) - \Delta t K_{ob}(B_p \hat{z}(t_k) + A_p x_p(t_k) + B_p u(t_k) + B_p K_{ob} x_p(t_k)).$$

4. When the next sampling time arrives, repeat Step 1.

The control law is derived in the context of disturbance rejection. The question arises as how the reference signals will enter the system. Bearing in mind that it is a state feedback control system, the state vector $x_p(t)$ is a deviation variable. Therefore, the state vector $x_p(t)$ has steady-state values for each of its components. The reference signals enter the state feedback control system via the steady-state values of the state variables. An example with MATLAB tutorial (see Tutorial 2.8) to demonstrate this approach is given for the wind turbine drive train control.

2.3.6 Food for Thought

1. If you knew that a disturbance input matrix B_d is different to B_p in (2.32), would you modify the system matrices $A,\ B,\ C$ to include this information?
2. Will the mismatch between the disturbance input matrix B_d and the system input matrix B_p affect the convergence of the estimated state vector $\hat{x}_p(t)$?

3. In the anti-windup implementation, is it essential for us to update the observer with the correct information of control signal saturation? If not, what might happen?
4. Is it possible to devise a control law by subtracting the disturbance only, which is

$$u(t) = -\hat{d}(t) \ ?$$

Would this control law be useful if the open-loop system is unstable?

2.4 Drive Train Control of a Wind Turbine

The application of state estimate feedback control with measured state variables is demonstrated with the control problem of wind turbine.

2.4.1 Modelling of Wind Turbine's Drive Train

A model for the drive train of a wind turbine is described by a two-mass system (Perdana [2008]) where two inertias are connected together through a spring that represents the low stiffness of the drive-train shaft. Damping is also used in the two-mass model (see Figure 2.11). This model has been used in power system stability analysis (Trudnowski et al. [2004]) and the validation of a two-mass model against an actual wind turbine is shown in Akhmatov et al. [2000]. More specifically, defining the angular velocities of the turbine and generator as ω_t and ω_r, and the angular positions as θ_t and θ_r, the mathematical model of a two-mass drive train is given as (Perdana [2008])

$$2H_t \frac{d\omega_t}{dt} = T_t - T_w - K_s(\theta_r - \theta_t) - D_m(\omega_r - \omega_t) \tag{2.59}$$

$$2H_g \frac{d\omega_r}{dt} = -T_e + T_L + K_s(\theta_r - \theta_t) + D_m(\omega_r - \omega_t) \tag{2.60}$$

$$\frac{d\theta_t}{dt} = \omega_t \tag{2.61}$$

$$\frac{d\theta_r}{dt} = \omega_r \tag{2.62}$$

where H_t and H_g are the inertial time constants for the turbine and generator, K_s is the shaft stiffness constant and D_m is the mutual damping coefficient. T_w is the input disturbance generated by wind changes and T_L represents the load disturbance reflecting the demand of power output.

The control system for the wind turbine drive train is part of a larger system. The manipulated variables are the turbine torque T_t and the motor torque T_e. In order to optimize the wind energy generation, the angular velocity of the generator

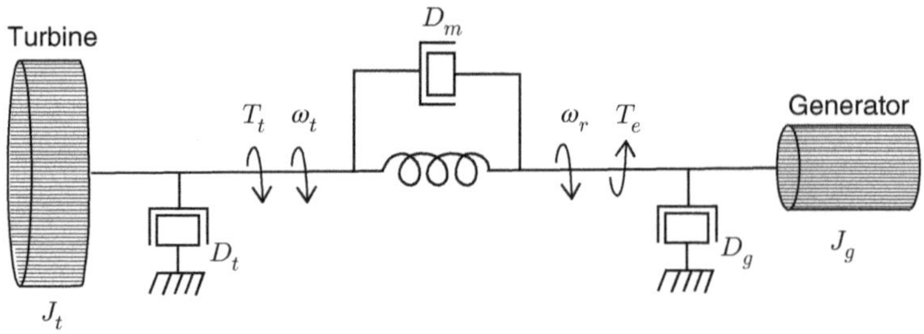

Figure 2.11 Illustration of two mass model for turbine (Perdana [2008]).

requires to be under control. Thus, the control objective is to achieve reference tracking of ω_r. To avoid causing mechanical stress to the wind turbine generator, the angular velocity of the turbine ω_t desires to be synchronized with that of the generator. Namely, in the steady-state operation, the control objective is $\omega_t = \omega_r$.

Applying Laplace transform to (2.59), with assumptions of zero initial conditions on ω_t and ω_r, we obtain the Laplace transform for $\Omega_t(s)$ as

$$\Omega_t(s) = \frac{s}{2H_t s^2 - D_m s - K_s}(T_t(s) - T_w(s)) - \frac{D_m s + K_s}{2H_t s^2 - D_m s - K_s}\Omega_r(s). \quad (2.63)$$

Similarly, we obtain the Laplace transform for $\Omega_r(s)$ as

$$\begin{aligned}\Omega_r(s) = &-\frac{s}{2H_g s^2 - D_m s - K_s}T_e(s) + \frac{s}{2H_g s^2 - D_m s - K_s}T_L(s)\\ &- \frac{D_m s + K_s}{2H_g s^2 - D_m s - K_s}\Omega_t(s).\end{aligned} \quad (2.64)$$

From (2.63) -(2.64), two observations follow. It is clear that the wind turbine drive train using the two mass model is an unstable system because some coefficients of the denominator of the transfer function are negative. There is a zero on the origin of the complex plane from both transfer functions relating T_t to Ω_t as well as relating T_e to Ω_r. At the first investigation, one might attempt to use two PID controllers for the two second order systems. However, PID controllers would not be suitable choices for the drive train model. First because of the zero in both transfer functions, integral action can not be used in the controller. The best one could do is to use two PD controllers. More importantly, because the interactive dynamics represented by the transfer functions between $\Omega_r(s)$ and $\Omega_t(s)$ are unstable, and the input disturbances T_w and T_L acting on the system exhibit unstable dynamics, the two PD controllers would have tremendous difficulties to stabilize the drive train separately as two individual controllers in the presence of disturbance and model uncertainty. A sound choice is to design a state feedback control to stabilize the wind turbine drive train.

Now, in order to perform state feedback controller design, the differential equations (2.59)-(2.62) are expressed in a compact matrix form:

$$\begin{bmatrix} \dot{\omega}_t(t) \\ \dot{\omega}_r(t) \\ \dot{\theta}_t(t) \\ \dot{\theta}_r(t) \end{bmatrix} = \overbrace{\begin{bmatrix} \frac{D_m}{2H_t} & -\frac{D_m}{2H_t} & \frac{K_s}{2H_t} & -\frac{K_s}{2H_t} \\ -\frac{D_m}{2H_g} & \frac{D_m}{2H_g} & -\frac{K_s}{2H_g} & \frac{K_s}{2H_g} \\ 1 & 0 & 0 & 0 \\ 0 & 1 & 0 & 0 \end{bmatrix}}^{A_p} \begin{bmatrix} \omega_t(t) \\ \omega_r(t) \\ \theta_t(t) \\ \theta_r(t) \end{bmatrix} + \overbrace{\begin{bmatrix} \frac{1}{2H_t} & 0 \\ 0 & -\frac{1}{2H_g} \\ 0 & 0 \\ 0 & 0 \end{bmatrix}}^{B_p} \begin{bmatrix} T_t \\ T_e \end{bmatrix}$$

$$+ \overbrace{\begin{bmatrix} -\frac{1}{2H_t} & 0 \\ 0 & \frac{1}{2H_g} \\ 0 & 0 \\ 0 & 0 \end{bmatrix}}^{B_d} \begin{bmatrix} T_w \\ T_L \end{bmatrix}. \tag{2.65}$$

For the current study, as in Perdana [2008], the turbine inertial constant $H_t = 2.6$ (sec), and the generator inertial constant $H_g = 0.22$ (sec), the stiffness constant $K_s = 141.0$ pu, and mutual damping coefficient $D_m = 3.0$ pu.

The rated speed for the generator is 42 rpm, which is 4.398 *rad/sec* and is used for the velocity reference signals of turbine and generator ω_t^* and ω_r^*. With $t_0 = 0$, the integration of the constant velocity reference gives the angular position reference signal for θ_t^* and θ_r^*.

The rated power for the generator is 200 KW. Thus, when the generator is operating at the normal operating condition, the power $P \approx T_e\omega_r$. Thus, load torque T_L used to indicate the power output drawn from the generator is calculated as

$$T_L = \frac{P}{\omega_r^*} = \frac{200000}{4.398} = 4.5473 \times 10^4$$

where the rated speed for the generator is used for the calculation of load torque in the simulation. T_w is the disturbance reflecting the changes of wind speed on the closed-loop control system, which is modelled as piece-wise constant.

With the physical parameters, the system matrices are

$$A_p = \begin{bmatrix} 0.577 & -0.577 & 27.115 & -27.115 \\ -6.818 & 6.818 & -320.455 & 320.455 \\ 1 & 0 & 0 & 0 \\ 0 & 1 & 0 & 0 \end{bmatrix}; \; B_p = \begin{bmatrix} 0.192 & 0 \\ 0 & -2.273 \\ 0 & 0 \\ 0 & 0 \end{bmatrix}$$

and the output matrix

$$C_p = \begin{bmatrix} 0 & 0 & 1 & 0 \\ 0 & 0 & 0 & 1 \end{bmatrix}.$$

The eigenvalues of the system are 22.7039, −15.3088, 0, 0, which confirm that the wind turbine drive train system is unstable and with two eigenvalues on the origin of the complex plane due to the integral relationship between the angular velocity and position.

2.4.2 Configuration of The Control System

To reject the low frequency disturbances represented by the variation of the wind torque T_w and the load torque T_L, integrators are required for the state feedback control system. Additionally, the reference tracking of the desired angular velocity signals ω_r^* and ω_t^* is required. An intuitive approach is to directly give references to the velocities with integral action. As an exercise, we can verify that the augmented system with integrators is not controllable. As a result, we will choose the turbine and the generator's angular positions θ_t and θ_r as the outputs. The constant reference signal for ω_r^* becomes a ramp signal with its slope determined by ω_r^* for a given initial time t_0:

$$\theta_r^* = \int_{t_0}^{t} \omega_r^* d\tau = \omega_r^*(t - t_0). \tag{2.66}$$

Figures 2.12(a)-(b) show the piece-wise constant reference signal ω_r^* and the corresponding reference signal θ_r^*.

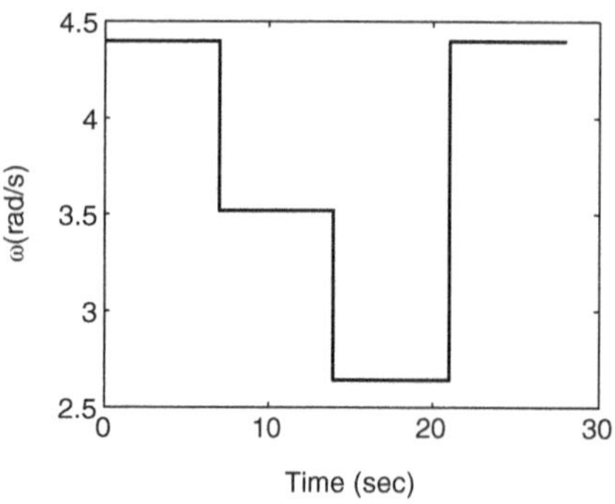

(a) Angular velocity signal

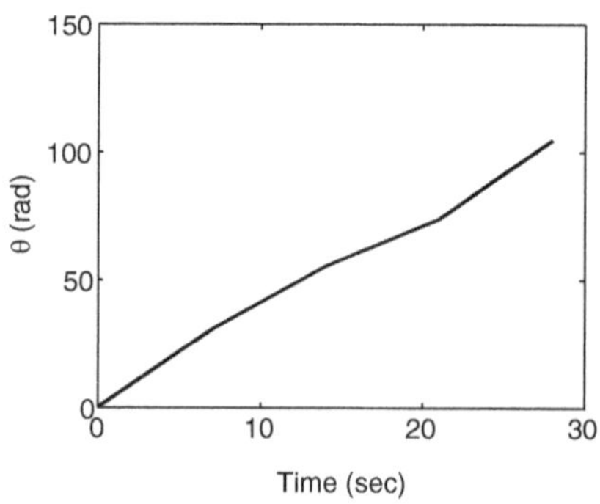

(b) Angular position signal

Figure 2.12 Reference signals for the angular velocity and position.

2.4.3 Design Method I

We assume that both the angular velocity and position of the wind turbine and generator are measured. Therefore, there is no need to use an observer.

By embedding integrators into the system model, we obtain the augmented state space model:

$$\begin{bmatrix} \ddot{x}_p(t) \\ \dot{y}(t) \end{bmatrix} = \begin{bmatrix} A_p & 0_{2\times2} \\ C_p & 0_{2\times2} \end{bmatrix} \begin{bmatrix} \dot{x}_p(t) \\ y(t) \end{bmatrix} + \begin{bmatrix} B_p \\ 0_{2\times2} \end{bmatrix} \begin{bmatrix} \dot{T}_t(t) \\ \dot{T}_e(t) \end{bmatrix}, \tag{2.67}$$

where $0_{2\times2}$ is a zero matrix with dimension 2×2, and the output $y(t)$ consists of the angular position $\theta_t(t)$ and $\theta_r(t)$. Note that the derivatives of the load torque T_L and the torque from wind T_w are very small due to the assumption of slow variation or piece wise constant, hence neglected from the model used in the design.

Now, for both $\theta_t(t)$ and $\theta_r(t)$, there is a double integrator in the augmented model where one is from the plant model and one is embedded through the augmentation. The outputs of the state feedback control system will track ramp signals without steady-state error. The embedded integrator in the controller will enable the closed-loop system to reject the low frequency torque disturbances from the wind variations and from the variations of power demand.

The augmented model has the following system matrices:

$$A = \begin{bmatrix} 0.577 & -0.577 & 27.115 & -27.115 & 0 & 0 \\ -6.818 & 6.818 & -320.455 & 320.455 & 0 & 0 \\ 1 & 0 & 0 & 0 & 0 & 0 \\ 0 & 1 & 0 & 0 & 0 & 0 \\ 0 & 0 & 1 & 0 & 0 & 0 \\ 0 & 0 & 0 & 1 & 0 & 0 \end{bmatrix}; \; B = \begin{bmatrix} 0.192 & 0 \\ 0 & -2.273 \\ 0 & 0 \\ 0 & 0 \\ 0 & 0 \\ 0 & 0 \end{bmatrix}.$$

We can use MATLAB function **ctrb.m** to calculate the controllability matrix for the pair A and B matrices and evaluate its rank by following the computation:

```
Lc=ctrb(A,B);
rank(Lc)
```

The rank of Lc matrix is 6, which equals the number of states in the augmented model. This means that the augmented model is controllable.

To find the state feedback controller, we can use MATLAB **lqr.m** function to find the controller gain. However, it is difficult to choose the suitable weighting matrices Q and R by only adjusting the diagonal elements in Q and R as one of the dominant closed-loop eigenvalues was at -0.3 and hard to move. In this situation, we deploy a constraint on the location of the closed-loop poles on the left of $-\alpha$ line and use the MATLAB **lqr.m** function to find the state feedback controller gain (see Section 1.5.4). For instance, if we select $\alpha = 6$, the MATLAB program used to achieve this design is given as follows.

```
Q=eye(6,6);
R=eye(2,2);
alpha=6;
K=lqr(A+alpha*eye(6,6),B,Q,R);
```

For this design, the closed-loop eigenvalues are

$$-34.792, -15.3296, -11.9268, -12.0261 \pm 0.0372i, -12.0000$$

and the state feedback control gain becomes

$$K = 10^3 \begin{bmatrix} 0.1788 & -0.0044 & 2.1301 & -0.1706 & 7.5450 & -0.1495 \\ 0.0523 & -0.0313 & 1.0583 & -0.7026 & 3.9024 & -3.4252 \end{bmatrix}.$$

In the simulation studies, the velocity reference signals of turbine and generator ω_t^* and ω_r^* are identically chosen as 4.398 rad/sec. There is a constant disturbance d_o acting on the turbine speed to model the effect of the wind where $d_o = 1.2 \times 4.398$ rad/sec. The generator's output is produced by adding an input disturbance T_L to the electrical torque when the generator produces the electrical power, where $T_L = -200 \times 10^3/(42 \times 2\pi/60)$ from the rated power output.

With the sampling interval $\Delta t = 0.001$ (sec), we simulate the closed-loop control performance. When the closed-loop control system for the wind turbine drive train starts up, the power output demand is assumed to be zero to avoid the disturbance caused by the transient response. A constant T_L is added as the load disturbance at one third of the simulation time. The simulation results are discussed as follows.

1. Figures 2.13(a)-(d) show the closed-loop responses of the angular position and velocity of the turbine and generator. It is clear that the closed-loop system outputs track the angular reference positions and the velocities. When the control system starts up, the wind speed is higher than the reference speed simulated as output disturbance acting on the system. It is seen that the initial responses go up and then come down to track the reference signals (see Figures 2.13(c)-(d)), showing that this disturbance is successfully rejected by the closed-loop controller.
2. At $t = 10$ sec, the input torque disturbance enters the system so the output power is produced. With this rated power, the velocity of the generator drops to -1152 rad/sec before it bounces back to 359 rad/sec and converges to the reference signal.

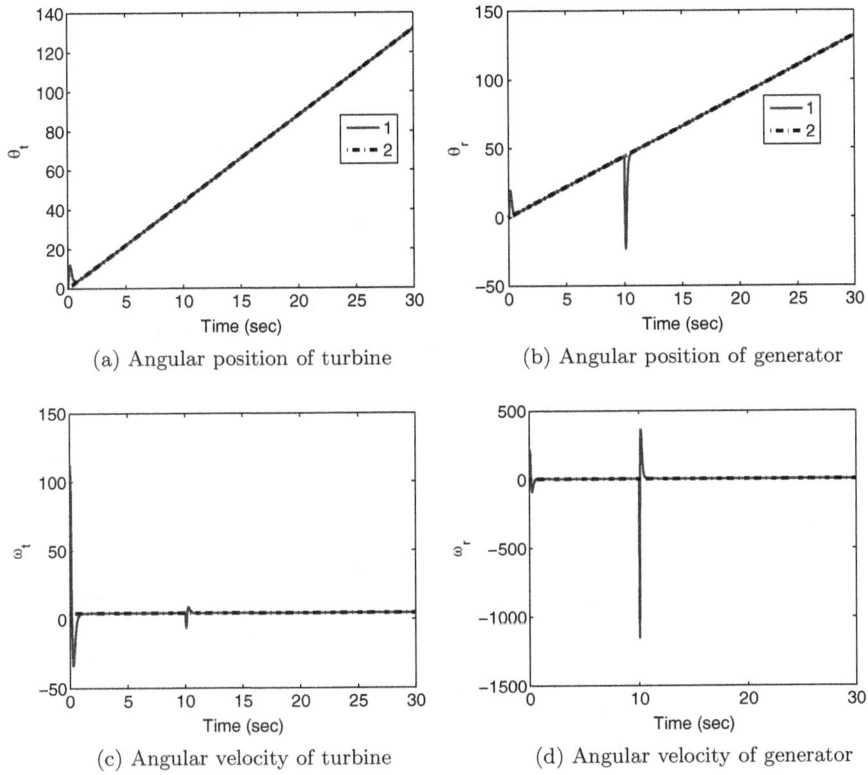

Figure 2.13 Closed-loop responses of turbine and generator. Key: line (1) output; line (2) reference.

3. Given that the power output is related to the angular speed of the generator, the large changes on the velocity will cause large changes in the output power that may cause instability to the power networks.
4. Because the open-loop system is unstable, reducing the feedback controller gain to reduce the large angular velocity variation in the disturbance rejection is not the best choice for the reason that such a reduction could destabilize the closed-loop system. Since the power output is modeled as load torque T_L, we can introduce a load demand filter in the form of

$$T_{fL}(s) = \frac{1}{\tau_f s + 1} T_L(s).$$

By suitably choosing the load demand filter, we can gradually introduce the load torque to the wind turbine drive train.
5. In this simulation, the filter time constant is chosen to be 1. Figures 2.14(a)-(d) show the closed-loop responses of the turbine and generator with the load

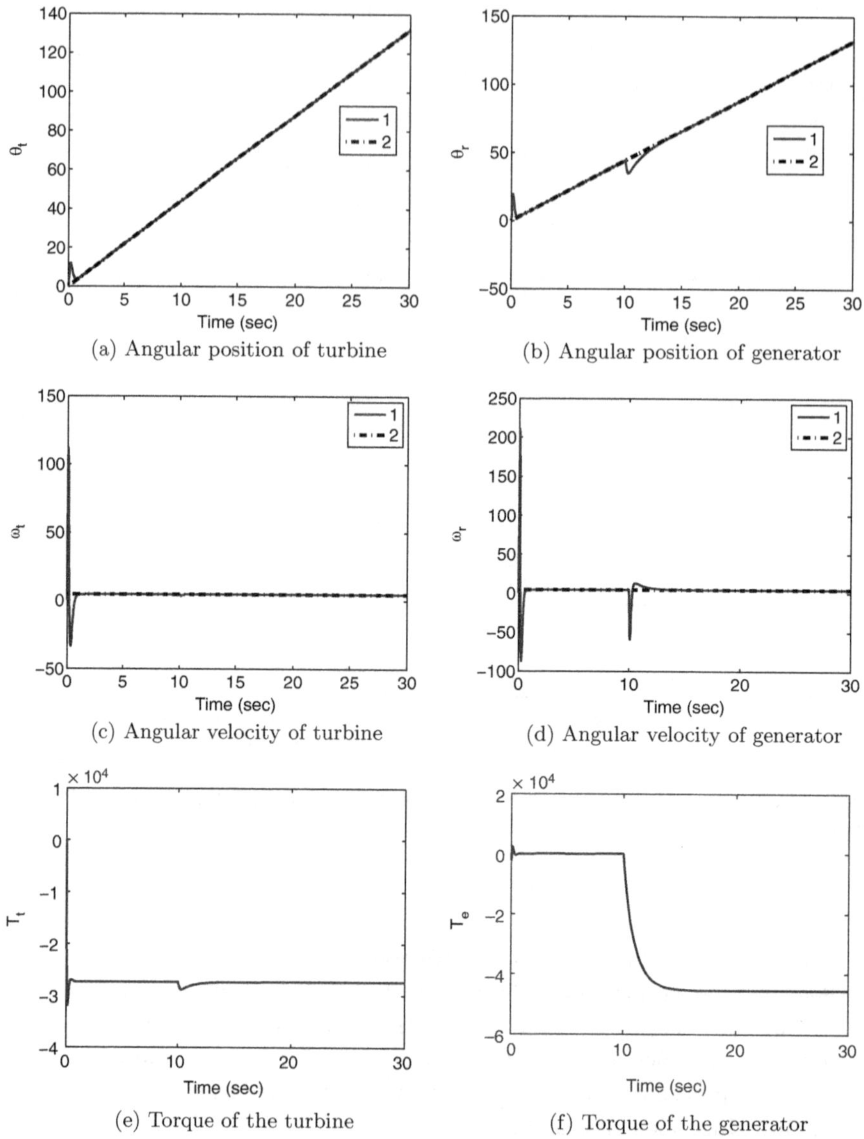

Figure 2.14 Closed-loop responses of turbine and generator with load demand filter. Key: line (1) output; line (2) reference.

demand filter. When the load torque is added to the system, the large drop in the generator's velocity has been drastically improved from −1152 *rad/sec* previously to −59.11 *rad/sec*. Also, the positive spike has been reduced from 359 *rad/sec* to 12.81 *rad/sec*.

6. Figures 2.14(e)-(f) show the control signals to the wind turbine drive train. It is seen from Figure 2.14(f) that the torque of the generator made a smooth transition from 0 to its steady-state value in the disturbance rejection.

2.4.4 Design Method II

The second approach will be based on the disturbance estimation using the measured state variables (see Section 2.3.5).

In this approach, we design the state feedback controller K_p using the pair of system matrices A_p and B_p. By specifying all desired closed-loop eigenvalues on the left of line −6, and the weighting matrices as identity matrices $Q = I_{4\times4}$ and $R = I_{2\times2}$, we find the state feedback controller gain

$$K_p = \begin{bmatrix} 116.3602 & -4.4290 & 769.7486 & -153.4586 \\ 52.3429 & -26.0123 & 466.2033 & -426.4333 \end{bmatrix}$$

and the corresponding closed-loop eigenvalues: −34.7920, −11.9457, −12.0334, −15.3296. One can verify as an exercise that one pair of eigenvalues in the closed-loop system are hard to move by only choosing the diagonal elements in Q and R matrices. Thus, we justify the use of prescribed degree of stability in the design.

To design the estimator for the input disturbance to generate integral action, the system matrices are

$$A_d = \begin{bmatrix} 0 & 0 \\ 0 & 0 \end{bmatrix}; \; B_p = \begin{bmatrix} 0.1923 & 0 \\ 0 & -2.2727 \\ 0 & 0 \\ 0 & 0 \end{bmatrix}.$$

Again, one can verify that one of the closed-loop eigenvalues is hard to move if we only try to adjust the diagonal elements of the weighting matrices. Thus, we choose $Q_d = I_{2\times2}$ and $R_d = I_{4\times4}$. With the desired closed-loop eigenvalues for the estimator specified to the left of −3 line on the complex plane, we find the estimator gain matrix as

$$K_{ob} = \begin{bmatrix} 31.2320 & 0 & 0 & 0 \\ 0 & -2.9760 & 0 & 0 \end{bmatrix}$$

with the resultant closed-loop eigenvalues as −6.7637, −6.0062.

The closed-loop control system using disturbance estimation is simulated with sampling interval $\Delta t = 0.001$ (sec) under the identical simulation conditions as discussed in Section 2.4.3, where a constant disturbance d_o acting on the turbine speed to model the effect of the wind where $d_o = 1.2 \times 4.398$ rad/sec while the generator's output is produced by adding an input disturbance T_L to the electrical torque. Again, the disturbance T_L is modelled as a filtered step signal to avoid large changes in the generator speed when the electrical load is introduced. The simulation results are discussed as follows.

1. Figures 2.15(a)-(b) show the closed-loop angular positions of the turbine and the generator, which track ramp signals. It is seen that both desired positions have been followed without steady-state errors in the presence of wind and torque disturbances.
2. Figures 2.15(c)-(d) show the closed-loop angular velocities of the turbine and generator, which are maintained to be constant signals. Both figures show that the wind and torque disturbances have been completely rejected without steady-state errors.
3. Figures 2.15(e)-(f) show the control signal responses, which are the torques of the turbine and generator, to the reference tracking and disturbance rejections. For the chosen set of control performance parameters, it is seen that the torque of the generator makes a smooth transition when the load disturbance kicks in.

The simulation results can be confirmed by following MATLAB Tutorial 2.8 presented in the next section.

2.4.5 MATLAB Tutorial on Design Method II

This tutorial is to show how to design and implement the disturbance-observer-based control system with measured state variables. We will use the wind turbine control as an example.

Tutorial 2.8 *Write a MATLAB program Ex4WindEstim.m that will design and simulate a disturbance-observer-based control system with known measured state variables.*

Step by Step

1. *Create a new file called Ex4WindEstim.m.*

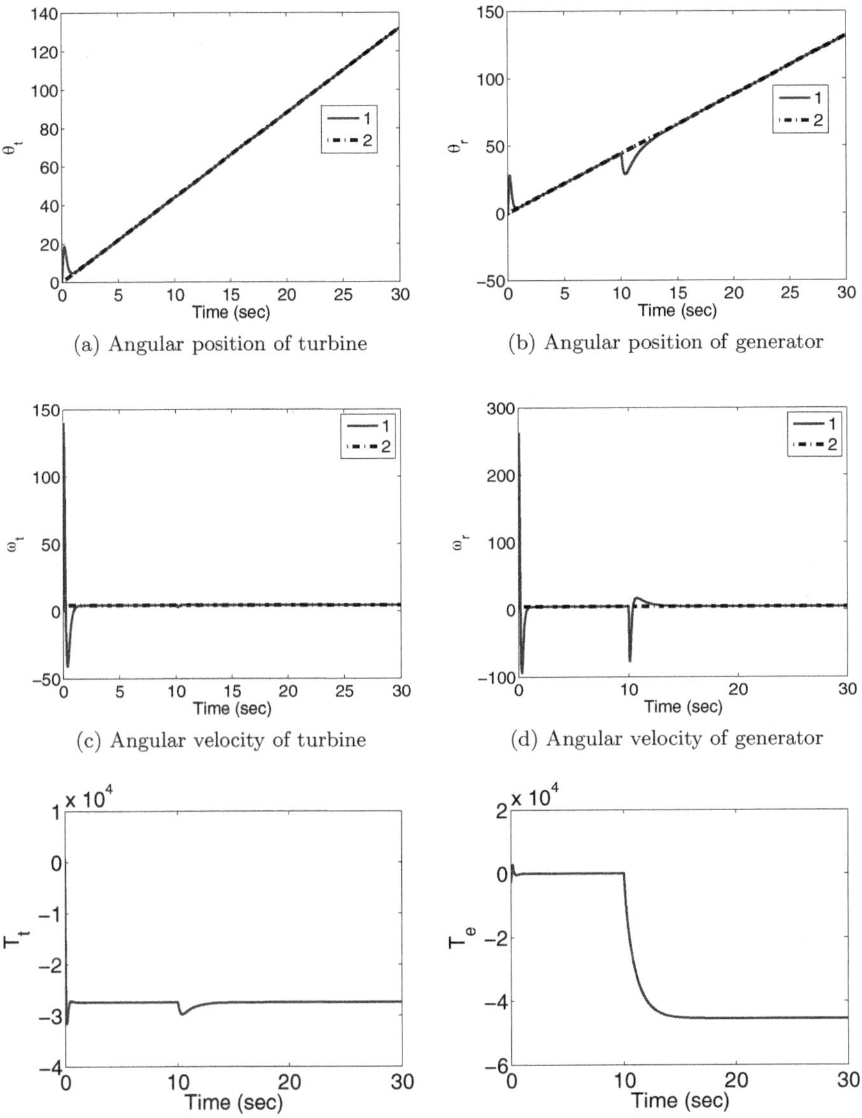

(a) Angular position of turbine

(b) Angular position of generator

(c) Angular velocity of turbine

(d) Angular velocity of generator

(e) Torque of the turbine

(f) Torque of the generator

Figure 2.15 Closed-loop responses of turbine and generator with load demand filter using disturbance estimation. Key: line (1) output; line (2) reference.

2. *Input the physical parameters of the wind turbine model and form the system matrices A_p, B_P and C_p (see (2.65)). The disturbance matrix B_d equals to $-B_p$. Enter the following program into the file:*

```
Ht=2.6;    %second
Hg=0.22;   %second
Ks=141.0;   %pu
Dm=3.0;    %pu
Ap1=[Dm/(2*Ht) -Dm/(2*Ht) Ks/(2*Ht)..
  -Ks/(2*Ht); -Dm/(2*Hg) Dm/(2*Hg)..
   -Ks/(2*Hg) Ks/(2*Hg)];
Ap=[Ap1;1 0 0 0;0 1 0 0];
Bp=[1/(2*Ht) 0; 0 -1/(2*Hg);..
  0 0; 0 0];
Cp=[0 0 1 0;0 0 0 1];
Bd=-Bp;
```

3. *Design the state feedback controller using the linear quadratic regulator design method, where all closed-loop poles are positioned at the left of −6 line on the complex plane. Continue entering the following program into the file:*

```
[n1,n_in]=size(Bp);
[m1,n1]=size(Cp);
Qc=eye(n1,n1);
Rc=eye(n_in,n_in);
alpha2=6;
K=lqr(Ap+alpha2*eye(n1,n1),Bp,Qc,Rc);
```

4. *Design the observer for estimating the disturbance using linear quadratic regulator design method, where all closed-loop poles of error system are positioned at −3 line on the complex plane. Continue entering the following program into the file:*

```
Aob=zeros(2,2);
Cob=Bp;
Qob=eye(2,2);
Rob=eye(4,4);
alpha1=3;
Kob=lqr(Aob'+3*eye(2,2),Cob',Qob,Rob)';
```

5. *Prepare closed-loop simulation for disturbance rejection and reference tracking.*
   (a) *Choose sampling interval and simulation time. Continue entering the following program into file:*

```
Tsim=30;
```

```
Deltat=0.001;
Nsim=Tsim/Deltat;
t=0:Deltat:(Nsim-1)*Deltat;
```

(b) *Enter the initial conditions for the simulation:*

```
xm=zeros(n1,1);
zhat=zeros(2,1);
y=zeros(2,1);
```

(c) *Define the reference signals to angular velocities of the turbine and generator:*

```
sp1=ones(1,Nsim)*42*2*pi/60;
sp2=ones(1,Nsim)*42*2*pi/60;
spw=[sp1;sp2];
```

(d) *Define the reference signals to angular positions of the turbine and generator, which are the integrated velocity signals. Continue entering the following program into the file:*

```
sp(:,1)=spw(:,1)*Deltat;
for kk=2:Nsim
sp(:,kk)=sp(:,kk-1)+spw(:,kk)*Deltat;
end
```

(e) *Define the disturbance from wind as 1.2 of the desired turbine speed, which will be directly added to the turbine speed:*

```
dwind=1.2*sp1;
```

(f) *Define the filtered load torque demand, which enters the simulation at one third of the simulation time. Continue entering the following program into the file:*

```
dm=200e3/(42*2*pi/60);
du2=-dm*[0*ones(1,Nsim/3) ones(1,2*Nsim/3)];
tauf=1;
[dnum,dend]=c2dm(1,[tauf 1],Deltat);
du2=filter(dnum,dend,du2);
```

(g) *One could also add a random low frequency disturbance to the turbine. Continue entering the following program into the file:*

```
e=randn(1,Nsim);
du1=0.5*filter(0.1,[1 -0.9999],e);
du=[du1;du2];
```

6. *Perform closed-loop simulation iteratively. Generate the state feedback variable with the reference signals to the state variables. Continue entering the following program into the file:*

```
for kk=1:Nsim;
    xr=[xm(1)-spw(1,kk);xm(2)-spw(2,kk);...
    xm(3)-sp(1,kk); xm(4)-sp(2,kk)];
```

7. *Generate the estimated disturbance signal. Continue entering the following program into the file:*

```
dhat=zhat+Kob*xr;
```

8. *Compute the control signal:*

```
u=-K*xr-dhat;
```

9. *At this step, impose constraints on the control signal if required.*
10. *Update the estimated intermediate disturbance vector with the control and state signals:*

```
zhat=zhat-Kob*Deltat*(Bp*zhat+Bp*Kob*xr+Ap*xr+Bp*u);
```

11. *Save the control signal, output signal and state signals for plots:*

```
u1(1:n_in,kk)=u;
y1(1:m1,kk)=y;
 xm1(kk)=xm(1);
 xm2(kk)=xm(2);
```

12. *Input the control signal to the system with the solution of the differential equation and complete one cycle of closed-loop simulation:*

```
xm=xm+(Ap*xm+Bp*u+Bd*du(:,kk))*Deltat;
xm(1,1)=xm(1,1)+dwind(1,kk);
y=Cp*xm;
 end
```

13. *Present the simulation results by plotting the input signal vector u_1 and output signal y_1 and state variables xm_1 and xm_2 and validate this closed-loop controller design and simulation by comparing the results shown in Section 2.4.4.*

Because the system is unstable, one must be very cautious about limits of the control signals. The constraints on the control signals could easily cause the closed-loop system to become unstable. An effective strategy as shown in the case study is to carefully design the reference trajectories and the load profiles in order to satisfy the physical constraints when controlling an unstable system.

2.4.6 Food for Thought

1. For the unstable system, if there are some errors in the physical parameters and we try to improve the robustness of the control system, should we try to reduce the controller gain or should we try to increase the controller gain?
2. Why is it very difficult and dangerous to limit the control signal amplitude if the physical system is unstable?
3. What strategies you might consider to reduce the control signal amplitude for control of an unstable system so that constraints can be met naturally?
4. Can you propose an alternative approach to reduce the effect of load disturbance on the network? Would you think that there are similarities between the reduction of load disturbance in transient response and the reduction of overshoot in reference tracking?

2.5 Summary

In this chapter, we have discussed state space design methods for reference tracking and disturbance rejection. In order for the control systems to track a step reference and reject a low frequency disturbance, integral action is required. Two approaches are discussed in this chapter to achieve the design requirement. One is to directly embed an integrator into the state feedback controller whilst the other is to estimate a constant input disturbance and subtract this disturbance from the control signal. Both methods are convenient for control system implementation and both have an anti-windup mechanism in the event of control signal reaching its saturation limits.

The other important aspects of this chapter are summarized as follows.

1. The traditional way of including integrator in the state space design is to introduce an integrated output variable as a feedback signal. This approach encounters problems in the implementation stage. In particular, when the control signal reaches its saturation limits, the event of integrator windup will occur.
2. We have taken a different approach to include integrator in the state space design. We have replaced the original state variables by the derivatives of the state variables in the state feedback and introduced the the output signal as the feedback signal. This approach will eliminate the effect of constant input disturbance on the state estimation. More importantly, the control algorithm can be directly implemented with an anti-windup mechanism when the control signal reaches its saturation limits.
3. We have designed the state estimate feedback control system to follow a constant reference signal and reject a low frequency disturbance by estimating a

constant input disturbance and subtracting this disturbance from the control signal. The control algorithm via disturbance estimation can be directly implemented, which has a naturally occurring anti-windup mechanism in the event of the control signal reaching its saturation limits.

4. When the state variables are measured, the disturbance estimation approach is modified to take the measured state variables into the feedback control. This proves to be useful for many applications in electrical and mechanical systems.

2.6 Further Reading

1. Detailed mathematical modelling of windmills was introduced in Perdana [2008].
2. State-space pole assignment controller design using output feedback was introduced in Seraji [1978], and with H_∞ design in Chilali and Gahinet [1996].
3. Pole-assignment controller design with extension to systems with time delay is proposed in Michiels et al. [2002].
4. Anti-windup mechanisms are introduced in Kapoor et al. [1998], Kothare et al. [1994], Peng et al. [1996], Tarbouriech and Turner [2009], Zaccarian and Teel [2011], Galeani et al. [2009], Wang [2016a], Wang [2016b].
5. disturbance-observer-based control is generalized to deal with uncertainties in Li et al. [2012].

Problems

Problem 2.1 Consider the practical controller I. Determine whether each of the systems below is controllable and whether their augmented system with embedded integrator is still controllable:

$$\dot{x}(t) = A_p x(t) + B_p u(t); y(t) = C_p x(t).$$

1. $A_p = \begin{bmatrix} 2 & -5 \\ -4 & 1 \end{bmatrix}$; $B_p = \begin{bmatrix} 1 \\ 2 \end{bmatrix}$; and $C_p = \begin{bmatrix} 1 & 1 \end{bmatrix}$;
2. $A_p = \begin{bmatrix} -2 & -3 \\ 1 & 0 \end{bmatrix}$; $B_p = \begin{bmatrix} 1 \\ 0 \end{bmatrix}$; and $C_p = \begin{bmatrix} 1 & 0 \end{bmatrix}$.
3. $A_p = \begin{bmatrix} -5 & -5 & 6 \\ 2 & -3 & -2 \\ -1 & -1 & -3 \end{bmatrix}$; $B_p = \begin{bmatrix} 1 & 0 \\ 0 & 2 \\ 1 & 1 \end{bmatrix}$; and $C_p = \begin{bmatrix} 1 & 1 & 0 \\ 2 & 1 & 2 \end{bmatrix}$.

4. $A_p = \begin{bmatrix} -5 & -5 & 6 \\ 2 & -3 & -2 \\ -1 & -1 & -3 \end{bmatrix}$; $B_p = \begin{bmatrix} 1 \\ 0 \\ 1 \end{bmatrix}$; and $C_p = \begin{bmatrix} 1 & 1 & 0 \\ 2 & 1 & 2 \end{bmatrix}$.
5. $A_p = \begin{bmatrix} -5 & -5 & 6 \\ 2 & -3 & -2 \\ -1 & -1 & -3 \end{bmatrix}$; $B_p = \begin{bmatrix} 1 & 0 \\ 0 & 2 \\ 1 & 1 \end{bmatrix}$; and $C_p = \begin{bmatrix} 2 & 1 & 2 \end{bmatrix}$.

For the system that is not controllable with the augmented system matrices A and B, calculate the original system transfer function using A_p, B_p, and C_p and the augmented system transfer function using A, B, and C. What are your observations?

Problem 2.2 A linearized model for a ball and plate balancing system that is described by the state space model (Wang [2020]):

$$\begin{bmatrix} \dot{x}_1(t) \\ \dot{x}_2(t) \end{bmatrix} = \begin{bmatrix} 0 & 0 \\ 1 & 0 \end{bmatrix} \begin{bmatrix} x_1(t) \\ x_2(t) \end{bmatrix} + \begin{bmatrix} -\frac{5g_0}{7} \\ 0 \end{bmatrix} \theta_x(t) \tag{2.68}$$

where g_0 is the gravity constant ($g_0 = 9.8m/(sec)^2$), the variable $x_1(t)$ is the velocity of the ball and $x_2(t)$ is the position of the ball on the plate. Only the position of the ball is measured. Design a state estimate feedback control system with integral action (via controller design) for this system.

1. The desired closed-loop characteristic polynomial for the controller is selected to be $(s + w_n)(s^2 + 2 \times 0.707w_ns + w_n^2)$ and the desired observer characteristic polynomial is selected to be $(s^2 + 4 \times 0.707w_ns + 4w_n^2)$. We will choose w_n equal to 3 and 5 respectively.
2. Simulate the closed-loop response with sampling interval $\Delta t = 0.001$ (sec) for the case when the ball makes a square wave movement. Here the input signal is a square wave signal with amplitude ± 0.075 and period of 10 *sec*. The initial conditions for the simulation are $x_1(0) = 0$ and $x_2(0) = 0.16$.
3. Find the peak of the control signal from the first simulation as U^{lim} and choose the parameters $u^{min} = -0.9U^{lim}$ and $u^{max} = 0.9U^{lim}$. Impose the limits in the closed-loop simulation with anti-windup mechanism and observe the closed-loop responses when the constraints become active.

Problem 2.3 Design a state estimate feedback control system with integral action via controller design for the plant described by the following state space model:

$$\dot{x}(t) = A_px(t) + B_pu(t); \; y(t) = C_px(t)$$

where $A_p = \begin{bmatrix} -5 & -5 & 6 \\ 2 & -3 & -2 \\ -1 & -1 & -3 \end{bmatrix}$; $B_p = \begin{bmatrix} 1 & 0 \\ 0 & 2 \\ 1 & 1 \end{bmatrix}$; and $C_p = \begin{bmatrix} 1 & 1 & 0 \\ 2 & 1 & 2 \end{bmatrix}$.

1. Use the MATLAB function **place.m** to find the controller K and observer gain K_{ob}. Choose the five closed-loop controller poles at -5, -5.1, -5.2, -5.3, -5.4 and the three observer poles at -7, -7.1, -7.2. Check if the MATLAB function has correctly assigned the closed-loop poles for this multi-input and multi-output system.
2. With sampling interval $\Delta t = 0.001$ (sec) and zero initial conditions, simulate the closed-loop system response. The reference signal to output y_1 is a square wave signal with amplitude of ± 1 and period of 4 seconds while the reference signal to y_2 is zero. A unit step input disturbance signal is added to the control signal $u_2(t)$ that enters the simulation at half of the simulation time.
3. Denote the peaks of the control signal from the first simulation as U_1^{lim} and U_2^{lim}. Choose $u_1^{min} = -0.85U_1^{lim}$, $u_1^{max} = 0.85U_1^{lim}$, and $u_2^{min} = -0.85U_2^{lim}$ $u_2^{max} = 0.85U_2^{lim}$. Impose the constraints in the closed-loop simulation with anti-windup mechanism and observer the effects of the control signal saturation on the closed-loop performance.

Problem 2.4 Continue from Problem 2.3. We use the MATLAB function **lqr.m** to find the controller gain K and observer gain K_{ob}. Choose both Q and R as identity matrices with appropriate dimensions, however constrain the closed-loop poles for the controller to -5 and those of observer to -7 with the prescribed degree of stability (see Section 1.5.4). Check the closed-loop pole locations for both control and observer systems.

1. With the same simulation conditions as outlined in Problem 2.3, simulate the closed-loop control system for the reference tracking and disturbance rejection.
2. Discuss the differences in closed-loop responses for the two designs with and without the control signal constraints.
3. With the same constraints on the control signals as in Problem 2.3, the closed-loop control performance becomes much worse. What modifications should we try in order to reduce both the controller and observer gains for improving the closed-loop performance in the presence of control signal constraints? What are your observations?

Problem 2.5 Design and simulate state estimate feedback control system with integral action via controller design for each of the following systems:

$$\dot{x}(t) = A_p x(t) + B_p u(t); y(t) = C_p x(t).$$

1. $A_p = \begin{bmatrix} 0 & 1 \\ -4 & 0 \end{bmatrix}$; $B_p = \begin{bmatrix} 0 \\ 1 \end{bmatrix}$; and $C_p = \begin{bmatrix} 1 & 0 \end{bmatrix}$.

2. $A_p = \begin{bmatrix} -1 & 0 & 6 \\ 2 & -3 & -2 \\ -1 & -1 & 0 \end{bmatrix}$; $B_p = \begin{bmatrix} 1 & 0 \\ 0 & 2 \\ 1 & 1 \end{bmatrix}$; and $C_p = \begin{bmatrix} 2 & 1 & 2 \\ 1 & 0 & 1 \end{bmatrix}$.

Use MATLAB **lqr.m** function to calculate the controller gain K with Q equal to C^TC and $R = I$ where C is the output matrix from (2.12) and **lqr.m** to calculate the observer gain K_{ob} with Q_{ob} being identity matrix and $R_{ob} = \epsilon I$ with $\epsilon = 0.1,\ 1,\ 10$ being considered for closed-loop performance. With appropriate choice of sampling interval Δt, evaluate the designs with closed-loop simulation studies for disturbance rejection and reference tracking where the reference signal and disturbance signals are step signals. What are your observations on the effect of ϵ on the closed-loop performance?

Problem 2.6 Determine whether each of the following system is observable and whether their augmented system with constant disturbance model is still observable:

$$\dot{x}(t) = A_p x(t) + B_p(u(t) + d(t));\ y(t) = C_p x(t).$$

1. $A_p = \begin{bmatrix} 0 & -2 \\ -1 & 1 \end{bmatrix}$; $B_p = \begin{bmatrix} 1 \\ 2 \end{bmatrix}$; and $C_p = \begin{bmatrix} 1 & 1 \end{bmatrix}$.
2. $A_p = \begin{bmatrix} -1 & -1 \\ 1 & 0 \end{bmatrix}$; $B_p = \begin{bmatrix} 0.5 \\ 0 \end{bmatrix}$; and $C_p = \begin{bmatrix} 1 & 0 \end{bmatrix}$.
3. $A_p = \begin{bmatrix} -2 & -1 & 3 \\ 1 & -2 & -1 \\ -1 & -1 & -3 \end{bmatrix}$; $B_p = \begin{bmatrix} 0 & 0 \\ 0 & 2 \\ 1 & 0 \end{bmatrix}$; and $C_p = \begin{bmatrix} 1 & 0 & 0 \\ 0 & 1 & 2 \end{bmatrix}$.
4. $A_p = \begin{bmatrix} -2 & -1 & 3 \\ 1 & -2 & -1 \\ -1 & -1 & -3 \end{bmatrix}$; $B_p = \begin{bmatrix} 1 \\ 1 \\ 1 \end{bmatrix}$; and $C_p = \begin{bmatrix} 1 & 1 & 0 \\ 0 & 1 & 2 \end{bmatrix}$.
5. $A_p = \begin{bmatrix} -2 & -1 & 3 \\ 1 & -2 & -1 \\ -1 & -1 & -3 \end{bmatrix}$; $B_p = \begin{bmatrix} 1 & 0 \\ 0 & 2 \\ 1 & 0 \end{bmatrix}$; and $C_p = \begin{bmatrix} 1 & 1 & 1 \end{bmatrix}$.

For the systems that are not observable with the augmented constant disturbance model, will we still be able to use the MATLAB function **place.m** or **lqr.m** to design the observer for the disturbance estimation?

Problem 2.7 A three spring and double mass system is illustrated in Figure 1.5 of Example 1.1. Its state space model is given as follows:

$$\begin{bmatrix} \dot{x}_1(t) \\ \dot{x}_2(t) \\ \dot{x}_3(t) \\ \dot{x}_4(t) \end{bmatrix} = \overbrace{\begin{bmatrix} 0 & 1 & 0 & 0 \\ -\frac{\alpha_1+\alpha_2}{M_1} & 0 & \frac{\alpha_2}{M_1} & 0 \\ 0 & 0 & 0 & 1 \\ \frac{\alpha_2}{M_2} & 0 & -\frac{\alpha_1+\alpha_2}{M_2} & 0 \end{bmatrix}}^{A_p} \begin{bmatrix} x_1(t) \\ x_2(t) \\ x_3(t) \\ x_4(t) \end{bmatrix} + \overbrace{\begin{bmatrix} 0 & 0 \\ \frac{1}{M_1} & 0 \\ 0 & 0 \\ 0 & \frac{1}{M_2} \end{bmatrix}}^{B_p} \begin{bmatrix} u_1(t) \\ u_2(t) \end{bmatrix}$$

$$\begin{bmatrix} y_1(t) \\ y_2(t) \end{bmatrix} = \overbrace{\begin{bmatrix} 1 & 0 & 0 & 0 \\ 0 & 0 & 1 & 0 \end{bmatrix}}^{C_p} \begin{bmatrix} x_1(t) \\ x_2(t) \\ x_3(t) \\ x_4(t) \end{bmatrix}$$

where the physical parameters are $M_1 = 2kg$, $M_2 = 4kg$, $\alpha_1 = 40N/m$ and $\alpha_2 = 100N/m$. Design and implement state estimate feedback control system via observer design for this system. Assume that only the distances $y_1(t)$ and $y_2(t)$ are measured.

1. Design the state feedback controller K using MATLAB **lqr.m** based on the A_p and B_p matrices, where $Q = I$ and R is chosen to be $0.1I$.
2. Design the observer with the augmented system matrices A and C for constant disturbances as in Section 2.3, where Q_{ob} and R_{ob} are identity matrices with appropriate dimensions. The locations of the closed-loop poles for the observer error system are constrained to be on the left of -3 line of the complex plane.
3. Choosing sampling interval $\Delta t = 0.0001$, simulate the closed-loop system response with a reference signal $r_1(t) = 0.1$ and $r_2(t) = 0.2$.

Part II

Discrete-time State Feedback Control

3

Introduction to Discrete-time Systems

3.1 Introduction

As we have learned from the previous chapters, the continuous-time control systems are very efficient in controlling the physical systems that have a fast sampling rate because they are designed in the continuous-time and implemented in the discrete-time, where an approximation due to the discretization is accepted in the implementation stage. Furthermore, the existence of a continuous-time model underpins the development of continuous-time control systems.

In the digital world, the physical models are often naturally presented in terms of sampled data. These systems commonly operate in a slower sampling environment, and it is difficult to describe them using differential equations in continuous-time due to information losses from the slow sampling rate. For these types of systems, it is natural to describe their dynamics using difference equations and design control systems in discrete-time. Designing the control systems in discrete-time presents advantages when continuous-time physical systems have a limited sampling rate. Because the discrete-time systems are more naturally connected to real world data, it is relatively easier to handle the systems with communication delays, missing data and multi-rate sampled data systems.

For those who are not familiar with discrete-time systems, the introductory materials presented in this chapter give a starting point. The chapter begins with the discretization of a continuous-time model (see Section 3.2). With a zero-order-hold, the discrete-time model is accurate at the sampling instant. Additionally, in this section, stability of discrete-time system is introduced based on the eigenvalues of the discrete-time system matrix and several examples are presented to illustrate how the sampling process leads to discrete-time models. In Section 3.3, input and output models in discrete-time are introduced, including the finite impulse response and step response models and the state

State Feedback Control and Kalman Filtering with MATLAB/Simulink Tutorials, First Edition.
Liuping Wang and Robin Ping Guan.

Companion website: www.wiley.com/go/feedback19

space models using a non-minimal state space (NMSS) realization. In Section 3.4 the z-transform is introduced as the discrete-time counterpart of the Laplace transform.

3.2 Discretization of Continuous-time Models

The control systems in discrete-time are in general designed using discrete-time models and implemented using the discrete-time control algorithms. If a physical model is given in continuous-time, the first step in the discrete-time control system design is to obtain a discrete-time model. Because the physical systems are associated with continuous-time, the bridges between the two environments are the analogue to digital (A-D) converter and the digital to analogue (D-A) converter. The former is to convert a continuous-time signal to a discrete-time signal, and the latter is to convert a discrete-time signal to a continuous-time signal. These two devices commonly exist either in the data acquisition equipment or in the micro controllers as part of the control systems. Thus, the discretization of a continuous-time model takes into consideration of these two devices as part of the modelling process. Figure 3.1 shows how these two devices bridge the environment of continuous-time to that of the discrete-time where $u(t_k)$ and $y(t_k)$ are the discrete-time input and output signals at the sampling instant t_k, and $u(t)$ and $y(t)$ are the corresponding continuous-time signals. The most commonly encountered D-A converter is to assume a constant between two samples of discrete-time data, which is called zero-order-hold in the modelling process.

3.2.1 Sampling of a Continuous-time Model

Assume that a continuous-time system is described by the continuous-time state space model:

$$\dot{x}(t) = A_c x(t) + B_c u(t) \tag{3.1}$$

$$y(t) = C_c x(t) + D_c u(t), \tag{3.2}$$

where $x(t)$ is the state vector with dimension $n \times 1$, $u(t)$ is the input signal with dimension $p \times 1$ and $y(t)$ is the output signal with dimension $m \times 1$. A_c, B_c, C_c and

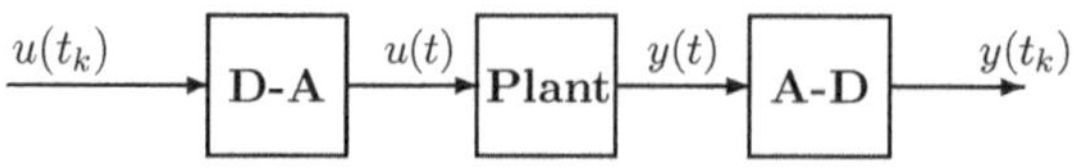

Figure 3.1 Illustration of the bridges between continuous-time and discrete-time signals.

D_c are the system matrices with corresponding dimensions. We use the subscript c in the system matrices to highlight that they are representing the parameters in a continuous-time model.

In order to derive a discrete-time model, we consider the solution of the differential equation (3.1) for a given initial state $x(t_0)$, which consists of the effect of $x(t_0)$ and that of the input signal $u(t)$ where $t \geq t_0$. Its analytical solution is given by:

$$x(t) = e^{A_c(t-t_0)}x(t_0) + \int_{t_0}^{t} e^{A_c(t-\gamma)}B_c u(\gamma)d\gamma, \tag{3.3}$$

where the first term on the right hand-side represents the effect of the initial condition $x(t_0)$ and the convolutional integral from the second term represents the influence of the input signal $u(t)$ for $t \geq t_0$. Now, let us consider a specific time interval $t_k \leq t < t_{k+1}$, where (3.3) is written as

$$x(t_{k+1}) = e^{A_c(t_{k+1}-t_k)}x(t_k) + \int_{t_k}^{t_{k+1}} e^{A_c(t_{k+1}-\gamma)}B_c u(\gamma)d\gamma. \tag{3.4}$$

With the assumption that there is a zero-order-hold in the D-A converter, $u(\gamma) = u(t_k)$ for $t_k \leq \gamma < t_{k+1}$, which is a constant, and as a result $u(\gamma)$ is taken out of the convolutional integral. This leads to

$$x(t_{k+1}) = e^{A_c(t_{k+1}-t_k)}x(t_k) + \left(\int_{t_k}^{t_{k+1}} e^{A_c(t_{k+1}-\gamma)}B_c d\gamma\right) u(t_k). \tag{3.5}$$

For a sampling interval Δt that is assumed to be a constant, we have $\Delta t = t_{k+1} - t_k$, thus,

$$e^{A_c(t_{k+1}-t_k)} = e^{A_c\Delta t}. \tag{3.6}$$

As for the convolutional integral in (3.5), a variable τ is chosen to be $\tau = t_{k+1} - \gamma$, which gives $d\tau = -d\gamma$. By substituting the limits of the original integral with the new limits corresponding to τ, also noting a sign change when the limits change positions, we obtain

$$\int_{t_k}^{t_{k+1}} e^{A_c(t_{k+1}-\gamma)}B_c d\gamma = \int_{0}^{\Delta t} e^{A_c\tau}B_c d\tau. \tag{3.7}$$

Combining (3.6) and (3.7), the solution of the differential equation (3.1) with given initial condition at $x(t_k)$ for $t_k \leq t < t_{k+1}$ is presented as

$$x(t_{k+1}) = e^{A_c\Delta t}x(t_k) + \int_{0}^{\Delta t} e^{A_c\tau}B_c d\tau u(t_k), \tag{3.8}$$

where we have used the assumption that the input signal is a constant between the sampling instants t_k and t_{k+1}.

There are two comments related.

1. Unlike the discretization we used for the implementation of a continuous-time observer in Chapters 1 and 2, there is no approximation involved in the derivation of (3.8) when we assume that there is a zero-order-hold in the D-A converter, meaning that the model is accurate at the sampling instant t_k.
2. The exponential matrix $e^{A_c\Delta t}$ and the integral matrix $\int_0^{\Delta t} e^{A_c\tau} B_c \, d\tau$ are functions of the continuous-time system matrices and the chosen sampling interval Δt, and these quantities are time-invariant for a constant Δt.

For notational simplicity, letting

$$A_m = e^{A_c\Delta t} \tag{3.9}$$

$$B_m = \int_0^{\Delta t} e^{A_c\tau} B_c d\tau. \tag{3.10}$$

Equation (3.8) becomes

$$x(t_{k+1}) = A_m x(t_k) + B_m u(t_k). \tag{3.11}$$

For the output equation (3.2), at the sampling time t_k, because there are no dynamics involved, it simply becomes

$$y(t_k) = C_m x(t_k) + D_m u(t_k), \tag{3.12}$$

where $C_m = C_c$ and $D_m = D_c$, hence the sampling process does not change the output equation. Equations (3.11) and (3.12) form the discrete-time state space model at the sampling time $t = 0,\ t_1, \dots,\ t_k\ \dots$, where $t_k = k\Delta t$, $k = 0,\ 1,\ 2, \dots$.

In the discrete-time environment, it is common practice to consider the sampling time t_k in terms of the sampling index k. Therefore, based on (3.11) and (3.12), the discrete-time state space model is expressed in the following difference equations:

$$x(k+1) = A_m x(k) + B_m u(k) \tag{3.13}$$

$$y(k) = C_m x(k) + D_m u(k). \tag{3.14}$$

The calculation of A_m matrix follows from the Maclaurin series (Kreyszig [2006]):

$$A_m = e^{A_c\Delta t} = I + A_c\Delta t + \frac{1}{2!}(A_c\Delta t)^2 + \frac{1}{3!}(A_c\Delta t)^3 + \dots \tag{3.15}$$

and the calculation of B_m takes the integration of the above series, leading to the following form:

$$B_m = \int_0^{\Delta t} e^{A_c \tau} d\tau B_c$$
$$= (\Delta t + \frac{1}{2} A_c \Delta t^2 + \frac{1}{3!} A_c^2 \Delta t^3 + \dots) B_c. \tag{3.16}$$

If A_c^{-1} exists, then $B_m = A_c^{-1}(e^{A_c \Delta t} - I) B_c$.

3.2.2 Stability of Discrete-time System

When the input signal $u(k) = 0$, for a given initial condition $x(0)$, the solution of difference equation (3.13) is easily computed iteratively as

$$x(1) = A_m x(0)$$
$$x(2) = A_m x(1) = A_m^2 x(0)$$
$$\vdots$$
$$x(k) = A_m x(k-1) = A_m^k x(0). \tag{3.17}$$

In order for the discrete-time system to be asymptotically stable, the eigenvalues of the A_m matrix must be strictly inside the unit circle of the complex plane: $|\lambda_i(A_m)| < 1$ for all $i = 1, 2, \dots, n$. Figure 3.2 illustrates the unit circle on the complex plane for a stable system. This unit circle is an important concept in the study of discrete-time systems. This ensures that each state in the state vector $x(k)$

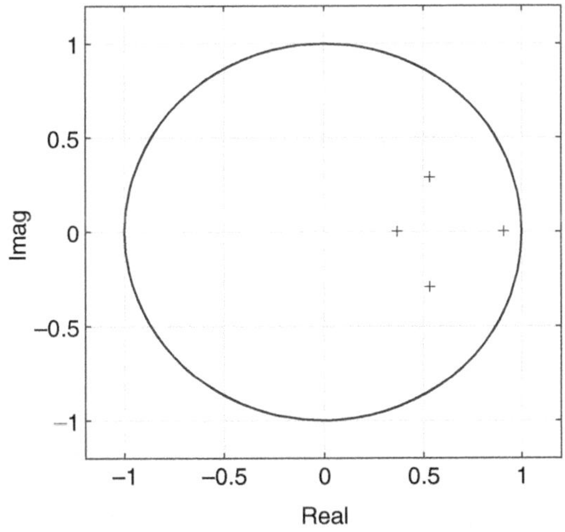

Figure 3.2 Illustration of the unit circle on the complex plane (solid line) and the eigenvalues of a stable system (+).

converges to zero as $k \rightarrow \infty$. When $u(k) \neq 0$, for a given initial condition $x(0)$, the state variables are computed iteratively as

$$
\begin{aligned}
x(1) &= A_m x(0) + B_m u(0) \\
x(2) &= A_m x(1) + B_m u(1) \\
&= A_m^2 x(0) + A_m B_m u(0) + B_m u(1) \\
&\vdots \\
x(k) &= A_m^k x(0) + \sum_{i=0}^{k-1} A_m^{k-i-1} B_m u(i)
\end{aligned} \tag{3.18}
$$

When all the eigenvalues of the A_m matrix are strictly inside the unit circle, with the assumption that the input signal $u(k)$ is bounded, then all the elements in the state vector $x(k)$ are bounded.

3.2.3 Examples of Discrete-time Models from Sampling

There are several examples presented in this section to illustrate how the sampling process leads to discrete-time models.

Example 3.1

Given the first order continuous-time system

$$
\begin{aligned}
\dot{x}(t) &= ax(t) + bu(t) \\
y(t) &= cx(t),
\end{aligned}
$$

where the parameter $a \neq 0$, find the discrete-time model for sampling interval Δt.

Solution

Since $a \neq 0$, based on (3.9) and (3.10), the discrete-time system model is

$$
\begin{aligned}
x(k+1) &= e^{a\Delta t}x(k) + \frac{b}{a}(e^{a\Delta t} - 1)u(k) \\
y(k) &= cx(k).
\end{aligned}
$$

The eigenvalue of the system is at $e^{a\Delta t}$.

It is worthwhile to emphasize that the parameters in the discrete-time model are dependent on the choice of sampling interval Δt. For the same continuous-time model, we could have a completely different discrete-time model if Δt changes. Additionally, if the continuous-time model is unstable with $a > 0$, then the discrete-time model is also unstable with the eigenvalue $e^{a\Delta t} > 1$.

Example 3.2
Considering the integrating system described by

$$\dot{y}(t) = u(t)$$

find the discrete-time model with sampling interval Δt.

Solution
The continuous-time state space model is

$$\dot{x}(t) = u(t)$$
$$y(t) = x(t).$$

For this continuous-time model, $A_m = e^{0\times\Delta t} = 1$, *and* $B_m = \int_0^{\Delta t} d\tau = \Delta t$. *Thus, the discrete- time model is*

$$x(k+1) = x(k) + \Delta t u(k); \; y(k) = x(k).$$

The eigenvalue of the discrete-time system is at 1, which is on the unit circle of the complex plane with imaginary part being zero.

As a special case when the input signal $u(t) = 0$, *the differential equation* $\dot{x}(t) = 0$ *describes the dynamic of a constant signal. The difference equation for the constant signal is*

$$x(k+1) = x(k)$$
$$y(k) = x(k).$$

Example 3.3
A double integrator system is described by the following differential equation:

$$\ddot{y}(t) = u(t).$$

Find the discrete-time model with sampling interval Δt.

Solution
Let $x_1(t) = y(t)$, $x_2(t) = \dot{x}_1(t)$ *and* $x(t) = \begin{bmatrix} x_1(t) & x_2(t) \end{bmatrix}^T$. *Then,* $\dot{x}_2(t) = u(t)$. *We obtain the following state space model in the continuous-time for the double integrator system:*

$$\begin{bmatrix} \dot{x}_1(t) \\ \dot{x}_2(t) \end{bmatrix} = \overbrace{\begin{bmatrix} 0 & 1 \\ 0 & 0 \end{bmatrix}}^{A_c} \begin{bmatrix} x_1(t) \\ x_2(t) \end{bmatrix} + \overbrace{\begin{bmatrix} 0 \\ 1 \end{bmatrix}}^{B_c} u(t) \tag{3.19}$$

$$y(t) = \overbrace{\begin{bmatrix} 1 & 0 \end{bmatrix}}^{C_c} \begin{bmatrix} x_1(t) \\ x_2(t) \end{bmatrix}. \tag{3.20}$$

To find the corresponding discrete-time model, we note that for $k = 2, 3, \ldots$,

$$A_c^k = \begin{bmatrix} 0 & 1 \\ 0 & 0 \end{bmatrix}^k = \begin{bmatrix} 0 & 0 \\ 0 & 0 \end{bmatrix}.$$

Therefore, based on (3.15) we have

$$e^{A_c \Delta t} = I + A_c \Delta t = \begin{bmatrix} 1 & \Delta t \\ 0 & 1 \end{bmatrix},$$

$$\int_0^{\Delta t} e^{A_c \tau} B_c d\tau = \begin{bmatrix} \frac{1}{2}\Delta t^2 \\ \Delta t \end{bmatrix}.$$

From these two expressions, we arrive at the discrete-time state space model:

$$\begin{bmatrix} x_1(k+1) \\ x_2(k+1) \end{bmatrix} = \overbrace{\begin{bmatrix} 1 & \Delta t \\ 0 & 1 \end{bmatrix}}^{A_m} \begin{bmatrix} x_1(k) \\ x_2(k) \end{bmatrix} + \overbrace{\begin{bmatrix} \frac{1}{2}\Delta t^2 \\ \Delta t \end{bmatrix}}^{B_m} u(k)$$

$$y(k) = \overbrace{\begin{bmatrix} 1 & 0 \end{bmatrix}}^{C_m} \begin{bmatrix} x_1(k) \\ x_2(k) \end{bmatrix}. \tag{3.21}$$

We note that the eigenvalues of the double integrator system are the solutions of the polynomial equation:

$$\det(\lambda I - A_m) = (\lambda - 1)^2 = 0$$

which are $\lambda_1 = \lambda_2 = 1$. It is worthwhile to emphasize that the eigenvalues of the double integrator in discrete-time are on the unit circle of the complex plane with imaginary component being 0 and they are not dependent on the sampling interval Δt, although some parameters in the discrete-time model are functions of Δt.

When the input signal $u(t) = 0$, the differential equation $\ddot{y}(t) = 0$ describes the dynamics of a ramp signal. Therefore, the state space model for a ramp signal is

$$\begin{bmatrix} x_1(k+1) \\ x_2(k+1) \end{bmatrix} = \begin{bmatrix} 1 & \Delta t \\ 0 & 1 \end{bmatrix} \begin{bmatrix} x_1(k) \\ x_2(k) \end{bmatrix}$$

$$y(k) = \begin{bmatrix} 1 & 0 \end{bmatrix} x(k). \tag{3.22}$$

Example 3.4

A continuous-time sinusoidal function with unknown amplitude a_m, phase ϕ, but known frequency ω_0, is described by the following expression:

$$y(t) = a_m \sin(\omega_0 t + \phi).$$

Find the corresponding discrete-time state space model with the sampling interval being Δt.

Solution

We choose $x_1(t) = a_m \sin(\omega_0 t + \phi)$ *and* $x_2(t) = a_m \cos(\omega_0 t + \phi)$. *Then,* $\dot{x}_1(t) = \omega_0 x_2(t)$, *and* $\dot{x}_2(t) = -\omega_0 x_1(t)$. *Thus, the continuous-time state space model is obtained for this choice of state variables:*

$$\begin{bmatrix} \dot{x}_1(t) \\ \dot{x}_2(t) \end{bmatrix} = \overbrace{\begin{bmatrix} 0 & \omega_0 \\ -\omega_0 & 0 \end{bmatrix}}^{A_c} \begin{bmatrix} x_1(t) \\ x_2(t) \end{bmatrix} \tag{3.23}$$

$$y(t) = \overbrace{\begin{bmatrix} 1 & 0 \end{bmatrix}}^{C_c} x(t). \tag{3.24}$$

To find the discrete-time state space model, we can verify that

$$A_c^2 = \begin{bmatrix} 0 & \omega_0 \\ -\omega_0 & 0 \end{bmatrix}^2 = \begin{bmatrix} -\omega_0^2 & 0 \\ 0 & -\omega_0^2 \end{bmatrix}$$

$$A_c^3 = \begin{bmatrix} 0 & -\omega_0^3 \\ \omega_0^3 & 0 \end{bmatrix}.$$

After calculating a few terms, we can show that

$$\begin{aligned} A_m &= I + A_c \Delta t + \frac{1}{2!}(A_c \Delta t)^2 + \frac{1}{3!}(A_c \Delta t)^3 + \dots \\ &= \begin{bmatrix} \cos(\omega_0 \Delta t) & \sin(\omega_0 \Delta t) \\ -\sin(\omega_0 \Delta t) & \cos(\omega_0 \Delta t) \end{bmatrix} \end{aligned} \tag{3.25}$$

where the sinusoidal functions are the representations of the following power series (Kreyszig [2006]):

$$\begin{aligned} \cos(\omega_0 \Delta t) &= 1 - \frac{1}{2!}(\omega_0 \Delta t)^2 + \frac{1}{4!}(\omega_0 \Delta t)^4 - \dots \\ &= \sum_{m=0}^{\infty} \frac{(-1)^m (\omega_0 \Delta t)^{2m}}{(2m)!} \\ \sin(\omega_0 \Delta t) &= \omega_0 \Delta t - \frac{1}{3!}(\omega_0 \Delta t)^3 + \frac{1}{5!}(\omega_0 \Delta t)^5 - \dots \\ &= \sum_{m=0}^{\infty} \frac{(-1)^m (\omega_0 \Delta t)^{2m+1}}{(2m+1)!}. \end{aligned}$$

With the expression of the A_m *matrix, we find the discrete-time state space model for the sinusoidal signal as*

$$\begin{aligned} \begin{bmatrix} x_1(k+1) \\ x_2(k+1) \end{bmatrix} &= \begin{bmatrix} \cos(\omega_0 \Delta t) & \sin(\omega_0 \Delta t) \\ -\sin(\omega_0 \Delta t) & \cos(\omega_0 \Delta t) \end{bmatrix} \begin{bmatrix} x_1(k) \\ x_2(k) \end{bmatrix} \\ y(k) &= \begin{bmatrix} 1 & 0 \end{bmatrix} x(k). \end{aligned} \tag{3.26}$$

The eigenvalues of the system matrix A_m are the solutions of the characteristic polynomial equation:

$$\det[\lambda I - A_m] = (\lambda - \cos(\omega_0 \Delta t))^2 + \sin^2(\omega_0 \Delta t)$$
$$= \lambda^2 - 2\cos(\omega_0 \Delta t)\lambda + 1 = 0 \quad (3.27)$$

where we have used the equality: $\sin^2(\omega_0 \Delta t) + \cos^2(\omega_0 \Delta t) = 1$. *The solutions of the second order polynomial equation (3.27) are given as*

$$\lambda_{1,2} = \cos(\omega_0 \Delta t) \pm j \sin(\omega_0 \Delta t).$$

Note that the continuous-time system matrix A_c from (3.23) has a pair of complex eigenvalues on the imaginary axis, $\pm j\omega_0$. After the discretization, the discrete-time system matrix also has a pair of complex eigenvalues, which are, however, located on the unit circle of the complex plane with real part being $\cos(\omega_0 \Delta t)$ *and imaginary part being* $\pm \sin(\omega_0 \Delta t)$.

The characteristic polynomial equation (3.27) is particularly important to us in the design of control systems for disturbance rejection and reference tracking of periodic signals (see Chapter 5).

From Example 3.4, we have seen in (3.26) that the discrete-time model for a continuous-time sinusoidal signal contains the parameters that are sinusoidal functions of the continuous-time frequency ω_0 and the sampling interval Δt. Because of the periodicity of the sinusoidal function, the parameters in (3.26) have the following relations:

$$\cos(\omega_0 \Delta t) = \cos\left(\omega_0\left(1 + \frac{2\pi}{\Delta t}m\right)\Delta t\right)$$
$$\sin(\omega_0 \Delta t) = -\cos\left(\omega_0\left(1 + \frac{\pi}{\Delta t}m\right)\Delta t\right),$$

where $m = 0, 1, 2, 3, \ldots,$ is an integer. This means that if $\omega_0 \Delta t > \pi$, the parameters in the state space model are no longer unique. Therefore, in order to preserve or capture all the information in the sinusoidal signal, the sampling interval Δt should be limited such that

$$\omega_0 \Delta t \leq \pi,$$

which is

$$\Delta t \leq \frac{\pi}{\omega_0}.$$

The limitation on the sampling interval Δt can also be revealed by the *Sampling Theorem*. The *Sampling Theorem* states that if the highest frequency component in a signal is f_{max}, then the signal should be sampled at the rate of at least $2f_{max}$ for the sampled data to describe the signal completely. This means that we have

$$\frac{1}{\Delta t} \geq 2f_{max},$$

where $\frac{1}{\Delta t}$ is the sampling frequency or sampling rate. For the sinusoidal signal here, $\omega_0 = 2\pi f_{max}$, which leads to the limit on Δt being $\frac{\pi}{\omega_0}$.

Example 3.5

Consider the continuous-time sinusoidal signal $y(t) = \sin(\omega_0 t)$ *where* $\omega_0 = 2$ *(rad/s). Find the discrete-time state space models that describe the sinusoidal signal with sampling intervals* $\Delta t = 0.25\frac{\pi}{\omega_0}$ *and* $\Delta t = 0.5\frac{\pi}{\omega_0}$, *respectively and examine the variations of their eigenvalues.*

Solution

When $\Delta t = 0.25\frac{\pi}{\omega_0}$, $\sin(\omega_0 \Delta t) = \cos(\omega_0 \Delta t) = 0.7071$. *Therefore, the state space model to describe the discrete-time sinusoidal signal becomes*

$$\begin{bmatrix} x_1(k+1) \\ x_2(k+1) \end{bmatrix} = \begin{bmatrix} 0.7071 & 0.7071 \\ -0.7071 & 0.7071 \end{bmatrix} \begin{bmatrix} x_1(k) \\ x_2(k) \end{bmatrix}$$

$$y(k) = \begin{bmatrix} 1 & 0 \end{bmatrix} x(k).$$

The eigenvalues are $0.7071 \pm j0.7071$. *When* $\Delta t = 0.5\frac{\pi}{\omega_0}$, $\sin(\omega_0 \Delta t) = 1$ *and* $\cos(\omega_0 \Delta t) = 0$. *The discrete-time state space model becomes,*

$$\begin{bmatrix} x_1(k+1) \\ x_2(k+1) \end{bmatrix} = \begin{bmatrix} 0 & 1 \\ -1 & 0 \end{bmatrix} \begin{bmatrix} x_1(k) \\ x_2(k) \end{bmatrix}$$

$$y(k) = \begin{bmatrix} 1 & 0 \end{bmatrix} x(k).$$

The eigenvalues are $\pm j1$.

As an illustration, when $\omega_0 = 2$ *(rad/s), Figure 3.3(a) shows the discrete-time sinusoidal signal for* $\omega_0 \Delta t = 0.25\pi$ *and Figure 3.3 (b) shows the case with* $\omega_0 \Delta t = 0.5\pi$. *It is seen that the choice of* $\omega_0 \Delta t = 0.25\pi$ *leads to the discrete-time sinusoidal signal*

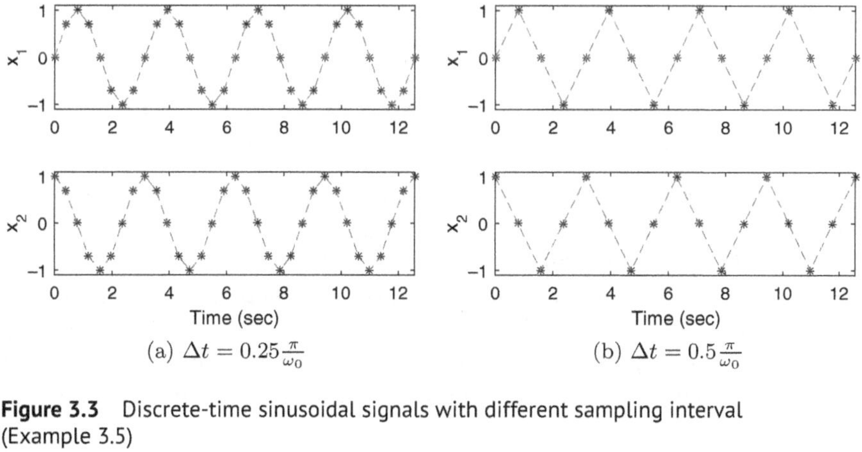

Figure 3.3 Discrete-time sinusoidal signals with different sampling interval (Example 3.5)

that has captured its original shape. However, with the case $\omega_0\Delta t = 0.5\pi$, the original sinusoidal signal becomes a triangular signal.

Example 3.6

Find the discrete-time state space model for the following system:

$$\dot{x}(t) = A_c x(t) + B_c u(t) \tag{3.28}$$

$$y(t) = C_c x(t), \tag{3.29}$$

where the system matrices are

$$A_c = \begin{bmatrix} -10 & 0.5 & 0 & 0 \\ -0.1 & -1 & 3 & 0 \\ 0.6 & 0 & -10 & -50 \\ 0 & 0 & 1 & 0 \end{bmatrix}; \; B_c = \begin{bmatrix} 1 \\ 0 \\ 1 \\ 1 \end{bmatrix};$$

$$C_c = \begin{bmatrix} 1 & 1 & 1 & 1 \end{bmatrix}$$

and the sampling interval $\Delta t = 0.01,\ 0.1,$ and 1 respectively. Observe the changes of eigenvalues of the discrete-time model.

Solution

We use the MATLAB function **c2dm.m** *to find the discrete-time state space model. After defining the continuous-time system matrices, Ac, Bc, Cc, and Δt, the discrete time state space model is found using the following code:*

```
[Am,Bm,Cm,Dm]=c2dm(Ac,Bc,Cc,0,Deltat,'zoh');
```

where the Dc parameter is zero, and we choose to use zero-order-hold in the discretization. As an illustration, we can verify that when $\Delta t = 0.1$, the following system matrices are obtained:

$$A_m = \begin{bmatrix} 0.3678 & 0.0298 & 0.0036 & -0.0074 \\ -0.0016 & 0.9048 & 0.1642 & -0.5119 \\ 0.0197 & 0.0007 & 0.2416 & -2.9080 \\ 0.0015 & 0.0000 & 0.0582 & 0.8231 \end{bmatrix}; \; B_m = \begin{bmatrix} 0.0632 \\ -0.0088 \\ -0.1173 \\ 0.0971 \end{bmatrix}.$$

There is not much pattern in the data structure of A_m and B_m matrices. However, the pattern becomes clearer with respect to the eigenvalues of the discrete-time system. The eigenvalues for the continuous-time system are $\lambda_1 = -9.9743$, $\lambda_{2,3} = -5.0088 \pm j4.9890$ and $\lambda_4 = -1.0080$. The eigenvalues of the system matrices of the discrete-time system are $\gamma_1 = e^{\lambda_1 \Delta t}$, $\gamma_{2,3} = e^{\lambda_{2,3}\Delta t}$ and $\gamma_4 = e^{\lambda_4 \Delta t}$. We observe that when $\Delta t = 0.01$, all four eigenvalues converge to the value 1 on the complex plane (see Figure 3.4(a) denoted by the o symbol). In contrast when $\Delta t = 1$, it is seen that the eigenvalues γ_1

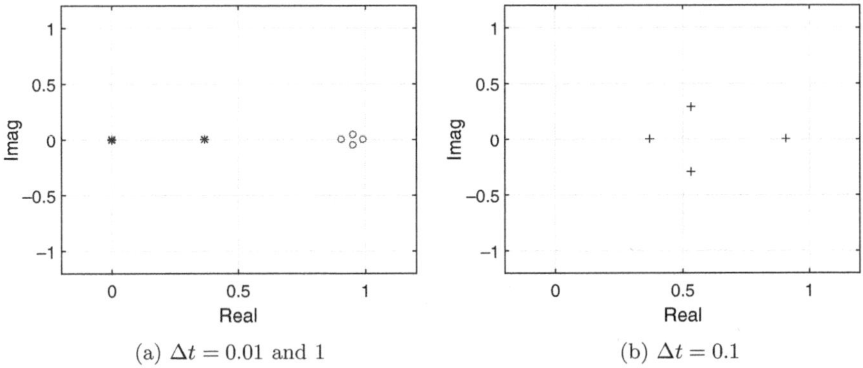

Figure 3.4 Eigenvalues of discrete-time system (Example 3.6). Key: * $\Delta t = 1$; o $\Delta t = 0.01$; + $\Delta t = 0.1$

and $\gamma_{2,3}$ converge to the origin of the complex plane (see Figure 3.4(a) denoted by the $*$ symbol). In comparison, when $\Delta t = 0.1$, Figure 3.4(b) shows that the four eigenvalues are distinct. Figures 3.4(a)-(b) show that when the sampling interval Δt is too small, the stable eigenvalues in the discrete-time system converge to the value 1, and when the sampling interval Δt is too large, the stable eigenvalues converge to the origin of the complex plane. Thus, the choice of sampling interval is very important in the discrete-time control system because it directly affects the dynamics of the system.

3.2.4 Food for Thoughts

1. What is the relationship between the sampling time t_k, sampling index k and sampling interval Δt?
2. How does a change in Δt affect the quantity $t_k - t_{k-1}$?
3. What is the key assumption on the input signal $u(t_k)$ used to derive the discrete-time state space model?
4. If the key assumption on the input signal is satisfied, is it correct to say that the discrete-time model is accurate at the sampling instant k?
5. Do you think that it is the simplest way to assume the zero-order-hold in the input signal when we do not have the information between the sampling instants?
6. Do you think that the discrete-time model will be more accurate if we replace the constant input between the two sampling instants with a linear function?
7. Assuming that a discrete-time system is described by $x(k+1) = -0.9x(k)$, is this system stable?
8. Assuming that $x(k+1) = 1.1x(k)$ and $x(0) = 1$, will $x(k) \to \infty$ as $k \to \infty$?

3.3 Input and Output Discrete-time Models

3.3.1 Input and Output Models

In the discrete-time, shift operators are defined, which will convert the operation of a difference equation to an algebraic operation. With the shift operators, higher order discrete-time models are naturally presented that will relate the input variables to the output variables.

The variable for the forward shift operator is defined as q, where for a discrete-time function $f(k)$, $k = \dots, -2, -1, 0, 1, 2, \dots$

$$qf(k) = f(k+1).$$

The inverse of the forward shift operator exists for the range of k where $k = \dots, -2, -1, 0, 1, 2, \dots$, and is defined as q^{-1} called the backward shift operator. For a discrete-time function $f(k)$,

$$q^{-1}f(k) = f(k-1).$$

The introduction of forward and backward shift operators (q, q^{-1}) gives us the tools to handle the difference equations using algebraic operations.

An immediate application of the shift operators is to convert the discrete-time state space models to models that only have input and output variables, *i.e.* without the state variables. With the forward shift operator q, the state variable $x(k+1) = qx(k)$, thus, the original state space model

$$x(k+1) = A_m x(k) + B_m u(k)$$

is written as

$$qx(k) = A_m x(k) + B_m u(k). \tag{3.30}$$

Noting that the difference operation is converted to the algebraic operation in terms of q, then, the state variable $x(k)$ is expressed as

$$x(k) = (qI - A_m)^{-1} B_m u(k). \tag{3.31}$$

Together with the output equation,

$$y(k) = C_m x(k) + D_m u(k)$$

$y(k)$ is expressed as

$$y(k) = [C_m (qI - A_m)^{-1} B_m + D_m] u(k). \tag{3.32}$$

The expression $[C_m (qI - A_m)^{-1} B_m + D_m]$ relates the discrete-time input variable $u(k)$ to the output variable $y(k)$. It is seen from (3.32) that the state variable $x(k)$ is eliminated and the model only contains the input and output variables.

Example 3.7

Given a continuous-time state space model,

$$\dot{x}(t) = A_c x(t) + B_c u(t); \; y(t) = C_c x(t)$$

where the system matrices are

$$A_c = \begin{bmatrix} -1 & 0 \\ 0 & -2 \end{bmatrix}; \; B_c = \begin{bmatrix} 1 \\ 2 \end{bmatrix}; \; C_c = \begin{bmatrix} 1 & 1 \end{bmatrix}$$

find the discrete-time model that relates input $u(k)$ to output $y(k)$ via the forward shift operator q. We assume that the initial condition $x(0) = 0$. The sampling interval Δt is chosen to be 0.1.

Solution

We will first find the discrete-time state space model:

$$x(k+1) = A_m x(k) + B_m u(k); \; y(k) = C_m x(k).$$

Because A_c is a diagonal matrix, $e^{A_c \Delta t}$ is also a diagonal matrix. Thus, we can find its analytical solution with two first order components as

$$\begin{aligned} x_1(k+1) &= e^{-\Delta t} x_1(k) - (e^{-\Delta t} - 1)u(k) \\ &= 0.9048 x_1(k) + 0.0952 u(k) \\ x_2(k+1) &= e^{-2\Delta t} x_2(k) - (e^{-2\Delta t} - 1)u(k) \\ &= 0.8187 x_2(k) + 0.1813 u(k) \\ y(k) &= x_1(k) + x_2(k). \end{aligned}$$

Letting $x_1(k+1) = qx_1(k)$ and $x_2(k+1) = qx_2(k)$, then we have the two state variables:

$$\begin{aligned} x_1(k) &= \frac{0.0952}{q - 0.9048} u(k) \\ x_2(k) &= \frac{0.1813}{q - 0.8187} u(k) \end{aligned}$$

leading to the expression of the output using the forward shift operator:

$$y(k) = \left(\frac{0.0952}{q - 0.9048} + \frac{0.1813}{q - 0.8187} \right) u(k) \tag{3.33}$$

$$= \frac{0.2764q - 0.2419}{q^2 - 1.7236q + 0.7408} u(k). \tag{3.34}$$

Furthermore, this relationship can be expressed as

$$(q^2 - 1.7236q + 0.7408) y(k) = (0.2764q - 0.2419) u(k)$$

Now, by using the definition of the operator $q^2 y(k) = y(k+2)$, $qy(k) = y(k+1)$, $qu(k) = u(k+1)$, we obtain the discrete-time model that relates the input signal to *the output signal without the state variables as*

$$y(k+2) - 1.723y(k+1) + 0.7408y(k) = 0.2764u(k+1) - 0.2419u(k).$$

Example 3.8

Continuing from Example 3.7, find the discrete-time model relating input to output using the backward shift operator.

Solution

We multiply the numerator and denominator in (3.34) by q^{-2}, leading to

$$\begin{aligned} y(k) &= \frac{q^{-2}(0.2764q - 0.2419)}{q^{-2}(q^2 - 1.7236q + 0.7408)}u(k) \\ &= \frac{0.2764q^{-1} - 0.2419q^{-2}}{1 - 1.7236q^{-1} + 0.7408q^{-2}}u(k). \end{aligned} \tag{3.35}$$

Equation (3.35) is expressed as

$$(1 - 1.7236q^{-1} + 0.7408q^{-2})y(k) = (0.2764q^{-1} - 0.2419q^{-2})u(k)$$

With the definition of the backward shift operator, $q^{-1}y(k) = y(k-1)$, $q^{-2}y(k) = y(k-2)$, $q^{-1}u(k) = u(k-1)$, $q^{-2}u(k) = u(k-2)$, we obtain the following difference equation:

$$y(k) - 1.723y(k-1) + 0.7408y(k-2) = 0.2764u(k-1) - 0.2419u(k-2).$$

From Examples 3.7 and 3.8, we can see that the shift operators are flexible in the descriptions of the system dynamics in discrete-time, and we can choose to use either the forward shift representation or the backward shift representation. In general, the backward shift representation is better suited if we need to handle real-time data because it uses the past values of the input and output data that are available to us at the sampling instant k.

3.3.2 Finite Impulse Response and Step Response Models

The Finite Impulse Response (FIR) model is widely used in system identification to obtain a mathematical model from experimental data. This model describes system dynamics using backward shift operator with the assumption that the discrete-time system is stable with all eigenvalues strictly inside the unit circle. To find the general expression of a finite impulse response model, we examine (3.32) where D_m is assumed to be zero:

$$\begin{aligned} y(k) &= C_m(qI - A_m)^{-1}B_m u(k) \\ &= C_m(I - A_m q^{-1})^{-1}q^{-1}B_m u(k) \\ &= [C_m B_m q^{-1} + C_m A_m B_m q^{-2} + \ldots + C_m A_m^{i-1} B_m q^{-i} + \ldots]u(k). \end{aligned} \tag{3.36}$$

Because the system is stable with $|\lambda_i(A_m)| < 1$ for all i, the coefficients from the power series expansion will converge to zero. Thus, as an approximation, there will be a finite number of terms in the discrete-time model (3.36). When the input signal $u(k)$ is a Kronecker delta function $\delta(k)$, which is the discrete- time impulse function defined by,

$$u(k-i) = \delta(k-i) = \begin{cases} 1 \text{ if } k = i \\ 0 \text{ if } k \neq i \end{cases}$$

the coefficients of the model (3.36) form the impulse response of the system.

In the same category, a step response model exists. When the input signal $u(k)$ is a step signal, which is defined by,

$$u(k-i) = \begin{cases} 1 \text{ if } k \geq i \\ 0 \text{ if } k < i \end{cases}$$

then the coefficients of the step response model consist of 0 ($k = 0$), C_mB_m ($k = 1$), $C_mB_m + C_mA_mB_m$ ($k = 2$), …, and the kth coefficient $\sum_{i=1}^{k} C_mA_m^{k-i}B_m$.

One of the advantages in using the FIR model or step response model is that they are transparent in providing information about the time delay in the system. A time delay describes the time interval between the time of the injection of an input signal in a system and the another time of its resulting response. With this understanding, we assume that a system has time delay τ_d, and if the sampling interval Δt is selected as

$$\Delta t = \frac{\tau_d}{n_d},$$

where n_d is an integer for simplicity, then the time delay information in the discrete-time FIR or step response model is captured by the number of zero coefficients. The FIR model for the time delay system becomes

$$\begin{aligned} y(k) &= [C_mA_m^{n_d-1}B_mq^{-n_d} + \ldots + C_mA_m^{i+n_d-1}B_mq^{-i-n_d} + \ldots]u(k) \\ &= C_mA_m^{n_d-1}B_mu(k-n_d) + \ldots + C_mA_m^{i+n_d-1}B_mu(k-n_d-i) + \ldots. \end{aligned}$$

Similarly, for the step response model, the first n_d coefficients are zero to reflect the time delay in the system.

Example 3.9

Considering the continuous-time system described by the Laplace transfer function model:

$$G(s) = \frac{1}{(5s+1)^2}e^{-2s}$$

find the coefficients of the discrete-time step response model and the impulse response model for $k = 0, 1, 2, \ldots, 59$. *We assume that sampling interval* $\Delta t = 0.5$.

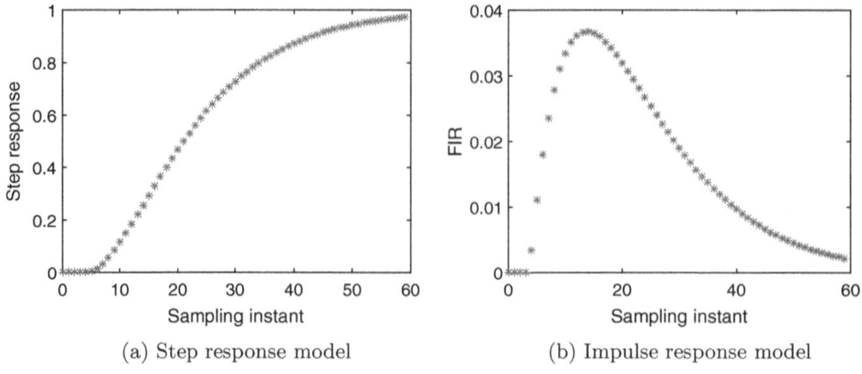

Figure 3.5 Discrete-time higher order model (Example 3.9)

Solution

The coefficients for the step response model g(k), $k = 0, 1, 2, \ldots, 59$ are generated using Simulink simulation where the sampling interval $\Delta t = 0.5$. The corresponding FIR model with coefficients h(k) is generated by taking difference operation of g(k), which is $h(k) = g(k+1) - g(k)$. Figure 3.5(a) shows the coefficients for the step response model and Figure 3.5(b) shows the coefficients for the FIR model. Indeed, there is a time delay in the system indicated by the fact that the first four coefficients of both the FIR model and step response model are zero where $n_d = \frac{\tau_d}{\Delta t} = \frac{2}{0.5} = 4$.

In general, assuming that a continuous-time system with time delay τ_d is contained in the input signal, which is

$$\dot{x}(t) = A_c x(t) + B_c u(t - \tau_d);$$
$$y(t) = C_c x(t),$$

if the sampling interval Δt is selected as $\Delta t = \frac{\tau_d}{n_d}$ where n_d is an integer, then the discrete-time model has the following form:

$$x(k+1) = A_m x(k) + B_m u(k - n_d) \tag{3.37}$$
$$y(k) = C_m x(k), \tag{3.38}$$

where $A_m = e^{A_c \Delta t}$ and $B_m = \left(\int_0^{\Delta t} e^{A_c \tau} d\tau\right) B_c$. This is based on the assumption that the number of delayed samples n_d is an integer, and as a result the input signal $u(k - n_d)$ is a constant with the zero-order-hold for each interval $t_k \leq t < t_{k+1}$. If $\frac{\tau_d}{\Delta t}$ is not an integer, (3.37) does not hold because the delayed signal will change between the sampling instants, thus a more comprehensive scheme is required to deal with this scenario (see Astrom and Wittenmark [1997]).

The following example is to show how to augment the state variables in order to take into account of the time delay in the state space model.

Example 3.10

Assume that a continuous-time system is described by the following state space model:

$$\dot{x}(t) = A_c x(t) + B_c u(t - \tau_d); \; y(t) = C_c x(t), \tag{3.39}$$

where the time delay $\tau_d = 3$ *(sec) and the system matrices are*

$$A_c = \begin{bmatrix} -0.1 & 0.2 \\ 0 & -0.5 \end{bmatrix}; \; B_c = \begin{bmatrix} 1 \\ 1 \end{bmatrix}; \; C_c = \begin{bmatrix} 1 & 1 \end{bmatrix}.$$

With the sampling interval $\Delta t = 1$ *(sec), find the discrete-time state space model that includes the time-delay in the state variables.*

Solution

We will first use the MATLAB function **c2dm.m** *to find the discrete-time system matrices:*

```
[Am,Bm,Cm,Dm]=c2dm(Ac,Bc,Cc,0,deltat,'zoh');
```

which is

$$x_m(k+1) = A_m x_m(k) + B_m u(k - n_d) \tag{3.40}$$

$$y(k) = C_m x_m(k), \tag{3.41}$$

where $n_d = 3$, A_m, B_m *and* C_m *are*

$$A_m = \begin{bmatrix} 0.9048 & 0.1492 \\ 0 & 0.6065 \end{bmatrix}; \; B_m = \begin{bmatrix} 1.0340 \\ 0.7869 \end{bmatrix} \; C_m = \begin{bmatrix} 1 & 1 \end{bmatrix}.$$

Now, we choose the augmented state vector $x(k)$ *to contain the original* $x_m(k)$ *and the past control signals as*

$$x(k) = \begin{bmatrix} x_m(k)^T & u(k-1) & u(k-2) & u(k-3) \end{bmatrix}^T$$

leading to the state space model:

$$x(k+1) = Ax(k) + Bu(k); \; y(k) = Cx(k), \tag{3.42}$$

where the augmented system matrices are

$$A = \begin{bmatrix} 0.9048 & 0.1492 & 0 & 0 & 0 \\ 0 & 0.6065 & 0 & 0 & 0 \\ 0 & 0 & 0 & 0 & 0 \\ 0 & 0 & 1 & 0 & 0 \\ 0 & 0 & 0 & 1 & 0 \end{bmatrix}; \; B = \begin{bmatrix} 1.0340 \\ 0.7869 \\ 1 \\ 0 \\ 0 \end{bmatrix}; \; C = \begin{bmatrix} 1 & 1 & 0 & 0 & 0 \end{bmatrix}.$$

Note that the inclusion of past control signals in the state variables has increased the model order of the system, where three eigenvalues of A matrix are at the origin of the complex plane.

3.3.3 Non-minimal State Space Realization

When the state variables are chosen to correspond to a set of plant input and output variables, the dimension of the states is increased and the state space model is no longer a minimal realization. This type of models is called Non-Minimal State Space (NMSS) model and is used in control system design. The following example is to show how to convert a shift operator model into an NMSS model.

Example 3.11
A discrete-time system is described by the following model:

$$y(k) = \frac{q - 1.1}{(q - 0.7)(q - 0.9)} u(k). \tag{3.43}$$

Convert this shift operator model to an NMSS model with the state variables corresponding to the measured input and output variables $u(k)$ and $y(k)$.

Solution
We will represent the shift operator model (3.43) using backward shift operator, which is

$$y(k) = \frac{q^{-1} - 1.1q^{-2}}{1 - 1.6q^{-1} + 0.63q^{-2}} u(k). \tag{3.44}$$

The corresponding difference equation is

$$y(k) = 1.6y(k-1) - 0.63y(k-2) + u(k-1) - 1.1u(k-2). \tag{3.45}$$

We choose the state vector $x_m(k) = \begin{bmatrix} y(k) & y(k-1) & u(k-1) \end{bmatrix}^T$ to obtain the following state space model:

$$\begin{bmatrix} y(k+1) \\ y(k) \\ u(k) \end{bmatrix} = \overbrace{\begin{bmatrix} 1.6 & -0.63 & -1.1 \\ 1 & 0 & 0 \\ 0 & 0 & 0 \end{bmatrix}}^{A_m} \begin{bmatrix} y(k) \\ y(k-1) \\ u(k-1) \end{bmatrix} + \overbrace{\begin{bmatrix} 1 \\ 0 \\ 1 \end{bmatrix}}^{B_m} u(k). \tag{3.46}$$

Note that the state vector $x_m(k)$ contains all the variables that are either measured or given at the sampling instant k, which means that $x_m(k)$ is available for a control application.

3.3.4 Food for Thought

1. Assuming the following state space model:

$$x(k+1) = x(k) + u(k); \quad y(k) = 2x(k)$$

   what is the input and output model in forward shift operator? What is the model in backward shift operator?

2. Following from the above question, using the model in backward shift operator, with the assumptions that $y(-1) = 0$, and $u(k)$ is a unit step signal, determine the values of $y(0), y(1), \dots, y(6)$. What are your observations about the trajectory of signal $y(k)$?
3. Repeat Question 2, however, with a unit impulse signal $u(k)$. What are your observations about the trajectory of $y(k)$? Compare the impulse response with the step response.
4. A discrete-time system is described by the following model:

$$y(k) = \frac{0.9q^{-3}}{1 - 0.1q^{-1}} u(k). \tag{3.47}$$

   What is the corresponding difference equation? Assuming that $u(k)$ is a unit step signal, $y(-1) = 0$, determine the values of $y(0), y(1), \dots, y(8)$. What are your observations about the trajectory of $y(k)$?
5. What is the mathematical model for (3.47) using the forward shift operator? What is the corresponding difference equation using the forward shift operator?
6. With the difference equation obtained using the forward shift operator, what initial conditions on $y(k)$ do we need to assume in order to obtain the same trajectory of $y(k)$ as in Question 4 above?
7. Would you say that it is likely more convenient to use the model with backward shift operator when we deal with real-time data?
8. A discrete-time system is described by the following shift operator model:

$$y(k) = \frac{0.3q^{-2} + 0.4q^{-3}}{(1 - 0.9q^{-1})(1 - 0.8q^{-1})} u(k).$$

   What is the NMSS state space representation for this system? How many eigenvalues does the system matrix have? Where are they?

3.4 z-Transforms

3.4.1 z-Transforms for Commonly Used Signals

The discrete-time counterpart of the Laplace transform is the z-transform. The z-transform maps the infinite time sequence $f(k)$ where $k = 0, 1, 2, \dots$, to a function $F(z)$ of complex variable z, where $z = e^{j\omega}$ with ω varying on the unit circle of the complex plane ($-\pi \leq \omega \leq \pi$). Hence, with the z-transform, we will have tools to perform frequency response analysis for discrete-time systems.

More specifically, for a discrete-time signal $f(k)$, its z-transform $F(z) = Z[f(k)]$ is defined as

$$F(z) = \sum_{k=0}^{\infty} f(k) z^{-k}. \tag{3.48}$$

The inverse of the z-transform is the following integration:

$$f(k) = \frac{1}{2\pi j} \oint F(z) z^{k-1} dz, \tag{3.49}$$

where the contour of the integration encloses all singularities of $F(z)$.

The following examples show how to obtain the z-transforms for the commonly used input signals.

Example 3.12

When the input signal is the Kronecker delta function $\delta(k)$, which is the discrete-time impulse function defined by,

$$\delta(k) = \begin{cases} 1 & if\ k = 0 \\ 0 & if\ k \neq 0 \end{cases}$$

find its z-transform.

Solution

From the definition of z-transform, we have

$$\begin{aligned} Z[\delta(k)] &= \sum_{k=0}^{\infty} \delta(k) z^{-k} \\ &= 1 \end{aligned} \tag{3.50}$$

because $\delta(0) = 1$, and $\delta(2) = \delta(3) = \ldots = 0$.

Example 3.13

A unit step input signal $u(k)$ is defined as

$$u(k) = \begin{cases} 1 & if\ k \geq 0 \\ 0 & if\ k < 0 \end{cases}$$

Find its z-transform.

Solution

From the definition of the z-transform, we have the infinite series:

$$Z[u(k)] = \sum_{k=0}^{\infty} z^{-k}$$

which can be written as

$$\begin{aligned} \sum_{k=0}^{\infty} z^{-k} &= 1 + z^{-1} + z^{-2} + \ldots + z^{-m} + \ldots \\ &= 1 + z^{-1} \sum_{k=0}^{\infty} z^{-k}. \end{aligned}$$

Therefore, $(1 - z^{-1}) \sum_{k=0}^{\infty} z^{-k} = 1$*, which leads to*

$$Z[u(k)] = \frac{1}{1 - z^{-1}}. \tag{3.51}$$

Example 3.14

When the input signal u(k) is a ramp signal where

$$u(k) = \begin{cases} k & \text{if } k \geq 0 \\ 0 & \text{if } k < 0 \end{cases}$$

find its z-transform.

Solution

From the definition of z-transform, we have the following infinite series:

$$Z[u(k)] = \sum_{k=0}^{\infty} kz^{-k} = 0 + z^{-1} + 2z^{-2} + 3z^{-3} + \ldots + \ldots$$

Note that for the infinite series above we have the following equality:

$$(1 - 2z^{-1} + z^{-2}) \sum_{k=0}^{\infty} kz^{-k} = z^{-1}.$$

Therefore, the z-transform for the ramp signal u(k) is given as

$$Z[u(k)] = \sum_{k=0}^{\infty} kz^{-k} = \frac{z^{-1}}{1 - 2z^{-1} + z^{-2}}. \tag{3.52}$$

Example 3.15

A sinusoidal input signal in discrete-time is given by

$$u(k) = a_m \cos(\omega_0 \Delta t k + \phi) = a_m \cos(\omega_d k + \phi),$$

where $\omega_d = \omega_0 \Delta t$ *is the discrete-time frequency. Find the z-transform of this sinusoidal signal.*

Solution

We consider the following equality:

$$a_m \cos(\omega_d k + \phi) = \frac{a_m}{2} \left[e^{j\phi} e^{j\omega_d k} + e^{-j\phi} e^{-j\omega_d k} \right].$$

Then, from the definition of the z-transform, we have

$$Z[u(k)] = \frac{a_m}{2} \left[e^{j\phi} \sum_{k=0}^{\infty} e^{j\omega_d k} z^{-k} + e^{-j\phi} \sum_{k=0}^{\infty} e^{-j\omega_d k} z^{-k} \right].$$

From the power series expansions (Kreyszig [2006]), we have

$$\sum_{k=0}^{\infty} e^{j\omega_d k} z^{-k} = \frac{1}{1 - e^{j\omega_d} z^{-1}}$$

$$\sum_{k=0}^{\infty} e^{-j\omega_d k} z^{-k} = \frac{1}{1 - e^{-j\omega_d} z^{-1}}$$

which leads to

$$Z[u(k)] = \frac{a_m}{2}\left[\frac{e^{j\phi}}{1 - e^{j\omega_d} z^{-1}} + \frac{e^{-j\phi}}{1 - e^{-j\omega_d} z^{-1}}\right]$$

$$= a_m \frac{\cos\phi - \cos(\omega_d - \phi) z^{-1}}{1 - 2\cos\omega_d z^{-1} + z^{-2}} \tag{3.53}$$

where we have used the equality:

$$2\cos\alpha = e^{j\alpha} + e^{-j\alpha}$$

which is based on the Euler formula,

$$e^{j\alpha} = \cos\alpha + j\sin\alpha.$$

For the case when $\phi = 0$,

$$Z[u(k)] = a_m \frac{1 - \cos\omega_d z^{-1}}{1 - 2\cos\omega_d z^{-1} + z^{-2}}.$$

When $\phi = \frac{\pi}{2}$, $u(k) = a_m \sin(\omega_d k)$, *then*

$$Z[a_m \sin(\omega_d k)] = a_m \frac{\sin\omega_d z^{-1}}{1 - 2\cos\omega_d z^{-1} + z^{-2}}. \tag{3.54}$$

3.4.2 *z*-Transfer Functions

Assume that a discrete-time system has a single input $u(k)$ and a single output $y(k)$, and their respective z-transforms are $U(z)$ and $Y(z)$. The z-transfer function for this system is written as

$$G(z) = \frac{Y(z)}{U(z)} = \frac{b_1 z^{n-1} + b_2 z^{n-2} + \ldots + b_n}{z^n + a_1 z^{n-1} + a_2 z^{n-2} + \ldots + a_n}. \tag{3.55}$$

Associated with the z-transfer function are the poles and zeros of the discrete-time system. The poles of the discrete-time system are the solutions of the following polynomial equation:

$$z^n + a_1 z^{n-1} + a_2 z^{n-2} + \ldots + a_n = 0 \tag{3.56}$$

which is based on the denominator of (3.55). The zeros of the discrete-time system are the solutions of the polynomial equation:

$$b_1 z^{n-1} + b_2 z^{n-2} + \ldots + b_n = 0 \tag{3.57}$$

which is based on the numerator of (3.55). The order of the z-transfer function is equal to the number of poles determined by the polynomial equation (3.56), which is expressed in a forward shift form.

The transfer function (3.55) is expressed in a forward shift form. By multiplying z^{-n} on both numerator and denominator of (3.55), we obtain the same transfer function with a backward shift form:

$$G(z) = \frac{Y(z)}{U(z)} = \frac{b_1 z^{-1} + b_2 z^{-2} + \ldots + b_n z^{-n}}{1 + a_1 z^{-1} + a_2 z^{-2} + \ldots + a_n z^{-n}}. \tag{3.58}$$

The section is completed with the following properties for z-transform.

1. **Linearity.** Assume that $w(k)$ and $v(k)$ have z- transforms $W(z) = Z[w(k)]$ and $V(z) = Z[v(k)]$ respectively. If $y(k) = c_1 w(k) + c_2 v(k)$, where c_1 and c_2 are constant, then

$$Z[y(k)] = c_1 W(z) + c_2 V(z). \tag{3.59}$$

2. **Backward shift property.** Assuming that $f(k)$ has z-transform $F(z)$ and $y(k) = f(k-m)$, where $m \geq 0$ is an integer, we have

$$Y(z) = z^{-m} F(z). \tag{3.60}$$

3. **Forward shift property.** Assuming that $f(k)$ has z-transform $F(z)$ and $y(k) = f(k+m)$, we have

$$Y(z) = z^m F(z) - z^m \sum_{i=0}^{m-1} f(i) z^{-i} \tag{3.61}$$

where $m \geq 0$ is an integer. For instance, say $m = 3$, then

$$Y(z) = z^3 F(z) - (z^3 f(0) + z^2 f(1) + z f(2)).$$

4. **Convolutional sum.** Assume that $f(k)$ and $g(k)$ have z-transforms $F(z)$ and $G(z)$, respectively. Then the z-transform of the convolutional sum is

$$Z\left[\sum_{i=0}^{k} f(i) g(k-i)\right] = F(z)G(z). \tag{3.62}$$

5. **Final value theorem.** Final value theorem provides us with an important tool for the discrete-time system analysis. Final value theorem in discrete-time system states that assuming that $f(k)$ has z-transform $F(z)$ and $(1 - z^{-1})F(z)$ does not have any poles on or outside of the unit circle, then

$$\lim_{k \to \infty} f(k) = \lim_{z \to 1} (1 - z^{-1}) F(z). \tag{3.63}$$

By comparing the z-transforms with the shift operators q introduced in Section 3.3.1, a few comments follow.

1. The z-transform is conceptually different from the shift operator q, where the variable z is defined as a complex variable: $z = e^{j\omega}$ with ω varying on the unit circle of the complex plane ($-\pi \leq \omega \leq \pi$) while the shift operator q represents a mathematical operation that converts a difference equation into an algebraic operation.
2. Note that in the z-transform, there is the backward shift property (see (3.60)). The backward shift property of z-transform says that if $y(k) = f(k - m)$, where $m \geq 0$ is an integer, we have

$$Y(z) = z^{-m}F(z).$$

In comparison, in terms of the backward shift operator, if $y(k) = f(k - m)$ and the backward shift operation is applied to $y(k)$, then

$$q[y(k)] = q^{-m}f(k).$$

When we are familiar with the discrete-time models, in practice, given a transfer function model, we often replace the z^{-1} by the shift operator q^{-1} directly, which then leads to difference equation in time domain. Conversely, given a shift operator model, we replace q^{-1} by z^{-1} directly, which then leads to frequency response analysis. This basically says that the discrete-time models offer a flexibility for us to interchange between the time domain and frequency domain without much complexity.
3. The model order, poles and zeros are calculated using the z-transfer function in forward shift form.

3.4.3 Food for Thought

1. Where is the pole of the z-transform for a step signal? Is it related to the amplitude of the signal?
2. Where are the poles of the z-transform for a ramp signal?
3. Where are the poles of the z-transform for a sinusoidal signal? What are they dependent on?
4. What are your observations on the pole locations for these commonly used signals?
5. What is the z-transform for the signal

$$y(k) = a^k?$$

6. Where is the pole location of the z-transform for $y(k) = a^k$? Assuming $a = 0.8$, find $y(k)$ for $k = 0, 1, 2, \dots, 6$. Repeat the same exercise for $a = 1.1$. What are

your observations with respect to the value of a, location of the pole and the behavior of the sequence in time domain?

7. Assuming that a discrete-time system is described by the z-transfer function:

$$G(z) = \frac{0.1}{z - 0.8}$$

what is the corresponding difference equation with respect to input signal $u(k)$ and output signal $y(k)$ with the forward shift operator?

8. The z-transfer function of a system is given by

$$G(z) = \frac{0.8}{z - 0.2}. \tag{3.64}$$

Supposing that the input signal $u(k)$ is a unit step, what is the final value for the output $y(k)$? Calculate the $y(k)$ for $k = 0, 1, \ldots, 6$. Do you think that the trajectory of $y(k)$ will converge to the final value?

9. If the input signal is a sinusoidal signal, for the same system given by (3.64), can we apply final value theorem to determine the value of $y(k)$ as $k \to$? If not, why?

3.5 Summary

In this chapter, we have discussed sampling of a continuous-time system, discrete-time state space models, discrete-time input and output models, and z-transforms. The important aspects of this chapter are summarized as follows.

1. The derivation of a discrete-time state space model from sampling of a continuous-time state space model is presented. The discrete-time model is accurate at the sampling instant k with the assumption that a zero-order-hold is used in the input signal.
2. The stability of a discrete-time model is determined by the locations of the eigenvalues of the system matrix A_m. In order for a discrete-time system to be stable, all eigenvalues must be strictly inside the unit circle of the complex plane.
3. Sampling of commonly encountered continuous-time signals is discussed. The discrete-time models for these signals will be used in the discrete-time control system design in the later chapters.
4. The eigenvalues of a discrete-time model are directly related to their continuous-time counterpart.
5. The concept of shift operators, both forward and backward shift, is important in discrete-time systems. The introduction of shift operators gives us the tools to handle the difference equations using algebraic operations.

6. Time-delay in a discrete-time system is calculated as the number of delayed samples n_d where $n_d = \frac{\tau_d}{\Delta t}$ is assumed an integer and τ_d is the time-delay in continuous-time. The effect of the time delay is revealed in the finite impulse response model or in the step response model.
7. The discrete-time counterpart of the Laplace transform in continuous-time is the z-transform. The z-transform gives us the tools to perform frequency analysis of a discrete-time system. With the z-transform, a z-transfer function for a discrete-time system is defined. The order of a discrete-time system is equal to the number of poles of the z-transfer function. The number of delayed samples corresponds to the number poles on the origin of the complex plane.

3.6 Further Reading

1. The textbooks in discrete-time systems include Astrom and Wittenmark [1997], Chen [1998], Bay [1999], Veloni and Miridakis [2017].
2. The topics about sampling of continuous-time signals and systems are discussed in Kalman and Bertram [1959], Feuer and Goodwin [1997], Astrom and Wittenmark [1997].
3. There are many books for the topics of system identification (see Ljung [1999], Soderstrom and Stoica [1989], Goodwin and Sin [1984], Young [2012], Soderstrom [2018]).
4. Books about linear algebra and matrix computations include Watkins [2004], Demmel [1997], Trefethen and Bau III [1997], Lax [2007].

Problems

Problem 3.1 Given a continuous-time state space model:

$$\dot{x}(t) = A_c x(t) + B_c u(t); \; y(t) = C_c x(t),$$

find the corresponding discrete-time state space model and determine its eigenvalues as well as stability of the discrete-time system. Discuss the relationships between the eigenvalues of the continuous-time and discrete-time models. MATLAB function **c2dm.m** is suggested for the discretization. Matrix D_c is assumed to be zero.

1.

$$A_c = \begin{bmatrix} -1 & 0 \\ 2 & -3 \end{bmatrix}; \; B_c = \begin{bmatrix} 1 & 1 \\ 0 & 2 \end{bmatrix}; \; C_c = \begin{bmatrix} 1 & 0 \end{bmatrix},$$

where sampling interval $\Delta t = 0.1$.

2.

$$A_c = \begin{bmatrix} 1 & 2 & 3 \\ 1 & 1 & 2 \\ 0 & 2 & -1 \end{bmatrix}; \ B_c = \begin{bmatrix} 1 \\ 2 \\ 1 \end{bmatrix};$$

$$C_c = \begin{bmatrix} 1 & 2 & 1 \\ 0 & 1 & 2 \end{bmatrix},$$

where sampling interval $\Delta t = 0.2$.

3.

$$A_c = \begin{bmatrix} -2 & 3 & 0 & 0 \\ 2 & -3 & 2 & 0 \\ 1 & 2 & -1 & 0 \\ 1 & 2 & 0 & -6 \end{bmatrix}; \ B_c = \begin{bmatrix} 2 & 1 \\ 1 & 0 \\ 2 & 2 \\ 0 & 3 \end{bmatrix};$$

$$C_c = \begin{bmatrix} 1 & 0 & 0 & 0 \\ 0 & 1 & 0 & 0 \end{bmatrix},$$

where sampling interval $\Delta t = 0.3$.

Problem 3.2 Given a continuous-time transfer function model $G(s)$, find the corresponding discrete-time state space model and determine its eigenvalues and stability of the discrete-time system. Discuss the relationships between the poles of the continuous-time transfer function and the eigenvalues of the discrete-time model. It is suggested to use (1) MATLAB function **tf.m** to enter the transfer function model; (2) MATLAB function **ssdata.m** to convert the transfer function model to state space model; (3) MATLAB function **c2dm.m** to obtain the discrete-time state space model.

1.

$$G(s) = \frac{1}{s^2 + 2\xi w_n s + w_n^2},$$

where $w_n = 2$, $\xi = 0.707$ and $\Delta t = 0.05$.

2.

$$G(s) = \begin{bmatrix} \frac{1}{s+1} & \frac{1}{s+3} \\ \frac{2}{s+4} & \frac{3(-s+3)}{(s+5)(s+10)} \end{bmatrix},$$

where sampling interval $\Delta t = 0.06$.

3.

$$G(s) = \begin{bmatrix} \frac{-s+2}{(s+1)^3} & 0 & \frac{0.1}{s+1} \\ \frac{-1}{(s+2)(s+0.5)} & \frac{0.5}{(s+3)^2} & 0 \\ 0 & 0 & \frac{s+2}{(s^2+0.1s+2)} \end{bmatrix},$$

where sampling interval $\Delta t = 0.05$.

Problem 3.3 Given a continuous-time transfer function model $G(s)$ that has a time delay where

$$G(s) = \frac{0.5e^{-3s}}{(s+0.1)(s+0.2)^2}$$

find the discrete-time state space model:

$$x(k+1) = A_m x(k) + B_m u(k - n_d); \; y(k) = C_m x(k)$$

and determine the eigenvalues of A_m matrix and the parameter n_d that is the number of samples in delay. The sampling interval Δt is chosen to be 0.2.

Problem 3.4 A continuous-time system is described by the transfer function:

$$G(s) = \frac{e^{-8s}}{(10s+1)^3}.$$

Build a Simulink program to compute its step response with sampling interval $\Delta t = 1$ (sec) and simulation time $T_{sim} = 200$ (sec).

1. Graphically present the step response coefficients for $t = 0, \Delta t, \dots, T_{sim}$.
2. With $\Delta t = 4$ (sec), repeat the computation of step response coefficients with the Simulink program. Graphically compare the step response using sampling interval $\Delta t = 4$ with the previous response. Discuss your observations in terms of the number of model coefficients and the values of the coefficients with respect to the variations of the sampling interval Δt.

Problem 3.5 Continue from the previous Problem 3.4. Compute the impulse response coefficients by taking difference operation of the step response coefficients. Graphically compare the two sets of impulse response coefficients where $\Delta t = 1$ and 4, respectively. Discuss your observations in terms of the number of model coefficients and the values of the impulse response coefficients with respect to the sampling interval Δt.

Problem 3.6 Given the finite impulse response model described by the difference equation:

$$y(k) = 0.6u(k-3) + 0.5u(k-4) + 0.2u(k-5) + 0.1u(k-6) + 0.01u(k-7)$$

find the state space model:

$$x(k+1) = A_m x(k) + B_m u(k);\ y(k) = C_m x(k)$$

where the state variables are chosen as,

$$x(k) = \left[u(k-1)\ \ u(k-2)\ \ \ldots\ \ u(k-7)\right]^T.$$

Problem 3.7 Assume that a discrete-time system with two inputs and two outputs is described by the finite impulse response model with the following z-transfer function:

$$G(z) = h_2 z^{-2} + h_3 z^{-3} + h_4 z^{-4} + h_5 z^{-5}$$

where the coefficient matrices are

$$h_2 = \begin{bmatrix} 0 & 0.1 \\ 0.2 & 0 \end{bmatrix};\ h_3 = \begin{bmatrix} 0.6 & -0.1 \\ 0 & 0 \end{bmatrix};\ h_4 = \begin{bmatrix} 0.1 & -0.05 \\ -0.1 & 0.4 \end{bmatrix};\ h_5 = \begin{bmatrix} 0.01 & 0 \\ 0 & 0.1 \end{bmatrix}.$$

Convert this z-transfer function into the state space model:

$$x(k+1) = A_m x(k) + B_m u(k);\ y(k) = C_m x(k)$$

where the state variables are chosen as,

$$x(k) = \left[u(k-1)^T\ \ u(k-2)^T\ \ \ldots\ \ u(k-5)^T\right]^T.$$

What are the advantages of the finite impulse response model when it is used to describe a multi-input and multi-output system with time-delay?

Problem 3.8 From the system

$$\begin{bmatrix} x_1(k+1) \\ x_2(k+1) \end{bmatrix} = \begin{bmatrix} 1 & 0 \\ 1 & 1 \end{bmatrix} \begin{bmatrix} x_1(k) \\ x_2(k) \end{bmatrix} + \begin{bmatrix} 1 \\ 0 \end{bmatrix} u(k)$$

$$y(k) = \begin{bmatrix} 0 & 1 \end{bmatrix} \begin{bmatrix} x_1(k) \\ x_2(k) \end{bmatrix}$$

the following values are obtained

$$y(1) = 0;\ y(2) = 1;\ u(1) = 1;\ u(2) = -1.$$

Determine the value of $x(k) = [x_1(k)\ x_2(k)]^T$ at $k = 3$.

Problem 3.9 Assuming that the input is a unit step, use final value theorem to determine $\lim_{k\to\infty} y(k)$ for the following systems:

1.

$$Y(z) = \frac{z - 0.6}{(z - 0.1)(z - 0.9)} z^{-3} U(z)$$

2.

$$x(k+1) = A_m x(k) + B_m u(k);\ y(k) = C_m x(k)$$

$$\text{where } A_m = \begin{bmatrix} 0.1 & 2 \\ 0 & 0.8 \end{bmatrix};\ B_m = \begin{bmatrix} 0.3 & 0 \\ 0.5 & 0.1 \end{bmatrix};\ C_m = \begin{bmatrix} 1 & 0.5 \\ 0 & 1 \end{bmatrix}.$$

Problem 3.10 Determine the poles, zeros and stability of the following systems:

1.

$$G(z) = \frac{1 - 0.3z^{-1}}{(1 - 0.4z^{-1})(1 - 0.5z^{-1})(1 - 1.1z^{-1})} z^{-3};$$

2.

$$x(k+1) = A_m x(k) + B_m u(k-3);\ y(k) = C_m x(k)$$

$$\text{where } A_m = \begin{bmatrix} 0.3 & \\ -0.1 & 0.8 \end{bmatrix};\ B_m = \begin{bmatrix} 1 & 0 \\ 0.5 & 0.1 \end{bmatrix};\ C_m = \begin{bmatrix} 1 & 0 \\ 0 & 1 \end{bmatrix}.$$

Problem 3.11 Compute frequency response of the following systems. MATLAB function **freqz.m** may be used in the computation.

1.

$$G(z) = \frac{1 - 0.3z^{-1}}{(1 - 0.6z^{-1})(1 - 0.9z^{-1})} z^{-3};$$

2.

$$x(k+1) = A_m x(k) + B_m u(k);\ y(k) = C_m x(k)$$

$$\text{where } A_m = \begin{bmatrix} 0.3 & \\ -0.1 & 0.8 \end{bmatrix};\ B_m = \begin{bmatrix} 1 & 0 \\ 0.5 & 0.1 \end{bmatrix};\ C_m = \begin{bmatrix} 1 & 0 \\ 0 & 1 \end{bmatrix}.$$

Problem 3.12 Assume that the discrete-time system:

$$y(k) - 1.0y(k-1) + 0.4y(k-2) = 0.2u(k-1) + 0.3u(k-2)$$

is controlled by $u(k) = K(r(k) - y(k))$, where $r(k)$ is the reference signal. Assume that $K = 0.8$. Is the closed-loop system stable? Find the answer by calculating the closed-loop poles. Verify your answer with closed-loop simulation by observing the output response. In your simulation, we assume $r(k) = 1$ for $k \geq 0$ and 0 for $k < 0$. The initial conditions of the control signal and output signal are all zero for $k < 0$. Use final value theorem to calculate the steady-state value of the output and compare the result with your simulation result.

4

Discrete-time State Feedback Control

4.1 Introduction

The discrete-time state feedback control systems share many similarities with their continuous-time counterpart. However, they also have their own distinguishable features.

The chapter begins with the introduction of state feedback control in discrete-time (see Section 4.2). Through a simple example, this section highlights the pole-assignment controller design in discrete-time and the selection of desired discrete-time poles. Additionally, controllability in discrete-time is explained in Section 4.2. In Section 4.3, the topics of observer and observability are covered. For a control system with multiple inputs or multiple outputs, the discrete-time linear quadratic regulator (DLQR) provides both optimal controller and observer design, and this topic is discussed in Section 4.4. The topic of DLQR with a prescribed degree of stability is presented in Section 4.5 with case studies to illustrate its applications.

4.2 Discrete-time State Feedback Control

4.2.1 Basic Ideas

Assume that a discrete-time state space model is described by the difference equation:

$$x(k+1) = Ax(k) + Bu(k), \tag{4.1}$$

where the state variable $x(k)$ has dimension $n \times 1$ and the input variable $u(k)$ has dimension $m \times 1$. Assuming that the the state vector $x(k)$ is measured, the state feedback control generates the control signal $u(k)$ with the controller K using

$$u(k) = -Kx(k),$$

State Feedback Control and Kalman Filtering with MATLAB/Simulink Tutorials, First Edition.
Liuping Wang and Robin Ping Guan.

Companion website: www.wiley.com/go/feedback19

where K is a gain matrix with dimension $m \times n$. For regulatory control, there is no explicit reference signal or output signal and the control objective is to use the control signal $u(k)$ to bring some initial state $x(0) \neq 0$ to zero, namely $x(k) \to 0$, as $k \to \infty$. The closed-loop system is formed by simply substituting the control signal $u(k)$ into the state space model (4.1), which gives

$$x(k+1) = (A - BK)x(k). \tag{4.2}$$

The objective of regulatory control is to establish a stable closed-loop system (4.2) with satisfactory dynamic performance. The closed-loop eigenvalues of the discrete-time system are the solutions of the characteristic equation:

$$\det(zI - (A - BK)) = 0. \tag{4.3}$$

The design objectives include the selection of K to ensure that the eigenvalues of the closed-loop system matrix $A - BK$ are strictly inside the unit circle and the closed-loop system has satisfactory dynamic performance.

In the applications of regulatory control, $x(k)$ is defined as $x(k) = X(k) - x^{ss}$, where $X(k)$ corresponds to the actual physical variables and x^{ss} corresponds to the steady-state operating conditions for the physical variables. Essentially, the regulatory control is to maintain the state variables in the vicinity of steady-state conditions as $X(k) \to x^{ss}$. A good example for regulatory control is the drive train control of a wind turbine with the steady-state operating conditions (see Section 2.4.4).

The state feedback control systems in discrete-time share many characteristics, properties, and design methods with the continuous-time state feedback control systems, which includes the same characteristic equation expression as in (4.3). The main difference is that the location of the desired closed-loop eigenvalues for the discrete-time systems must be inside the unit circle. For those who are familiar with the continuous-time control systems, it is an advantage because the new knowledge can be gained quickly when we focus on the differences.

Example 4.1

From Example 3.3, the following discrete-time model for a double integrator was obtained:

$$\begin{bmatrix} x_1(k+1) \\ x_2(k+1) \end{bmatrix} = \begin{bmatrix} 1 & \Delta t \\ 0 & 1 \end{bmatrix} \begin{bmatrix} x_1(k) \\ x_2(k) \end{bmatrix} + \begin{bmatrix} \frac{1}{2}\Delta t^2 \\ \Delta t \end{bmatrix} u(k)$$
$$y(k) = \begin{bmatrix} 1 & 0 \end{bmatrix} x(k). \tag{4.4}$$

Find the discrete-time state feedback controller K that will lead to a closed-loop system that has eigenvalues at α_1 and α_2, where $0 \leq \alpha_1 < 1$, $0 \leq \alpha_2 < 1$. Discuss the controller K in relation to the sampling interval Δt and the location of the closed-loop eigenvalues.

Solution
The control signal is expressed as

$$u(k) = -\begin{bmatrix} k_1 & k_2 \end{bmatrix} \begin{bmatrix} x_1(k) \\ x_2(k) \end{bmatrix}$$

Thus, the closed-loop system matrix becomes

$$\begin{aligned} A - BK &= \begin{bmatrix} 1 & \Delta t \\ 0 & 1 \end{bmatrix} - \begin{bmatrix} \frac{1}{2}\Delta t^2 \\ \Delta t \end{bmatrix} \begin{bmatrix} k_1 & k_2 \end{bmatrix} \\ &= \begin{bmatrix} 1 - \frac{1}{2}\Delta t^2 k_1 & \Delta t - \frac{1}{2}\Delta t^2 k_2 \\ -\Delta t k_1 & 1 - \Delta t k_2 \end{bmatrix}. \end{aligned} \tag{4.5}$$

Now, with the desired closed-loop polynomial being chosen as

$$(z - \alpha_1)(z - \alpha_2) = z^2 - (\alpha_1 + \alpha_2)z + \alpha_1\alpha_2,$$

we equate the actual closed-loop polynomial to the desired closed-loop polynomial, that is

$$\det(zI - (A - BK)) = z^2 - (\alpha_1 + \alpha_2)z + \alpha_1\alpha_2$$

where the matrix $A - BK$ is given by (4.5). By comparing the coefficients of z, we find that the solutions for the controller gains are

$$k_1 = \frac{\alpha_1\alpha_2 - (\alpha_1 + \alpha_2) + 1}{\Delta t^2} \tag{4.6}$$

$$k_2 = \frac{1.5 - 0.5(\alpha_1 + \alpha_2) - 0.5\alpha_1\alpha_2}{\Delta t}. \tag{4.7}$$

There are three comments as follows.

1. *The discrete-time controller gains are inversely proportional to the sampling interval Δt. It means that if the system is sampled at a faster rate, the resulting discrete-time controller gains are larger.*
2. *In the case of $\alpha_1 = \alpha_2 = 0$, which is the deadbeat controller with the desired closed-loop eigenvalues at the origin of the complex plane, $k_1 = \frac{1}{\Delta t^2}$ and $k_2 = \frac{1.5}{\Delta t}$.*
3. *For the double integrator system, the open-loop eigenvalues are at 1. If we place the desired closed-loop eigenvalues closer to 1, the controller gain is reduced.*

To avoid having very large controller gains, when the sampling interval Δt is small, we should choose the desired closed-loop poles α_1 and α_2 in relation to the sampling interval Δt. An intuitive way is to use the desired closed-loop pole location in continuous-time as a benchmark in selecting the desired closed-loop poles in discrete-time. For instance, if we choose the desired closed-loop poles in continuous-time as p_1 and p_2 (Real(p_1) < 0, Real(p_2) < 0), then with a selected Δt, the desired closed-loop poles become $\alpha_1 = e^{p_1 \Delta t}$ and $\alpha_2 = e^{p_2 \Delta t}$.

Example 4.2

We will continue from Example 4.1 by demonstrating how to relate the desired closed-loop poles in discrete-time to those in the continuous-time.

Solution

For simplicity, we choose two identical continuous-time poles where $p_1 = p_2 = -1$. The desired closed-loop time constants are equal to 1 (sec). If we wish to capture the desired closed-loop response time T_s with the number of samples N_s in discrete-time, then the sampling interval Δt is selected as

$$\Delta t = \frac{T_s}{N_s}. \tag{4.8}$$

For this particular choice of desired closed-loop poles at −1, it can be verified through simulation that the closed-loop response time is roughly estimated as $T_s = 6$ (sec). If we choose the number of samples $N_s = 60$ to capture the system dynamics in discrete-time, then $\Delta t = \frac{6}{60} = 0.1$ (sec). With this choice of Δt, the desired closed-loop poles are $\alpha_1 = \alpha_2 = e^{-0.1} = 0.9048$. The resulting controller gains are $k_1 = 0.9056$ and $k_2 = 1.8580$. Figure 4.1(a) shows the closed-loop state response ($x_1(k)$) to the state initial conditions $x_1(0) = x_2(0) = 1$ and Figure 4.1(b) shows the control signal used to generate this response.

Now, supposing that the system is in a fast sampling environment with $\Delta t = 0.01$, maintaining the same poles at −1 in continuous-time, the desired closed-loop poles in the discrete-time are $\alpha_1 = \alpha_2 = e^{-0.01} = 0.99$. The controller gains are $k_1 = 0.9901$, and $k_2 = 1.9851$. We can verify that it takes 600 samples for the closed-loop state variables to reach zero.

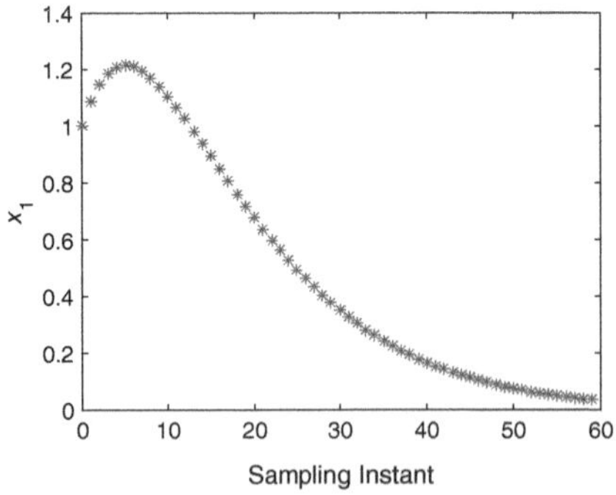

(a) State

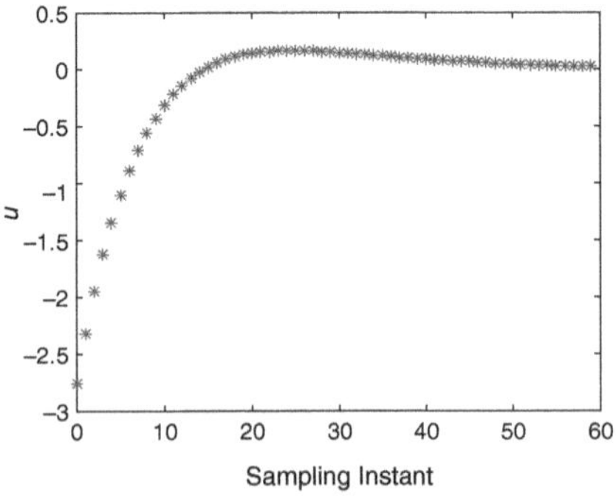

(b) Control

Figure 4.1 Closed-loop response (Example 4.2).

In contrast, if we did not consider the smaller sampling interval in the selection of the desired closed-loop poles, for example, by fixing the desired closed-loop poles at $\alpha_1 = \alpha_2 = 0.5$, the controller gains for $\Delta t = 0.01$ are $k_1 = 2500$ and $k_2 = 87.5$. The immediate consequence for this poor choice is a very large control signal. For example, if $x_1(0) = x_2(0) = 1$, the control signal at the initial response is

$$u(0) = -k_1(0)x_1(0) - k_2x_2(0) = -2587.5$$

Many problems would be encountered with this large controller gain, one of them being the possibility of a severe violation of the operational constraints.

If we wish to use a pair of complex poles in the continuous-time where $p_{1,2} = -p_r \pm jp_i$, then the desired closed-loop poles in discrete-time are:

$$\alpha_{1,2} = e^{-p_r\Delta t}(\cos(p_i\Delta t) \pm j\sin(p_i\Delta t)).$$

4.2.2 Controllability in Discrete-time

Controllability in discrete-time is defined as follows.

Controllability Assuming that the state space model is described by

$$x(k+1) = Ax(k) + Bu(k) \tag{4.9}$$

$$y(k) = Cx(k), \tag{4.10}$$

the system is controllable if for any initial state $x(0)$ and any final state $x(n_f)$, there exists a control sequence, $u(0)$, $u(1)$, …, $u(n_f)$ that transfers the initial state $x(0)$ to the final state $x(n_f)$ in a finite time.

The definition in discrete-time is focused on the input sequence, $u(0)$, $u(1)$, …, $u(n_f)$, which is capable of moving any initial state $x(0)$ to any other final state $x(n_f)$ in a finite time. However, it does not require a specific trajectory for the movement and there is no constraint imposed on the input signal.

In discrete-time, we can check whether a system is controllable by calculating the rank of the controllability matrix. The controllability matrix L_c is defined as

$$L_c = \begin{bmatrix} B & AB & A^2B & \dots & A^{n-1}B \end{bmatrix},$$

where n is the number of state variables. The discrete-time system is controllable if and only if the controllability matrix L_c has rank of n. If the system has only one input where L_c has dimension $n \times n$, then L_c having rank of n implies that its determinant is not equal to 0, which means that L_c^{-1} exists if the system is controllable.

We construct the following illustration of controllability in relation to the L_c matrix. From the state space model (4.9), with a given initial state $x(0)$, we can calculate the state variables sequentially as,

$$x(1) = Ax(0) + Bu(0)$$
$$x(2) = A^2x(0) + Bu(1) + ABu(0)$$
$$x(3) = A^3x(0) + Bu(2) + ABu(1) + A^2Bu(0)$$
$$\vdots$$
$$x(n) = A^nx(0) + Bu(n-1) + ABu(n-2) + \ldots + A^{n-1}Bu(0). \tag{4.11}$$

Let $U = \begin{bmatrix}u(n-1)^T & u(n-2)^T & \ldots & u(0)^T\end{bmatrix}^T$ and $L_c = \begin{bmatrix}B & AB & \ldots & A^{n-1}B\end{bmatrix}$. We write (4.11) in matrix form:

$$x(n) = A^nx(0) + L_cU, \tag{4.12}$$

which is equivalent to

$$L_cU = x(n) - A^nx(0). \tag{4.13}$$

If there is only one input signal and if the system is controllable, then the controllability matrix L_c is invertible. For this case, we can find the control sequence, $u(n-1), u(n-2), \ldots, u(0)$ using

$$U = L_c^{-1}(x(n) - A^nx(0)). \tag{4.14}$$

With respect to the concept of controllability, (4.14) says that if we choose the final state $x(n_f) = x(n)$, the vector U that contains $u(0), u(1), \ldots, u(n)$ will transfer the initial state $x(0)$ to the final state $x(n_f)$ in a finite step n.

If the system has multiple inputs and if the system is controllable, with the rank of L_c being n, then there are n independent columns in the L_c matrix. With the independent columns, we could find the control sequence that would move any arbitrary initial state $x(0)$ to any final state $x(n)$ in n step. A relaxed definition to controllability is stabilizability.

Stabilizability. A system is stabilizable if its uncontrollable modes, if any, are stable. Its controllable modes may be stable or unstable.

To check if a system is controllable, MATLAB program **ctrb.m** can be used to form the L_c matrix, followed with the application of **rank.m** program. For instance, after defining A and B matrices, we perform the following computation:

```
Lc=ctrb(A,B);
q=rank(Lc);
```

If the parameter q equals n that is the number of state variables, then the system is controllable.

Because the discrete-time characteristic polynomial has the same expression (see (4.3)) as that of the continuous-time system, the MATLAB programs written for pole-assignment controller design in continuous-time, such as **place.m** or **T2place.m**, can be directly used for the discrete-time state feedback controller design. In the discrete-time controller design, the desired closed-loop poles are chosen to be within the unit circle.

Controllability in continuous-time is explained with detailed examples in Section 1.3.2 if one wishes to study more on this topic.

4.2.3 Food for Thought

1. In a discrete-time control system design, should we consider the effect of sampling interval Δt when selecting the desired closed-loop poles? Why?
2. Given a discrete-time system with the state space model:

$$x(k+1) = 0.9x(k) + 0.1u(k) \tag{4.15}$$

   what is the state feedback controller K if we choose the desired closed-loop pole at 0.6?
3. If we wish to have a zero feedback controller gain ($K = 0$) for the same system (4.15), what desired closed-loop pole should we select?
4. Assume that a discrete-time system has the state space model:

$$x(k+1) = 0.7x(k) + 0.3u(k) + c$$

   where the constant disturbance $c \neq 0$ is unknown. Do you think that we can find the state feedback controller $K < \infty$ so that $x(k) \to 0$ as $k \to \infty$?
5. A discrete-time system is given by the transfer function

$$G(z) = \frac{z - 0.3}{(z - 0.5)(z - 0.3)}. \tag{4.16}$$

   If we write this second order system in a state space model, do you think that it is controllable? If not, what might be the cause? Is this system stabilizable?

4.3 Discrete-time Observer Design

4.3.1 Basic Ideas about Discrete-time Observer

For the state feedback control introduced in Section 4.2, the state vector $x(k)$ is available from measurement. If $x(k)$ is not measured, however the output $y(k)$ is measured, then we can construct an observer to estimate $x(k)$ to form a state estimate feedback control system.

Assuming that the state space model is given by

$$x(k+1) = Ax(k) + Bu(k) \tag{4.17}$$

$$y(k) = Cx(k) \tag{4.18}$$

the idea of a discrete-time observer is to estimate the state vector $x(k)$ using the following equation:

$$\hat{x}(k+1) = \overbrace{A\hat{x}(k) + Bu(k)}^{I} + \overbrace{K_{ob}(y(k) - C\hat{x}(k))}^{II}. \tag{4.19}$$

The first part of this observer equation (see Part I) is based on the original dynamic model, and the second part (see Part II) is to introduce a compensation of the error with the observer gain K_{ob}, which is between the measured and the estimated outputs from the sampling instant k. We use (4.19) for the implementation of the observer, which utilizes the measurement $y(k)$ to compute the estimate $\hat{x}(k+1)$.

For the design of an observer, we will form the following observer error system. Define the state error:

$$\tilde{x}(k) = x(k) - \hat{x}(k).$$

The dynamic equation for the state error is formed by subtracting (4.19) from (4.17):

$$\begin{aligned}\tilde{x}(k+1) &= A\tilde{x}(k) - K_{ob}(y(k) - C\hat{x}(k)) \\ &= (A - K_{ob}C)\tilde{x}(k) \end{aligned} \tag{4.20}$$

where the output equation (4.18) is used to replace $y(k)$ in order to obtain (4.20).

In the observer design, we will ensure that the observer error system (4.20) is stable with desired dynamic response. This means that the eigenvalues of the error system matrix $A - K_{ob}C$ are strictly inside the unit circle with desired locations. The stability of the state error system ensures that the state error variable $\tilde{x}(k)$ converges to zero as $k \to \infty$, and as a consequence, the estimated state $\hat{x}(k)$ converges to the true state $x(k)$.

Example 4.3

A sinusoidal signal with frequency $\omega_0 = 0.1$ *rad/sec has a period of* $T_0 = \frac{2\pi}{\omega_0} =$ 62.8319 *seconds. This signal is measured with a sensor and the sensor has measurement noise. Thus, we are only able to have the measured signal that is corrupted with the measurement noise as shown in Figure 4.2. In this example, we will construct a discrete-time observer that will estimate the true signal from the measurement data where the sampling interval* $\Delta t = 0.01$ *(sec).*

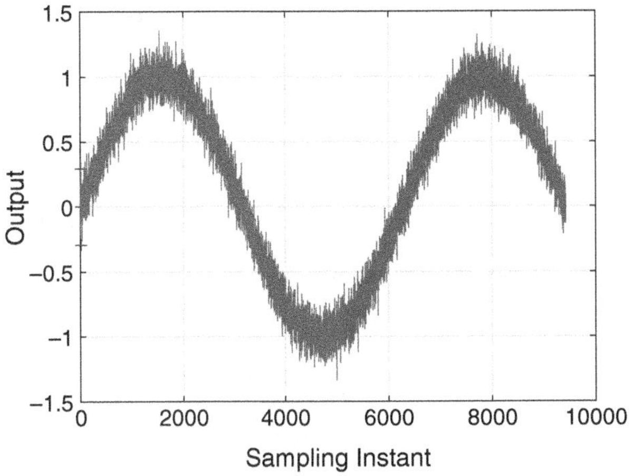

Figure 4.2 Measurement data (Example 4.3).

Solution

From Example 3.4, the discrete-time state space model for a sinusoidal signal corrupted with measurement noise $\epsilon(k)$ is given by

$$\overbrace{\begin{bmatrix} x_1(k+1) \\ x_2(k+1) \end{bmatrix}}^{x(k+1)} = \overbrace{\begin{bmatrix} \cos(\omega_0 \Delta t) & \sin(\omega_0 \Delta t) \\ -\sin(\omega_0 \Delta t) & \cos(\omega_0 \Delta t) \end{bmatrix}}^{A} \overbrace{\begin{bmatrix} x_1(k) \\ x_2(k) \end{bmatrix}}^{x(k)} \tag{4.21}$$

$$y(k) = \overbrace{\begin{bmatrix} 1 & 0 \end{bmatrix}}^{C} x(k) + \epsilon(k). \tag{4.22}$$

Based on (4.21) and (4.22), the observer that will estimate the state $x_1(k)$ and $x_2(k)$ with the effect of noise filter has the following form:

$$\hat{x}(k+1) = A\hat{x}(k) + K_{ob}(y(k) - C\hat{x}(k)), \tag{4.23}$$

where the input signal $u(k) = 0$, and $\hat{x}(k) = \begin{bmatrix} \hat{x}_1(k) & \hat{x}_2(k) \end{bmatrix}^T$. The closed-loop error system is

$$\tilde{x}(k+1) = (A - K_{ob}C)\tilde{x}(k) - K_{ob}\epsilon(k) \tag{4.24}$$

where $\tilde{x}(k) = x(k) - \hat{x}(k)$. Note that in comparison with (4.20), the measurement noise $\epsilon(k)$ has an effect on the error state $\tilde{x}(k)$.

Assuming that the closed-loop poles for the observer error system are chosen to be $0 \leq \alpha_1 < 1$ and $0 \leq \alpha_2 < 1$, the observer gain K_{ob} is chosen such that the eigenvalues *of the system matrix $A - K_{ob}C$ are equal to α_1 and α_2. For notational simplicity, we define $a_1 = \cos(\omega_0 \Delta t)$ and $b_1 = \sin(\omega_0 \Delta t)$ from the state space model (4.21). Letting $K_{ob} = [k_1\ k_2]^T$, we can verify that the following characteristic polynomial:*

$$\det(zI - (A - K_{ob}C)) = \det \begin{bmatrix} z - a_1 + k_1 & -b_1 \\ b_1 + k_2 & z - a_1 \end{bmatrix}$$

$$= z^2 + (-2a_1 + k_1)z + a_1^2 - k_1 a_1 + b_1^2 + b_1 k_2. \tag{4.25}$$

The desired closed-loop polynomial with the poles at α_1 and α_2 is

$$A_{cl}(z)^d = (z - \alpha_1)(z - \alpha_2) = z^2 - (\alpha_1 + \alpha_2)z + \alpha_1\alpha_2. \tag{4.26}$$

Equating (4.25) with (4.26) gives the solutions of the observer gain as

$$k_1 = -(\alpha_1 + \alpha_2) + 2a_1;\ k_2 = \frac{1}{b_1}(\alpha_1\alpha_2 + a_1^2 - a_1(\alpha_1 + \alpha_2) - b_1^2).$$

The next step is to choose the desired poles for the observer error system. If for simplicity we choose $\alpha_1 = \alpha_2 = 0.9$, then we have

$$k_1 = 0.2;\ k_2 = 9.9989.$$

To validate how the observer works in a noisy environment, we add a normally distributed noise with variance 0.01 to a sinusoidal signal (see Figure 4.2). Figure 4.3(a) compares $\hat{x}_1(k)$ to $x_1(k)$ (the true output). It is clearly seen that $\hat{x}_1(k)$ quickly estimates the true state $x_1(k)$, however, there is estimation error associated with $\hat{x}_1(k)$.

Now, we wish to reduce the effect of the estimation error in $\hat{x}_1(k)$ by reducing the observer gain. In a fast sampling environment, we propose using a pair of desired poles in the continuous-time and then converting them into

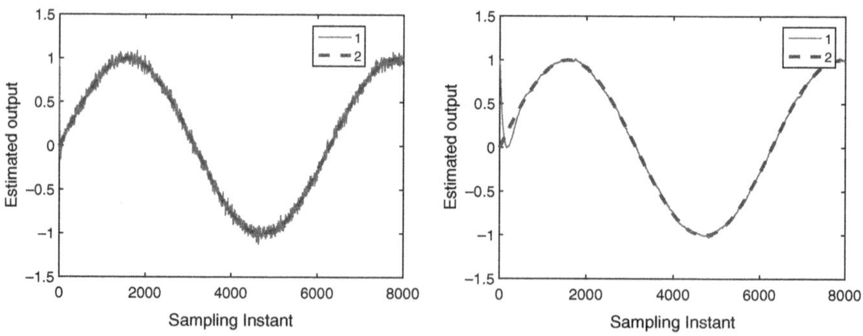

(a) $\alpha_1 = \alpha_2 = 0.9$ (b) $w_n = 1,\ \xi = 0.707$

Figure 4.3 Comparison between $\hat{x}_1(k)$ and $x_1(k)$ (Example 4.3). Key: line (1) $\hat{x}_1(k)$; line (2) $x_1(k)$.

discrete-time to reflect the influence of sampling interval. With a damping coefficient $\xi = 0.707$ and natural frequency $w_n = 1$, a pair of complex poles becomes $p_{1,2} = -\xi w_n \pm j\xi w_n = -0.707 \pm j0.707$ in the continuous-time, which gives the desired closed-loop poles in the discrete-time as

$$\alpha_{1,2} = e^{-\xi w_n \Delta t}(\cos(\xi w_n \Delta t) \pm j \sin(\xi w_n \Delta t)) = 0.9929 \pm j0.0070.$$

With this choice of desired closed-loop poles for the error system, we obtain

$$k_1 = 0.0141; \; k_2 = 0.0980.$$

Clearly the observer gain becomes much smaller compared to the previous case where $\alpha_{1,2} = 0.9$. Figure 4.3(b) compares the estimated state $\hat{x}_1(k)$ to the true state $x_1(k)$. It is seen that the effect of noise is significantly reduced, however, it takes a longer time for the estimated state to converge to the true state $x_1(k)$. In general, larger observer gain will lead to amplification of measurement noise.

4.3.2 Observability in Discrete-time

The concept of observability is associated with an observer design, which has the following definition.

Observability Assuming that a discrete-time system has the state space model:

$$x(k+1) = Ax(k) + Bu(k); \; y(k) = Cx(k)$$

the system is observable if there is a finite k such that the knowledge of the inputs $u(0), u(1), \ldots, u(k-1)$ and outputs $y(0), y(1), \ldots, y(k-1)$ is sufficient to determine the initial state $x(0)$ of the system.

The definition requires only that the initial state $x(0)$ be determined in a finite time with the given input and output data, but how the initial state is determined is not specified.

The observability matrix is defined as

$$L_o = \begin{bmatrix} C \\ CA \\ CA^2 \\ \vdots \\ CA^{n-1} \end{bmatrix}$$

where n is the dimension of the state vector $x(k)$. The system is observable if and only if the matrix L_o has rank n. To illustrate this, we construct the following example. For simplicity, we assume that the input signal $u(k) = 0$ for all k. We can express the output measurement in relation to the initial condition $x(0)$ sequentially as

$$y(0) = Cx(0)$$
$$y(1) = CAx(0)$$
$$y(2) = CA^2x(0)$$
$$\vdots$$
$$y(n-1) = CA^{n-1}x(0)$$

which can be written in vector form:

$$Y = L_o x(0) \tag{4.27}$$

where the vector Y contains the measurement sequence $y(0), y(1), \ldots, y(n-1)$. If the system has only one output, then L_o has dimension $n \times n$. Therefore, rank of L_o being n implies that det $L_o \neq 0$. For this case, we can determine $x(0)$ with the following equation:

$$x(0) = L_o^{-1}Y.$$

If the system has more than one output, then rank of L_o being n means that there are n independent rows in the observability matrix. We can choose the n independent rows to form $\overline{L}_o$ and the corresponding $\overline{Y}$. Then, the initial condition $x(0)$ can be determined through

$$x(0) = \overline{L}_o^{-1}\overline{Y}.$$

A more relaxed definition to observability is detectability.

Detectability. A system is detectable if its unobservable modes, if any, are stable. Its observable modes may be stable or unstable.

To check if a system is observable, we can use the MATLAB program **ctrb.m**, noting that the observability matrix L_o is a transpose to the controllability matrix L_c if we replace matrix B by C^T and A by A^T. For example, after introducing the matrices A and C, we can calculate L_o and determine its rank as

```
Lo=ctrb(A',C')';
q=rank(Lo);
```

If the variable q equals the number of states n, then the system is observable.

Observability in continuous-time is illustrated with detailed examples in Section 1.6.3 if one wishes to study more on this topic.

Note that the observer error system has the characteristic equation:

$$\det(zI - (A - K_{ob}C)) = 0 \tag{4.28}$$

which is equal to

$$\det(zI - (A^T - C^T K_{ob}^T)) = 0. \tag{4.29}$$

The characteristic equation (4.29) is identical to the one that has been used for the design of state feedback controller K (see (4.3), except that in the observer design case, we replace A by A^T, B by C^T and K by K_{ob}^T.

When systems are more complex and their analytical solutions for the observer gain are difficult to find, we resort to using numerical approaches to find K_{ob}. The MATLAB programs **place.m** or **T2place.m** written for the continuous-time control system can used by assuming that the system is observable. As an illustration, assuming that the A and C matrices are defined and the desired poles for the observer error system are chosen as a vector P_{ob}, we calculate K_{ob} using the following program:

```
Kob=place(A',C',Pob)';
```

4.3.3 Food for Thought

1. A signal is known to behave like a piece-wise constant and is corrupted by noise. Design an observer to filter the noise. If the pole of the observer error system is selected to be 0.8, what is the observer gain K_{ob}?
2. Continue from 1. If we want to reduce the effect of noise further, is it correct for us to choose a desired pole for the observer error system closer to the origin of the complex plane, say, 0.6?
3. Is it correct to say that measurement noise is one of the key factors to be considered when selecting the desired poles for the observer error system?
4. Assume a discrete-time system with one sample of measurement delay is described by the following difference equations:

   $$x(k+1) = ax(k) + bu(k); \; y(k) = x(k-1),$$

   where a and b are scalars. What state space model would you propose to include the measurement delay for the observer design?
5. Continue from 4. What if the parameter a changed after some time, say from $a = 0.8$ to $a = 0.6$, will the observer still give the correct estimate after the change?
6. A discrete-time system with one sample of actuator delay is described by the difference equations:

   $$x(k+1) = Ax(k) + Bu(k-1); \; y(k) = Cx(k).$$

   Do we need to consider the effect of actuator delay in the observer design? Why?

4.4 Discrete-time Linear Quadratic Regulator (DLQR)

4.4.1 Objective Function for DLQR

For a discrete-time system described by the difference equation:

$$x(k+1) = Ax(k) + Bu(k)$$

where $u(k) = -Kx(k)$, the discrete-time linear quadratic regulator is to find the state feedback controller K by minimizing the objective function:

$$J = \sum_{k=0}^{\infty} \left(x(k)^T Qx(k) + u(k)^T Ru(k)\right) \tag{4.30}$$

where both Q and R are symmetric with Q being semi-positive definite and R being positive definite. Here, we assume that the pair of A, B matrices is controllable and the pair of A, Γ matrices is observable, where $Q = \Gamma^T \Gamma$. Both Q are R matrices are used in the performance specification, which will be discussed later with examples.

The optimization control problem has two aspects. First, the controller K is required to stabilize the discrete-time system. Second, the solution should be the result of minimizing the objective function J. We will concentrate on illustrating the first part of the problem and as for the proof of optimality, other books books will provide detailed information (Anderson and Moore [1979], Anderson and Moore [1971], Bay [1999], Goodwin et al. [2000]).

4.4.2 Optimal Solution

For a given initial condition $x(0)$, the analytical solution of state $x(k)$ in closed-loop control has the following expression:

$$x(k) = (A - BK)^k x(0). \tag{4.31}$$

Thus, by substituting (4.31) into the objective function (4.30), we obtain:

$$\begin{aligned} J &= x(0)^T \sum_{k=0}^{\infty} \left((A - BK)^k\right)^T Q(A - BK)^k x(0) \\ &\quad + x(0)^T \sum_{k=0}^{\infty} \left((A - BK)^k\right)^T K^T RK(A - BK)^k x(0) \\ &= x(0)^T \sum_{k=0}^{\infty} \left((A - BK)^k\right)^T (Q + K^T RK)(A - BK)^k x(0). \end{aligned} \tag{4.32}$$

Let P denote the quantity that represents the infinite sum in (4.32), which is

$$P = \sum_{k=0}^{\infty} \left((A - BK)^k\right)^T (Q + K^T RK)(A - BK)^k. \tag{4.33}$$

We can verify that P satisfies the following algebraic equation:

$$(A - BK)^T P(A - BK) - P = -(Q + K^T RK), \tag{4.34}$$

where the subtraction on the left-hand side of the equation cancels all the terms except the first, which is $-(Q + K^T RK)$. The algebraic equation (4.34) is the

Lyapunov equation in discrete-time (Chen [1998]). The Lyapunov theorem states that all eigenvalues of $A - BK$ have magnitudes less than 1 if and only if for the given positive definite symmetric matrix $Q + K^T RK$, the discrete Lyapunov equation (4.34) has a unique symmetric solution P and P is positive definite.

Because the matrix P and the controller gain K are both unknown in (4.34), it is not feasible to solve the algebraic equation directly. To find the solutions, the state feedback controller gain K is chosen as the following function of P:

$$K = (R + B^T PB)^{-1} B^T PA. \tag{4.35}$$

Substituting this K into the algebraic equation (4.34) leads to

$$A^T \{P - PB(R + B^T PB)^{-1} B^T P\} A + Q - P = 0. \tag{4.36}$$

This is the algebraic Riccati equation in discrete-time.

In summary, the first effort is to find the unique symmetric solution P that is positive definite. Then, the controller K is found using (4.35) with the P matrix from the solution of the algebraic equation (4.36). Based on the Lyapunov theorem, as long as P is unique, symmetric, and positive definite, the closed-loop system matrix $A - BK$ has all eigenvalues strictly inside the unit circle. Therefore, the closed-loop system is stable.

The optimality of K, which minimizes the objective function (4.30), follows from the definition of the objective function with finite horizon,

$$J_N = x(N)^T P_N x(N) + \sum_{k=i}^{N-1} \left(x(k)^T Qx(k) + u(k)^T Ru(k) \right). \tag{4.37}$$

The minimization at each time step k gives (Bay [1999])

$$K_k = (R + B^T P_{k+1} B)^{-1} B^T P_{k+1} A \tag{4.38}$$

$$P_k = (A - BK_k)^T P_{k+1} (A - BK_k) + Q + K_k^T RK_k. \tag{4.39}$$

At the steady-state $k \to \infty$, $P_k = P_{k+1} = P$, $K_k = K$. Then equation (4.39) becomes identical to (4.34), and equation (4.38) becomes identical to (4.35). Therefore, the solution of the algebraic Riccati equation (4.36) gives the controller K that minimizes the objective function (4.30), where its minimum is

$$J_{min} = x(0)^T Px(0). \tag{4.40}$$

The MATLAB program **dlqr.m** can be used to solve the algebraic Riccati equation (4.36) and calculate the controller gain K. Upon choosing Q and R matrices, with given A and B matrices, the K, P and the closed-loop eigenvalues E are calculated using the following MATLAB code:

```
[K,P,E]=dlqr(A,B,Q,R);
```

The selection of weighting matrices Q and R is illustrated with several examples for the linear quadratic regulator in continuous-time (see Section 1.5.3 of Chapter 1).

4.4.3 Observer Design using DLQR

The DLQR design can be applied to state estimation. This topic will be discussed in the later chapters on Kalman filters (see Chapters 7 and 8). At present, we will focus on the idea of observer design. In the observer design, we assume that the state space model is described by

$$x(k+1) = Ax(k) + Bu(k);\ y(k) = Cx(k)$$

and the estimated state $\hat{x}(k)$ is obtained through the observer equation:

$$\hat{x}(k+1) = A\hat{x}(k) + Bu(k) + K_{ob}(y(k) - C\hat{x}(k)) \tag{4.41}$$

In order to find the observer gain K_{ob}, we will first solve the algebraic Riccati equation,

$$A\left[P - PC^T(R + CPC^T)^{-1}CP\right]A^T + Q - P = 0 \tag{4.42}$$

and find the observer gain K_{ob} using the P matrix from the solution of the algebraic Riccati equation:

$$K_{ob} = APC^T\left(R + CPC^T\right)^{-1}. \tag{4.43}$$

Comparing the algebraic Riccati equation (4.42) with the Riccati equation (4.36) from the controller design, we notice that they are identical, except that the matrix A is replaced by A^T and B by C^T. The same is true for the observer gain K_{ob} by comparing (4.43) with the controller gain K (see (4.35)) in addition to the transpose relationship between these two.

We can use MATLAB controller design program **dlqr.m** for the observer design. After defining the A and C matrices, and the selection of Q and R matrices, we can calculate K_{ob} using the following program:

```
[K1,P,E]=dlqr(A',C',Q,R);
Kob=K1';
```

4.4.4 Food for Thought

1. Is it correct to say that the minimum of the objective function J is equal to the following quantity:

   $$J_{min} = x(0)^T Px(0)$$

   where P is the Riccati solution and $x(0)$ is the initial state?

2. For a second order system, could we use any of the following Q matrices in the controller design:

$$Q = \begin{bmatrix} 1 & 3 \\ 4 & 2 \end{bmatrix}; Q = \begin{bmatrix} 1 & 0 \\ 0 & -1 \end{bmatrix}; Q = \begin{bmatrix} 0 & 0 \\ 0 & 0 \end{bmatrix}?$$

Why?

3. If a system has two inputs, which R matrix we could use:

$$R = \begin{bmatrix} 1 & 2 \\ 0 & 1 \end{bmatrix}; R = \begin{bmatrix} 1 & 0 \\ 0 & 1 \end{bmatrix}; R = \begin{bmatrix} 0 & 0 \\ 0 & 0 \end{bmatrix}?$$

4. Assuming that a discrete-time system has the following state space model:

$$x(k+1) = \begin{bmatrix} 0.9 & 0 \\ 0 & 0.8 \end{bmatrix} x(k) + \begin{bmatrix} 1 \\ 0 \end{bmatrix} u(k)$$

if we choose Q to be the identity matrix and $R = 1$, can we find the controller K using the MATLAB **dlqr.m**? Why do you think so?

5. Given a system with the following state space model:

$$x(k+1) = \begin{bmatrix} 0.7 & 0 \\ 0 & 1 \end{bmatrix} x(k) + \begin{bmatrix} 1 \\ 1 \end{bmatrix} u(k);\ y(k) = \begin{bmatrix} 1 & 0 \end{bmatrix} x(k),$$

if we choose Q to be the identity matrix and $R = 1$, can we find the observer gain K_{ob} using the MATLAB **dlqr.m**?

6. What are the key differences between the above two systems?

4.5 Discrete-time LQR with Prescribed Degree of Stability

Closed-loop performance with the discrete-time LQR is determined by the selection of weighting matrices Q and R. This task is often performed via a trial-and-error process. For instance, we can select the pair of initial Q and R matrices to start the state feedback controller design, followed by evaluation of the closed-loop eigenvalues and simulation of the closed-loop responses. This process can be quite time consuming if the system has many state variables and input variables.

4.5.1 Basic Ideas about a Prescribed Degree of Stability

A classic approach in Anderson and Moore [1979], Anderson and Moore [1971] introduced a constraint on the closed-loop eigenvalues such that they are all strictly inside the α-circle of the complex plane where $\alpha < 1$. Therefore, by choosing an α value, we can effectively change the locations of the closed-loop eigenvalues to achieve a desired dynamic performance. This approach is also called DLQR with a prescribed degree of stability.

Assume that a state space model is given by

$$x(k+1) = Ax(k) + Bu(k). \tag{4.44}$$

In the discrete-time LQ regulator design with a prescribed degree of stability, the objective function is selected as

$$J = \sum_{k=0}^{\infty} \left(\alpha^{-2k}x(k)^T Qx(k) + \alpha^{-2k}u(k)^T Ru(k)\right), \tag{4.45}$$

where $\alpha < 1$. Because $\alpha < 1$, the time-varying weight factor α^{-2k} increases as k increases. The discrete-time LQR design is to find K such that the objective function J (4.45) is minimized. At first glance, it is not obvious how the new optimal control could be solved with the time-varying weight factor α^{-2k}. The optimization problem is transformed by defining the pair of intermediate variables:

$$x_\alpha(k) = \alpha^{-k}x(k)$$

$$u_\alpha(k) = \alpha^{-k}u(k).$$

The objective function (4.45) is re-written in terms of $x_\alpha(k)$ and $u_\alpha(k)$:

$$J = \sum_{k=0}^{\infty} \left(x_\alpha(k)^T Qx_\alpha(k) + u_\alpha(k)^T Ru_\alpha(k)\right). \tag{4.46}$$

The original state space model (4.44) is also transformed with the intermediate variables, where the time-varying factor $\alpha^{-(k+1)}$ is multiplied on both sides of the equation:

$$\alpha^{-(k+1)}x(k+1) = \alpha^{-(k+1)}Ax(k) + \alpha^{-(k+1)}Bu(k)$$

leading to

$$x_\alpha(k+1) = A_\alpha x_\alpha(k) + B_\alpha u_\alpha(k) \tag{4.47}$$

where the transformed system matrices A_α and B_α are defined as

$$A_\alpha = \frac{A}{\alpha}, \; B_\alpha = \frac{B}{\alpha}.$$

We will find the feedback control signal $u_\alpha(k) = -Kx_\alpha(k)$ to minimize the objective function (4.46). For this purpose, the DLQR problem with the time-varying weight factor α^{-2k} is formulated as the problem that minimizes the objective function J (see (4.46)) subject to the state space model (4.47). This formulation is identical to that of the original LQR design where it does not have the time-varying weight factor α^{-2k}. Therefore, the optimal solutions to the feedback controller K and the

P matrix are identical to those given by (4.35) and (4.36), which are

$$K = (R + B_\alpha^T P B_\alpha)^{-1} B_\alpha^T P A_\alpha \tag{4.48}$$

$$A_\alpha^T \{P - PB_\alpha (R + B_\alpha^T P B_\alpha)^{-1} B_\alpha^T P\} A_\alpha + Q - P = 0. \tag{4.49}$$

To show how the solutions given by (4.48) and (4.49) lead to a prescribed degree of stability, the following closed-loop system is formulated based on state space model (4.47):

$$x_\alpha(k+1) = (A_\alpha - B_\alpha K) x_\alpha(k). \tag{4.50}$$

When the unique and positive definite matrix P is found, the solution of the algebraic Riccati equation (4.49) ensures that all closed-loop eigenvalues of $(A_\alpha - B_\alpha K)$ are strictly inside the unit circle. Note that in relation to the original state $x(k)$, we have

$$x_\alpha(k+1) = \alpha^{-(k+1)} x(k+1); \; x_\alpha(k) = \alpha^{-k} x(k)$$

with the initial condition $x_\alpha(0) = x(0)$. The closed-loop expression (4.50) is equivalent to

$$x(k+1) = \alpha(A_\alpha - B_\alpha K) x(k). \tag{4.51}$$

The closed-loop eigenvalues of (4.51) are the solutions of the polynomial equation:

$$\begin{aligned} \rho(z) &= \det(zI - \alpha(A_\alpha - B_\alpha K)) \\ &= \alpha \det\left(\frac{z}{\alpha} I - (A_\alpha - B_\alpha K)\right) = 0. \end{aligned} \tag{4.52}$$

Since all the closed-loop eigenvalues of $(A_\alpha - B_\alpha K)$ are strictly inside the unit circle, (4.52) means that they are all scaled by the parameter $\alpha < 1$ where α is chosen by the user. This result is practically significant. Although the DLQR solutions do not give the locations of the closed-loop eigenvalues and some of them may be very close to the unit circle due to the choice of Q and R matrices, by simply choosing an $\alpha < 1$, we design the controller K with the pair $\frac{A}{\alpha}$, $\frac{B}{\alpha}$, which results in that the closed-loop system matrix $A - BK$ has all eigenvalues strictly inside the α-circle.

The MATLAB program **dlqr.m** can be used to achieve the prescribed degree of stability without modification. For instance, for a pair of system matrices A and B, with a choice of Q, R, and α, to ensure all the closed-loop eigenvalues are within the α-circle, we use the following code:

```
[K,P,E]=dlqr(A/alpha, B/alpha, Q, R);
```

The closed-loop eigenvalues are calculated as

```
E=alpha*E;
```

4.5.2 Case Studies

The following example is used to illustrate vibration control using linear quadratic regulator.

Example 4.4
A system in civil engineering is described by the following continuous-time state space model:

$$\dot{x}(t) = A_p x(t) + B_p u(t); \; y(t) = C_p x(t), \tag{4.53}$$

where the system matrices are

$$A_p = \begin{bmatrix} -1.5 & 0.6467 & 0 & -717 & 260.4 & 0 \\ 0.6467 & -1.26 & 0.6133 & 260.4 & -493.5333 & 233.1333 \\ 0 & 0.6133 & -0.6133 & 0 & 233.1333 & -233.1333 \\ 1 & 0 & 0 & 0 & 0 & 0 \\ 0 & 1 & 0 & 0 & 0 & 0 \\ 0 & 0 & 1 & 0 & 0 & 0 \end{bmatrix};$$

$$B_p = \begin{bmatrix} -0.0333 & 0.0333 & 0 \\ 0 & -0.0333 & 0.0333 \\ 0 & 0 & -0.0333 \\ 0 & 0 & 0 \\ 0 & 0 & 0 \\ 0 & 0 & 0 \end{bmatrix}; \; C_p = \begin{bmatrix} 0 & 0 & 0 & 1 & 0 & 0 \\ 0 & 0 & 0 & 0 & 1 & 0 \\ 0 & 0 & 0 & 0 & 0 & 1 \end{bmatrix}.$$

The system is oscillatory if the initial condition $x(0) \neq 0$. The control objective is to use state feedback control to reduce the vibration.

Solution
There are three vibration modes in this system, which is evident from the locations of the eigenvalues of system matrix A_p: $-1.0622 \pm j30.2335$, $-0.5497 \pm j21.5111$, $-0.0748 \pm j8.0874$. With these eigenvalues, we can re-construct the three characteristic polynomial equations to determine their natural frequencies and damping coefficients. For example, for the first pair of open-loop eigenvalues, we have:

$$s^2 + 2\xi_1 w_{n1} s + w_{n1}^2 = (s + 1.0622 + j30.2335)(s + 1.0622 - j30.2335) = 0$$

which gives the natural frequency $w_{n1} = 30.2522$ and the damping coefficient $\xi_1 = 0.0351$. The remaining natural frequencies are $w_{n2} = 21.5181$ and $w_{n3} = 8.0878$, and the damping coefficients are $\xi_2 = 0.0255$, $\xi_3 = 0.0093$. To capture the three oscillatory modes in the discrete-time, we select the sampling interval $\Delta t = 0.01$, which

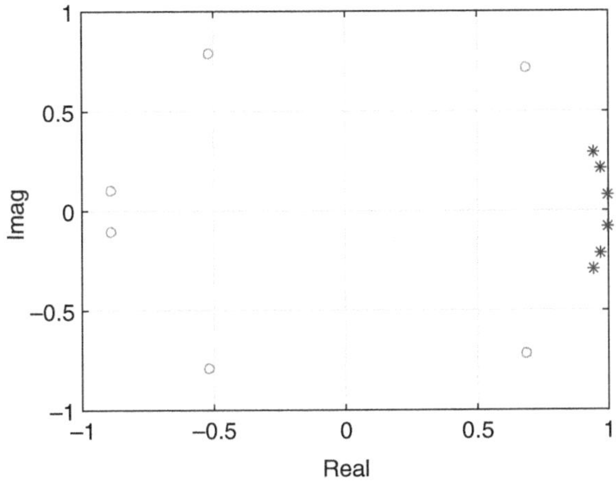

Figure 4.4 Illustration of eigenvalues in discrete-time (Example 4.4). Key: star $\Delta t = 0.01$, circle $\Delta t = 0.1$.

gives the three pairs of discrete-time eigenvalues as,

$$0.9446 \pm j0.2946,\ 0.9716 \pm j0.2123;\ 0.9960 \pm j0.0807$$

which are illustrated (see the stars) in Figure 4.4. In contrast, if we had chosen $\Delta t = 0.1$, *two of the three pairs of eigenvalues moved across the imaginary axis to have negative real parts (see the circles in Figure 4.4), which is an indication that the sampling interval* $\Delta t = 0.1$ *was too large.*

To design the state feedback control to produce sufficient damping for the system, the weighting matrix is $Q = C_p^T C_p$ *that has the last three diagonal elements as* 1 *and the rest* 0, *with more emphasis placed to reduce the errors on the last three state variables. The weighting matrix* R *is selected as* $R = \epsilon I$ *with* ϵ *adjustable. Apparently, with the choices of* Q *and* R *matrices, it is difficult to change the closed-loop eigenvalues with the variations of* ϵ. *Figure 4.5 shows the clusters of the closed-loop eigenvalues for* $\epsilon = 1, 0.1, 0.01$. *For instance, when* $\epsilon = 0.01$, *the closed-loop eigenvalues are* $0.9446 \pm j0.2946$, $0.9959 \pm j0.0807$, $0.9716 \pm j0.2123$. *When comparing the eigenvalues from the open-loop system, they have only negligible changes. This means that in order to achieve sufficient vibration control, we would need to consider the off-diagonal elements in* Q *and* R *matrices. But this process is based on trial-and-error and is time consuming.*

Now, assuming that we choose $\alpha = 0.99$, *using the DLQR design with a prescribed degree of stability, the controller is re-designed and the closed-loop eigenvalues become:* $0.9445 \pm j0.2946$, $0.9776 \pm j0.0792$, $0.9628 \pm j0.2104$. *In comparison, the second and third pairs of the closed-loop eigenvalues are changed by introducing the prescribed degree of stability. Figure 4.6 shows the locations of the closed-loop eigenvalues in relation to the* α*-circle.*

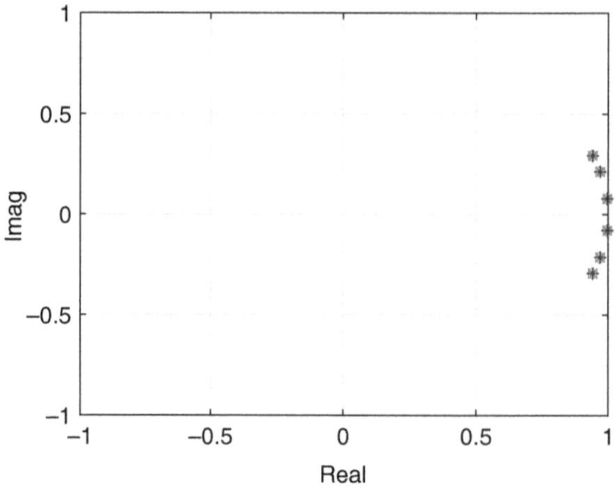

Figure 4.5 Illustration of closed-loop eigenvalues with $\epsilon = 1, 0.1, 0.01$ (Example 4.4).

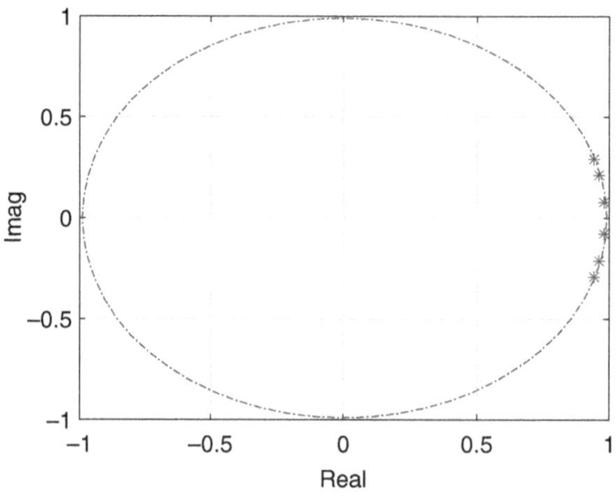

Figure 4.6 Illustration of closed-loop eigenvalues with $\epsilon = 0.01$ and $\alpha = 0.99$ (Example 4.4). Key: * the eigenvalues, dash-dot: the α-circle.

To compare the closed-loop responses, the initial conditions $x_1(0) = x_2(0) \ldots = x_6(0) = 1$ are chosen in the simulation studies. Figure 4.7(a) shows that the oscillatory closed-loop response of $x_6(k)$ to the initial states without the prescribed degree of stability ($\alpha = 1$). In contrast, Figure 4.7(b) shows that the re-designed controller significantly reduces the vibrations within three hundred samples. From an engineering design point of view, we need to examine how the control signal behaves with much improved vibration control. Figure 4.8(a) shows that control signal response without the prescribed degree of stability ($\alpha = 1$), which is seen to have amplitude variations within ± 0.5. In contrast, with the prescribed degree of stability where the vibrations have been significantly reduced, Figure 4.8(b) shows that the control (u_1) amplitude variations are between -1000 and 500. How fast

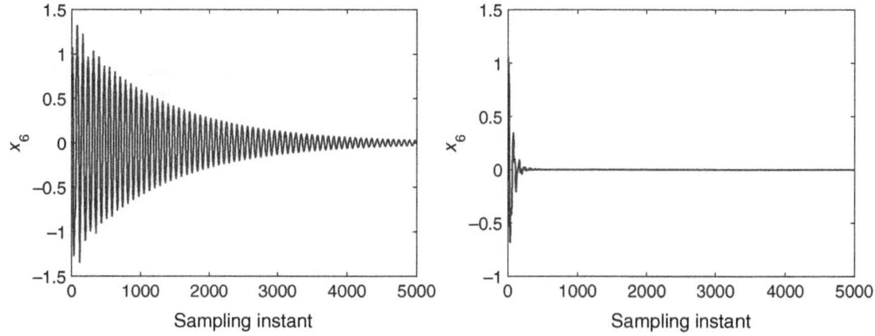

(a) $\alpha = 1$ (b) $\alpha = 0.99$

Figure 4.7 Comparison of closed-loop state response with and without prescribed degree of stability (Example 4.4).

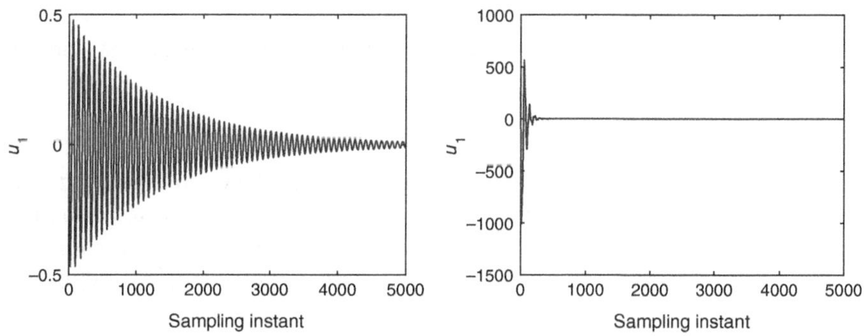

(a) $\alpha = 1$ (b) $\alpha = 0.99$

Figure 4.8 Comparison of closed-loop control response with and without prescribed degree of stability (Example 4.4).

we can reduce the vibrations is dependent on the capacity of the actuators as larger control signals will be required for the better performance. Therefore, the α value needs to be selected carefully with regard to actuators' constraints.

With Dp being zero, the following MATLAB program is used to obtain the results with discretization of the continuous-time model.

```
[A,B,C,D]=c2dm(Ap,Bp,Cp,Dp,Deltat,'zoh');
Q=zeros(6,6);
Q(4:6,4:6)=eye(3);
R=0.01*eye(3);
alpha=0.99;
[K,P,E]=dlqr(A/alpha,B/alpha,Q,R);
E=alpha*E;
```

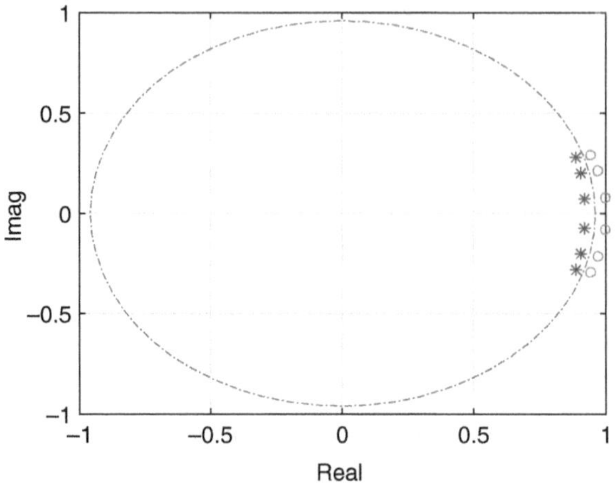

Figure 4.9 Illustration of closed-loop eigenvalues of the error system (Example 4.5). Key: $*$ $\alpha = 0.96$; o $\alpha = 1$; dash-dot α-circle.

The following example is to show how to use the prescribed degree of stability in the design of observer.

Example 4.5

Consider the oscillatory system given in Example 4.4. For this example, we will design an observer to estimate the state variables $x_1(k)$, $x_2(k)$, and $x_3(k)$ with the output measurements $y_1(k)$, $y_2(k)$, and $y_3(k)$. We assume that there is no process input, which is $u(k) = 0$, and the system has the same initial states $x_1(0) = x_2(0) \ldots = x_6(0) = 1$ with sampling interval $\Delta t = 0.01$. The measurement noises are independent and have zero mean and variance of 0.01. Design an observer to estimate the state vector $x(k)$.

Solution

We will choose $R = 0.01I$ and Q to be a zero matrix, where R is based on the variance of the output noise and Q is selected to be zero because there is no noise on state variables as we assumed. The closed-loop eigenvalues for the observer error system are

$$0.9446 \pm j0.2946, \ 0.9716 \pm j0.2123, \ 0.9960 \pm j0.0807.$$

Figure 4.9 shows the locations of the closed-loop eigenvalues (see the circles). Clearly, we anticipate that the convergence rate for the observer errors is slow because the closed-loop eigenvalues are very close to the unit circle. To speed up the convergence of the errors, we choose $\alpha = 0.96$, which gives a new K_{ob}. The closed-loop eigenvalues are

$$0.8892 \pm j0.2773, \ 0.9053 \pm j0.1978, \ 0.9193 \pm 0.0745.$$

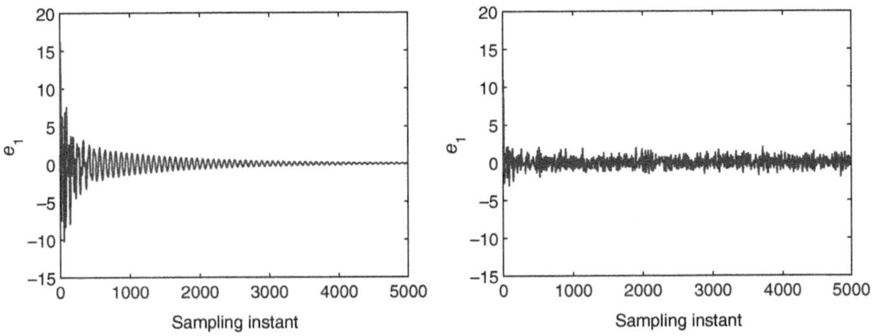

(a) $\alpha = 1$ (b) $\alpha = 0.96$

Figure 4.10 Comparison of error response with and without prescribed degree of stability (Example 4.5).

Figure 4.9 shows the new locations of the closed-loop eigenvalues (the stars) in comparison with those from the previous case (the circle). It is seen that all the new eigenvalues are within the α-circle.

To evaluate the performance of the observers, we generate the output data by solving the difference equation with the initial conditions specified. Three independent noises are added to the outputs.

Figure 4.10(a) shows the response of the error $e_1(k) = x_1(k) - \hat{x}_1(k)$ ($x_1(k)$ is unknown in applications) without the prescribed degree of stability ($\alpha = 1$). It is seen that it takes about 3500 samples for the error to converge to near zero, and the error is quite large at the initial stage of estimation with variations between −10 and 16.

In comparison, Figure 4.10(b) shows the response of the error $e_1(k)$ with this new K_{ob} where the prescribed degree of stability is used in the design. It is seen that the convergence of error is achieved within 100 samples, however, there is a noise component in the estimated variable $\hat{x}_1(k)$. Additionally, the initial error response is reduced to the range between −3 and 11.

Overall, we need to choose the parameter α carefully in balancing the relationship between the dynamic response of the errors and the attenuation of the measurement noise. As α reduces, the effect of measurement noise increases.

With the discrete-time A and C matrices defined, the following MATLAB program is used to generate the design:

```
alpha=0.96;
Q=zeros(6,6);
R=0.01*eye(3);
[K1,P,E]=dlqr(A'/alpha,C'/alpha,Q,R);
```

```
Kob=K1';
E=alpha*E;
```

4.5.3 Food for Thought

1. What objective function do we use to find the controller K when using the prescribed degree of stability method? What happens to this objective function when $\alpha = 1$?
2. Assume that a discrete-time system is described by the following state space model:

$$x(k+1) = \begin{bmatrix} 0.8 & 0 \\ 0 & 0.9 \end{bmatrix} x(k) + \begin{bmatrix} 1 \\ 0 \end{bmatrix}.$$

   With $Q = I$ and $R = 1$, can we use the MATLAB program **dlqr.m** to find the controller K? Now, using the method of prescribed degree of stability where $\alpha = 0.9$, can we use the same function to find K? Why?
3. Assume the same discrete-time system as in 2, however with Q being a zero matrix and $R = 1$. Do you think that the MATLAB program **dlqr.m** would find the controller K when $\alpha = 1$? How about $\alpha = 0.96$? Why?
4. If we wish the maximum time constant of a desired closed-loop system to be 10 seconds, with the sampling interval $\Delta t = 0.5$ seconds, what α value would you propose to use so that all the desired closed-loop constants are within the limit?
5. Assume that a discrete-time system is described by the following state space model:

$$x(k+1) = \begin{bmatrix} 0.8 & 0 \\ 1 & 0 \end{bmatrix} x(k) + \begin{bmatrix} 1 \\ 0 \end{bmatrix}.$$

   With the choice of Q being a zero matrix and $R = 1$, do you think that the MATLAB function **dlqr.m** would find the controller gain K? Can we do the same for $\alpha = 0.6$?
6. Does the method of prescribed degree of stability tell us exactly where the closed-loop poles are?

4.6 Summary

This chapter has discussed discrete-time state feedback controller design and observer design. Assuming that the system matrices A, B, and C are given for the discrete-time model, the controller design is based on the A and B matrices while the observer design is based on the A and C matrices. The controller design is designed to achieve stability and desired dynamic response speed of the closed-loop feedback control system $A - BK$. The observer design is designed to achieve the same objectives for the observer error system $A - K_{ob}C$.

The commonly used design methods for both controllers and observers are the pole-assignment controller design and the design of a discrete-time linear quadratic regulator (DLQR).

The other important aspects of this chapter are summarized as follows.

1. When using a pole-assignment controller design method for discrete-time, the desired closed-loop eigenvalues (or poles) are specified within the unit circle of the complex plane. The desired eigenvalues in discrete-time could be selected by converting desired continuous-time eigenvalues with the choice of sampling interval Δt.
2. Observers are used to estimate the state vector $x(k)$ when $x(k)$ is unknown. Observers are designed using the A and C matrices to achieve stability and desired dynamic response of the observer error system:

   $$\tilde{x}(k+1) = (A - K_{ob}C)\tilde{x}(k).$$

3. We can use the controller design programs for observer design where the matrices A^T and C^T are used in substitution of A and B matrices as we would do for the controller design.
4. In the implementation of the observer, the following observer equation is used:

   $$\hat{x}(k+1) = A\hat{x}(k) + Bu(k) + K_{ob}(y(k) - C\hat{x}(k))$$

   In contrast to its continuous-time counterpart (see Section 1.6 of Chapter 1), the observer implementation does not involve numerical approximation. This is one of the advantages when using the discrete-time control system design.
5. Discrete-time linear quadratic regulator (DLQR) design provides an effective tool for both controller design and observer design of multi-input and multi-output systems. The MATLAB program **dlqr.m** offers the optimal solution of the controller and observer. In the DLQR design, the selection of weighting matrices Q and R determines the optimal controller and optimal observer. Afterwards, the closed-loop eigenvalues are computed.
6. The discrete-time linear quadratic regulator with prescribed degree of stability constrains all closed-loop eigenvalues inside the α-circle of the complex plane where $\alpha < 1$. This is performed by scaling the system matrices used in the MATLAB program **dlqr.m** and is effective when the task of selecting Q and R matrices becomes too complex.

4.7 Further Reading

1. Linear optimal control and filter books include Kwakernaak and Sivan [1972], Anderson and Moore [1979], Goodwin and Sin [1984].
2. Discrete-time control using polynomial based design is presented in Kucera [1980].

Problems

Problem 4.1 Assume that a discrete-time system is described by the following state space model:

$$x(k+1) = Ax(k) + Bu(k)$$

where $A = \begin{bmatrix} \alpha_1 & \alpha_2 \\ 0 & 1 \end{bmatrix}$ and $B = \begin{bmatrix} \beta_1 \\ \beta_2 \end{bmatrix}$. Find the state feedback controller K such that the closed-loop eigenvalues are positioned at 0.8 and 0.9.

Problem 4.2 Use MATLAB function **place.m** to find the state feedback controller K for the state space model:

$$x(k+1) = Ax(k) + Bu(k)$$

1. The desired closed-loop poles eigenvalues are specified as 0.3, 0.4 and 0.5, the A and B matrices are

$$A = \begin{bmatrix} 0.5 & 0 & 0.1 \\ 0.1 & 0.7 & 1 \\ 0 & 0 & 1 \end{bmatrix}; \ B = \begin{bmatrix} 1 \\ 1.2 \\ 1.3 \end{bmatrix}.$$

2. The desired closed-loop eigenvalues are specified as 0.6, 0.61, 0.62, 0.63, and the A and B matrices are

$$A = \begin{bmatrix} 1 & 1.2 & 0 & 1.3 \\ 0 & 1.5 & 0.5 & -0.6 \\ -2 & 0 & 0 & 0.9 \\ 0 & 2 & 0 & 0.5 \end{bmatrix}; \ B = \begin{bmatrix} 2 \\ 1.5 \\ 2.1 \\ 0.9 \end{bmatrix}.$$

Problem 4.3 A continuous-time system is described by the following differential equation:

$$\dot{x}(t) = A_c x(t) + B_c u(t),$$

where the system matrices A_c and B_c are given as

$$A_c = \begin{bmatrix} -0.5 & 1 & 0 \\ 0 & -0.8 & 1 \\ 0 & 0 & 0 \end{bmatrix}; \ B_c = \begin{bmatrix} 1 \\ -1 \\ 1 \end{bmatrix}.$$

The sampling interval Δt for the system is chosen to be 0.2 (sec).

1. Find the corresponding discrete-time system matrices A and B using MATLAB program **c2dm.m** with zero-order-hold, where C_c and D_c matrices have appropriate dimensions, but with arbitrary values.

2. Determine if the discrete-time system is controllable.
3. Design the state feedback controller K using MATLAB program **place.m** where the desired closed-loop eigenvalues are 0.7, 0.71, 0.72.
4. Simulate the state feedback control control system with the initial conditions $x_1(0) = 1, x_2(0) = 2$ and $x_3(0) = 3$.

Problem 4.4 Continue from Problem 4.3. Instead of using the set of desired closed-loop poles in discrete-time as in Problem 4.3, we propose to choose the desired closed-loop poles in continuous-time, which gives us an insight into the closed- loop response time.

1. Assuming that all the desired closed-loop eigenvalues in the continuous-time are selected as −1, find the discrete-time desired closed-loop eigenvalues and the state feedback controller gain K where the sampling interval $\Delta t = 0.1$, 0.2 and 0.4.
2. Discuss the variations of the controller K with respect to the changes in the sampling interval Δt.

Problem 4.5 Check controllability and observability of the following state space model:

$$\begin{bmatrix} x_1(k+1) \\ x_2(k+1) \\ x_3(k+1) \\ x_4(k+1) \end{bmatrix} = \begin{bmatrix} 0 & 1 & 0 & 0 \\ 3\alpha^2 & 0 & 0 & 2\alpha \\ 0 & 0 & 0 & 1 \\ 0 & -2\alpha & 0 & 0 \end{bmatrix} \begin{bmatrix} x_1(k) \\ x_2(k) \\ x_3(k) \\ x_4(k) \end{bmatrix} + \begin{bmatrix} 0 \\ 1 \\ 0 \\ 0 \end{bmatrix} u(k)$$

$$y(k) = \begin{bmatrix} 0 & 0 & 1 & 0 \end{bmatrix} \begin{bmatrix} x_1(k) \\ x_2(k) \\ x_3(k) \\ x_4(k) \end{bmatrix}$$

where $\alpha \neq 0$.

Problem 4.6 Suppose that a two input and two output system is described by the difference equation:

$$\begin{bmatrix} x_1(k+1) \\ x_2(k+1) \end{bmatrix} = \begin{bmatrix} 1 + \frac{\beta}{a_1}\Delta t & \frac{\beta}{a_1}\Delta t \\ \frac{\beta}{a_2}\Delta t & 1 + \frac{\beta}{a_2}\Delta t \end{bmatrix} \begin{bmatrix} x_1(k) \\ x_2(k) \end{bmatrix} + \begin{bmatrix} \frac{1}{a_1}\Delta t & 0 \\ 0 & \frac{1}{a_2}\Delta t \end{bmatrix} \begin{bmatrix} u_1(k) \\ u_2(k) \end{bmatrix} \tag{4.54}$$

where $a_1 \neq 0, a_2 \neq 0$.

1. Determine if the system is controllable if $\beta = 0$.
2. We assume that we only have one input available for control purpose, say that the first input $u_1(k)$ is available. Determine if the system is still controllable if $\beta = 0$. Additionally, design a pole-assignment controller for this single input system and investigate the effect of $\beta = 0$ on the closed- loop control system.

Problem 4.7 Consider the dynamic system with the state space model:

$$\begin{bmatrix} x_1(k+1) \\ x_2(k+1) \end{bmatrix} = \begin{bmatrix} 1 + \frac{\beta}{a_1}\Delta t & \frac{\beta}{a_1}\Delta t \\ \frac{\beta}{a_2}\Delta t & 1 + \frac{\beta}{a_2}\Delta t \end{bmatrix} \begin{bmatrix} x_1(k) \\ x_2(k) \end{bmatrix} + \begin{bmatrix} \frac{1}{a_1}\Delta t & 0 \\ 0 & \frac{1}{a_2}\Delta t \end{bmatrix} \begin{bmatrix} u_1(k) \\ u_2(k) \end{bmatrix}$$

$$\begin{bmatrix} y_1(k) \\ y_2(k) \end{bmatrix} = \begin{bmatrix} \lambda & 0 \\ 0 & 1 \end{bmatrix} \begin{bmatrix} x_1(k) \\ x_2(k) \end{bmatrix}$$

where $a_1 \neq 0$, $a_2 \neq 0$, and the sampling interval $\Delta t > 0$.

1. Assuming that $\lambda \neq 0$, $\beta = 0$, determine if the system is observable.
2. Assuming that $\lambda = 0$, $\beta \neq 0$, determine if the system is observable.
3. Assuming that $\lambda = 0$ and $\beta = 0$, determine if the system is observable. Design an observer using pole-assignment technique and discuss what happens when $\lambda = 0$ and $\beta = 0$.

Problem 4.8 A continuous-time sinusoidal signal $y_0(t)$ is severely corrupted by noise, where $y_0(t)$ is described by the following discrete-time function

$$y_0(t) = a_m sin(\omega_0 t + \theta)$$

$$y(t) = y_0(t) + \epsilon(t)$$

where $y(t)$ is the measured continuous-time signal and $\epsilon(t)$ is the noise, the frequency parameter $\omega_0 > 0$ is known, but a_m and θ are unknown.

1. Write down the continuous-time state space model that could lead to filtering the noise.
2. Given a sampling interval Δt, design a discrete-time observer to estimate the noise-free signal $y_0(k)$, where all the closed-loop observer poles are chosen to be $e^{-3\omega_0 \Delta t}$.
3. Validate your design with the following simulation studies. Assuming that $\omega_0 = 1$, and $a_m = 2$ and $\theta = \frac{\pi}{4}$ and $\Delta t = 0.01$, generate the noise-free sinusoidal signal $y_0(k)$ and add the white noise $\epsilon(k)$ with standard deviation of 0.2. Calculate the estimated signal $\hat{y}(k)$ using the discrete-time observer designed where the initial states of the observer are selected as $\hat{x}_1(0) = 0$ and $\hat{x}_2(0) = 0$.

Problem 4.9 A double integrator system in discrete-time is described by the following difference equation:

$$\begin{bmatrix} x_1(k+1) \\ x_2(k+1) \end{bmatrix} = \begin{bmatrix} 1 & \Delta t \\ 0 & 1 \end{bmatrix} \begin{bmatrix} x_1(k) \\ x_2(k) \end{bmatrix} + \begin{bmatrix} \frac{1}{2}\Delta t^2 \\ \Delta t \end{bmatrix} u(k)$$

where $\Delta t = 1$. Design a state feedback controller using the discrete-time LQR design method to bring the initial states $x_1(0) = x_2(0) = 1$ to zero. Demonstrate the selection of Q and R matrices on the effect of closed-loop response.

1. Choosing $R = 1$ and assuming that Q is a diagonal matrix for a simple case, what is the objective function J in the discrete-time quadratic regulator? Experiment variations of Q matrix with the following three cases.
   (a) With $Q(1,1) = Q(2,2) = 1$, what are the controller gain K, Riccati solution P, and the closed-loop eigenvalues E? What is J_{min}?
   (b) With $Q(1,1) = 100$ and $Q(2,2) = 1$, what are the controller gain K, Riccati solution P, and the closed-loop eigenvalues E? What is J_{min}?
   (c) With $Q(1,1) = 1$ and $Q(2,2) = 100$, what are the controller gain K, Riccati solution P, and the closed-loop eigenvalues E? What is J_{min}?
   (d) Simulate the closed-loop state feedback control for the above three cases where the initial conditions $x_1(0) = x_2(0) = 1$. Compare the state variables and the control signals. What are your observations?
2. Choosing Q as the identity matrix ($Q(1,1) = Q(2,2) = 1$), experiment variations of R with the following three cases.
   (a) With $R = 0.01$, what are the controller gain K, Riccati solution P and the closed-loop eigenvalues E? What is J_{min}?
   (b) With $R = 0.1$, what are the controller gain K, Riccati solution P and the closed-loop eigenvalues E? What is J_{min}?
   (c) With $R = 10$, what are the controller gain K, Riccati solution P and the closed-loop eigenvalues E? What is J_{min}?
   (d) Simulate the closed-loop state feedback control for the above three cases where the initial conditions $x_1(0) = x_2(0) = 1$. Compare the state variables and the control signals. What are your observations?

Problem 4.10 A three-springs-double mass system has the following state space model:

$$\begin{bmatrix} \dot{x}_1(t) \\ \dot{x}_2(t) \\ \dot{x}_3(t) \\ \dot{x}_4(t) \end{bmatrix} = \begin{bmatrix} 0 & 1 & 0 & 0 \\ -\frac{\alpha_1+\alpha_2}{M_1} & 0 & \frac{\alpha_2}{M_1} & 0 \\ 0 & 0 & 0 & 1 \\ \frac{\alpha_2}{M_2} & 0 & -\frac{\alpha_1+\alpha_2}{M_2} & 0 \end{bmatrix} \begin{bmatrix} x_1(t) \\ x_2(t) \\ x_3(t) \\ x_4(t) \end{bmatrix} + \begin{bmatrix} 0 & 0 \\ \frac{1}{M_1} & 0 \\ 0 & 0 \\ 0 & \frac{1}{M_2} \end{bmatrix} \begin{bmatrix} u_1(t) \\ u_2(t) \end{bmatrix}$$

$$\begin{bmatrix} y_1(t) \\ y_2(t) \end{bmatrix} = \begin{bmatrix} 1 & 0 & 0 & 0 \\ 0 & 0 & 1 & 0 \end{bmatrix} \begin{bmatrix} x_1(t) \\ x_2(t) \\ x_3(t) \\ x_4(t) \end{bmatrix}$$

where the two input variables are the two applied forces u_1 and u_2 and two output variables are the distances of the mass block movement $y_1 = x_1(t)$ corresponding to block one and $y_2 = x_3(t)$ corresponding to block two. The state variables $x_2(t) = \dot{y}_1(t)$ and $x_4(t) = \dot{y}_2(t)$. The physical parameters are $M_1 = 3kg$, $M_2 = 6kg$, $\alpha_1 = 40N/m$ and $\alpha_2 = 100N/m$. The sampling interval $\Delta t = 0.1$ (sec) is used to discretize the continuous-time system.

1. Find the discrete-time state space model with the sampling interval $\Delta t = 0.1$ (sec).
2. Determine if the discrete-time model is controllable and observable.
3. Design discrete-time linear quadratic regulator K with $R = I$ and investigate three cases of Q matrix: (i) $Q(1,1) = Q(3,3) = 1$, the rest of elements in Q are zero; (ii) $Q(2,2) = Q(4,4) = 1$, the rest of elements in Q are zero; (iii) Q is an identity matrix. Can you find the controller K for the three choices of Q matrix? If so, where are the closed-loop eigenvalues?
4. Simulate the closed-loop state feedback control system with initial state variables $x_1(0) = x_3(0) = 1, x_2(0) = x_4(0) = 0$ for the cases that a state feedback controller K is designed. Compare the control signals and the state variables for the different choices of Q matrix. What are your observations?
5. Find an observer gain K_{ob} using MATLAB **dlqr.m**, where $Q_{ob} = I$ and $R_{ob} = I$. Where are the closed-loop eigenvalues for the observer error system?
6. Simulate the closed-loop state estimate control system with initial state variables $x_1(0) = x_3(0) = 1$, $x_2(0) = x_4(0) = 0$, and all initial states for the observer are chosen to be zero. Compare the control signals and estimated state variables for the different choices of Q matrix from the controller design, maintaining the same observer gain K_{ob}. What are your observations?

Problem 4.11 Continue from Problem 4.10 and design state feedback controller and observer using the DLQR method with prescribed degree of stability.

1. For the controller design, we wish to constrain the maximum closed-loop time constant to be 20 seconds. With R chosen as identity matrix and $Q(1,1) = Q(3,3) = 1$, the rest of elements in Q being zero, find the state feedback controller gain K using MATLAB **dlqr.m** function.
2. For the observer design, we wish to constrain the maximum time constant for the observer error system to be 10 seconds. With both Q_{ob} and R_{ob} chosen as

identity matrices, find the observer gain K_{ob} using MATLAB function **dlqr.m** function.

3. Simulate the closed-loop state estimate control system with initial state variables $x_1(0) = x_3(0) = 1$, $x_2(0) = x_4(0) = 0$ and all initial states for the observer are chosen to be zero. Present the control signals and estimated state variables with respect to the discrete-time $t = 0,\ \Delta t,\ 2\Delta t,\ \ldots,$. What are your observations?
4. Change the maximum time constant for the controller to be 10 second and for the observer to be 5 second, what are the new state feedback controller K and observer K_{ob}? Repeat the simulation studies with the new controller and observer. Compare the control signals and the estimated state variables with those obtained from the previous case. What are your observations in terms of closed-loop response time?

5

Disturbance Rejection and Reference Tracking via Observer Design

5.1 Introduction

Beginning with this chapter, reference signals and disturbance signals are introduced in the discrete-time state feedback control. For the discrete-time control system design, the reference and disturbance signals include step signal, ramp signal, sinusoidal signals, or any arbitrary combinations between them. Thus, the discrete-time design allows the disturbance rejection and reference tracking of complex signals in a multi-input and multi-output system and offers a much broader scope in various control system applications than their continuous-time counterpart.

The objective of this chapter is to introduce disturbance rejection and reference tracking with a disturbance- observer-based approach. The chapter begins with the general discussions on disturbance signals and their characterizations (see Section 5.2). After having been familiar with the commonly used disturbance models, Section 5.3 proposes how to compensate the disturbances in the state estimation. In Section 5.4, a disturbance-observer-based state feedback control is introduced. In Section 5.5, the disturbance-observer-based control system is analyzed. With this disturbance-observer-based approach, an anti-windup mechanism is naturally introduced for the controller implementation (see Section 5.6). Both MATLAB and Simulink tutorials are given in this chapter for the design and implementation of this state feedback controller.

5.2 Disturbance Models

In general, a disturbance is associated with a slowly changing variable, and in the frequency domain it has pre-dominant low frequency contents. The disturbance

State Feedback Control and Kalman Filtering with MATLAB/Simulink Tutorials, First Edition.
Liuping Wang and Robin Ping Guan.

Companion website: www.wiley.com/go/feedback19

signals are often part of the process input and output signals and are not measured independently unless special sensors are deployed for the purpose of their measurement.

5.2.1 Commonly Encountered Disturbance Signals

Constant disturbance. The most commonly encountered disturbance is a constant with varied amplitude for a deterministic case. For instance, the electrical torque load is modelled as a piece-wise constant. As illustrated in Chapter 3, the dynamic equation for a constant signal $\mu(k)$ is described as

$$\mu(k+1) = \mu(k)$$

with initial condition $\mu(0) \neq 0$, which can also be described by the forward shift operator as

$$D(q)\mu(k) = 0$$

where $D(q) = q - 1$. The polynomial model $D(q)$ is called disturbance model from here on.

For a stochastic case, there is a white noise input to the disturbance process and $\mu(k)$ satisfies the following difference equation:

$$\mu(k+1) = \mu(k) + \epsilon(k) \tag{5.1}$$

and with the forward shift operator:

$$D(q)\mu(k) = \epsilon(k)$$

where $\epsilon(k)$ is white noise with zero mean and variance σ^2. The disturbance $\mu(k)$ can be expressed in terms of $D(q)$ as

$$\mu(k) = \frac{\epsilon(k)}{D(q)}.$$

Or in backward shift operator:

$$\mu(k) = \frac{q^{-1}\epsilon(k)}{D(q^{-1})}$$

The signal $\mu(k)$ from the stochastic case is often refereed to as random walk signal. In the simulation studies, we need to generate the disturbance signal. In order for the results to be repeatable, a seed for the random signal generator should be selected. With $\sigma = 0.1$, Figure 5.1(a) shows the random walk signal with seed being 0, and Figure 5.1(b) shows the random walk signal with seed being 1. From these

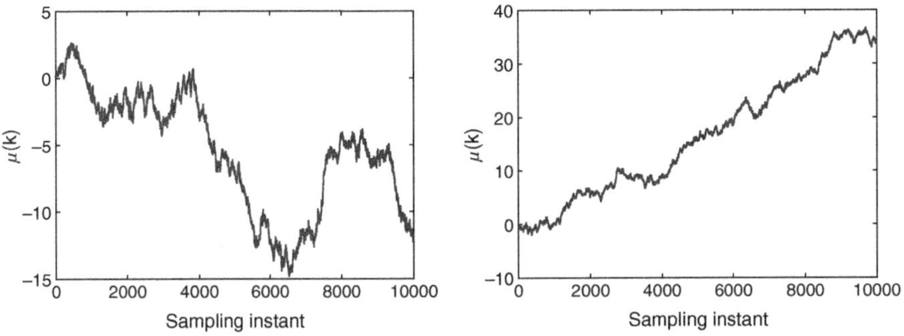

(a) Seed=0 (b) Seed=1

Figure 5.1 Random walk signal.

two signals, it is clearly seen that they are very different when the seed changes. The following program is used to generate the disturbance signal:

```
Nsim=10000;
randn('seed',1)
noi=0.1*randn(1,Nsim);
numd-1;
dend=[1 -1];
mu=filter(numd,dend,noi);
```

In Sections 5.3 and 5.4, we will see that the assumption of input constant disturbance or random walk disturbance will lead to the feedback control systems that have integral action when they are designed to overcome the disturbances.

Harmonic disturbance. The disturbance $\mu(k)$ is a sinusoidal signal with known discrete frequency ω_d satisfying the difference equation for the deterministic case:

$$\mu(k+2) - 2\cos\omega_d \mu(k+1) + \mu(k) = 0,$$

or in the forward shift operator:

$$(q^2 - 2\cos\omega_d q + 1)\mu(k) = 0.$$

For the stochastic case, the difference equation is

$$\mu(k+2) - 2\cos\omega_d \mu(k+1) + \mu(k) = \epsilon(k),$$

where $\epsilon(k)$ is white noise with zero mean and variance σ^2. For the harmonic disturbance case, the disturbance model is

$$D(q) = q^2 - 2\cos\omega_d q + 1.$$

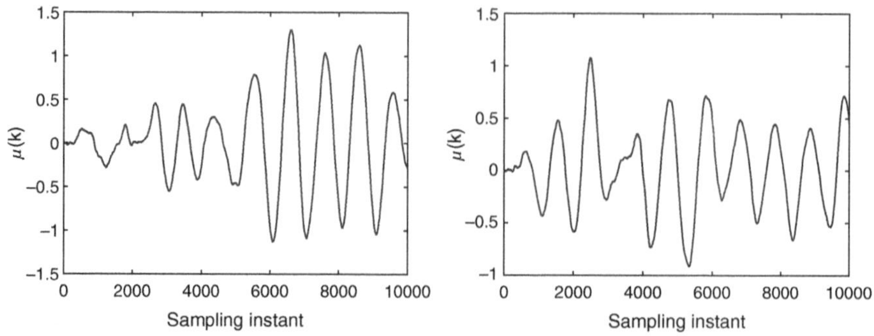

(a) Seed=0 (b) Seed=1

Figure 5.2 Sinusoidal disturbance.

With $\sigma = 0.001$, Figure 5.2(a) shows the harmonic disturbance with frequency $f = 0.1$ Hz, sampling interval $\Delta t = 0.01$ where the seed is chosen to be 0, and Figure 5.2(b) shows the harmonic disturbance generated with seed being 1. To generate the sinusoidal signals, the discrete-time frequency ω_d is determined as $\omega_d = 2\pi f \Delta t$.

Ramp disturbance. When the disturbance $\mu(k)$ is a ramp signal in the deterministic case, it satisfies the difference equation:

$$\mu(k+2) - 2\mu(k+1) + \mu(k) = 0.$$

In the stochastic case, it has the following form:

$$\mu(k+2) - 2\mu(k+1) + \mu(k) = \epsilon(k)$$

where $\epsilon(k)$ is white noise with zero mean and variance σ^2. For a ramp disturbance, the disturbance model is

$$D(q) = q^2 - 2q + 1$$

and $\mu(k)$ can be expressed as

$$\mu(k) = \frac{\epsilon(k)}{D(q)}.$$

To overcome the ramp disturbance at the steady-state, the control system will be designed to have double integral action as shown in Section 5.5.

A system can have a combination of the above disturbances. For instance, if the disturbance contains both random walk and a sinusoidal disturbance with frequency ω_d. Then, $\mu(k)$ is modelled with the forward shift operator:

$$(q-1)(q^2 - 2\cos\omega_d q + 1)\mu(k) = \epsilon(k)$$

which is

$$\mu(k) = \frac{\epsilon(k)}{D(q)}$$

where $D(q) = (q-1)(q^2 - 2\cos\omega_d q + 1)$.

In Section 6.5.2 of Chapter 6, we will show how to find a disturbance model $D(q)$ for a given complex periodic signal with frequency decompositions.

5.2.2 State Space Model with Input Disturbance

Assuming that an input disturbance is $\mu(k)$, the state space model that has the disturbance components becomes

$$x_m(k+1) = A_m x_m(k) + B_m u(k) + B_d \mu(k) \tag{5.2}$$

$$y(k) = C_m x_m(k). \tag{5.3}$$

For the majority of the applications, the exact information about B_d is not readily available, and it may not be necessary to have the information for the control system design as shown in the sequel. So we simply assume that they are part of the process input and output. By doing so, the state space model is simplified to

$$x_m(k+1) = A_m x_m(k) + B_m(u(k) + \mu(k)) \tag{5.4}$$

$$y(k) = C_m x_m(k). \tag{5.5}$$

Because there may be more than one source of input disturbance, it is assumed that the ith component of the disturbance vector is described by

$$\mu_i(k+1) = \frac{\epsilon_i(k)}{D(q^{-1})} \tag{5.6}$$

where $\epsilon_i(k)$ is a white noise sequence with zero mean and variance σ_i^2, $1 \leq i \leq m$. Here $D(q^{-1})$ is expressed in backward shift operator and is comprised of the least common factors of the disturbance models from all sources. For instance, supposing that the system has a random walk disturbance acting on $u_1(k)$ and a sinusoidal disturbance with frequency ω_d acting on $u_2(k)$, the disturbance model $D(q^{-1})$ is expressed as

$$D(q^{-1}) = (1 - q^{-1})(1 - 2\cos\omega_d q^{-1} + q^{-2}).$$

The same $D(q^{-1})$ is applied to each component in $\mu(k)$. In general, with the assumption that $D(q^{-1})$ is of degree γ, it can be written using the backward shift operator as

$$D(q^{-1}) = 1 + d_1 q^{-1} + d_2 q^{-2} + d_3 q^{-3} + \ldots + d_\gamma q^{-\gamma}. \tag{5.7}$$

5.2.3 Food for Thought

1. What is the dynamic model for a random walk using the backward shift operator?
2. What is the z-transform of a constant disturbance? With $z = e^{j\omega}, 0 \leq \omega \leq \pi$, can you calculate the magnitude of its frequency response for the initial condition $\mu(0) = 1$?
3. What is the z-transform of a sinusoidal disturbance with initial condition $\mu(0) = \mu(1) = 1$? Is it correct to say that the magnitude of the frequency response of the sinusoidal signal is zero every where except when $\omega = \pm\omega_d$? What is its value at $\omega = \pm\omega_d$?
4. What are your observations for the magnitudes of the frequency responses of the constant signal and sinusoidal signal?
5. A discrete-time system is described by the following difference equations:

$$x(k+1) = 0.8x(k) + 0.6(u(k) + \mu(k)); \;\; y(k) = x(k).$$

What is the key difference between the plant input signal $u(k)$ and the input disturbance $\mu(k)$?

5.3 Compensation of Input and Output Disturbances in Estimation

In this section, our objective in state estimation is to correctly estimate the state vector $x_m(k)$ in the presence of input disturbance and output disturbance.

5.3.1 Motivational Example

To demonstrate the effect of disturbance on the state estimation, we examine the following example.

Example 5.1

In Example 4.5, we have demonstrated the design of observer for an oscillatory system with measurement noise. Now, for the same system, there is a sinusoidal disturbance with frequency equal to 0.1 *Hz acting on* $x_1(k)$. *Estimate the state variables* $x_1(k)$, $x_2(k)$ *and* $x_3(k)$ *using the same observer as described in Example 4.5. For convenience, the mathematical model is given as below, which is sampled with* $\Delta t = 0.01$ *(sec),*

$$\dot{x}(t) = A_p x(t) + B_p u(t); \;\; y(t) = C_p x(t) \tag{5.8}$$

where the system matrices are

$$A_p = \begin{bmatrix} -1.5 & 0.6467 & 0 & -717 & 260.4 & 0 \\ 0.6467 & -1.26 & 0.6133 & 260.4 & -493.5333 & 233.1333 \\ 0 & 0.6133 & -0.6133 & 0 & 233.1333 & -233.1333 \\ 1 & 0 & 0 & 0 & 0 & 0 \\ 0 & 1 & 0 & 0 & 0 & 0 \\ 0 & 0 & 1 & 0 & 0 & 0 \end{bmatrix};$$

$$B_p = \begin{bmatrix} -0.0333 & 0.0333 & 0 \\ 0 & -0.0333 & 0.0333 \\ 0 & 0 & -0.0333 \\ 0 & 0 & 0 \\ 0 & 0 & 0 \\ 0 & 0 & 0 \end{bmatrix}; \; C_p = \begin{bmatrix} 0 & 0 & 0 & 1 & 0 & 0 \\ 0 & 0 & 0 & 0 & 1 & 0 \\ 0 & 0 & 0 & 0 & 0 & 1 \end{bmatrix}.$$

Solution

We generate the disturbance signal $\mu(k)$ as

$$\mu(k) = \frac{0.0005q^{-1}}{1 - 2\cos\omega_d q^{-1} + q^{-2}}\epsilon(k)$$

where $\omega_d = 2\pi f \Delta t$ with $f = 0.1$ (Hz), $\Delta t = 0.01$ (sec). In the simulation, $\epsilon(k)$ is a white noise with $\sigma = 0.1$. The disturbance sequence $\mu(k)$ is added to the first state variable $x_1(k)$ when generating the simulated output data. Figures 5.3(a)–(b) show two of the measured output signals. We can see from these figures that the sinusoidal

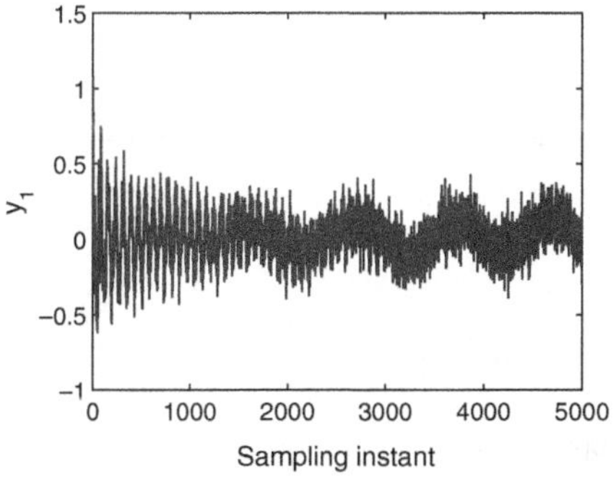

(a) Output y_1

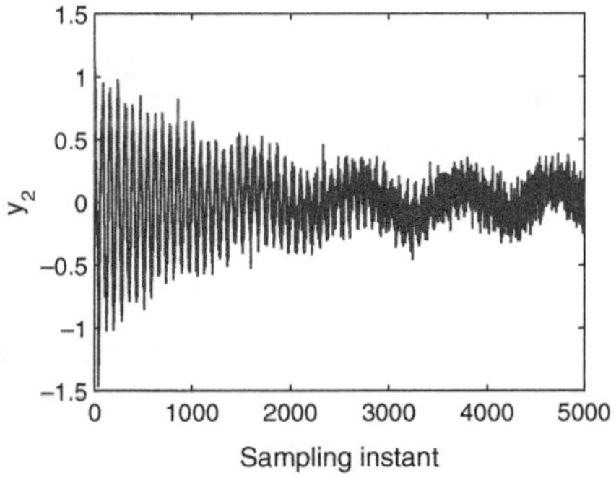

(b) Output y_2

Figure 5.3 Output signals in presence of sinusoidal disturbance (Example 5.1).

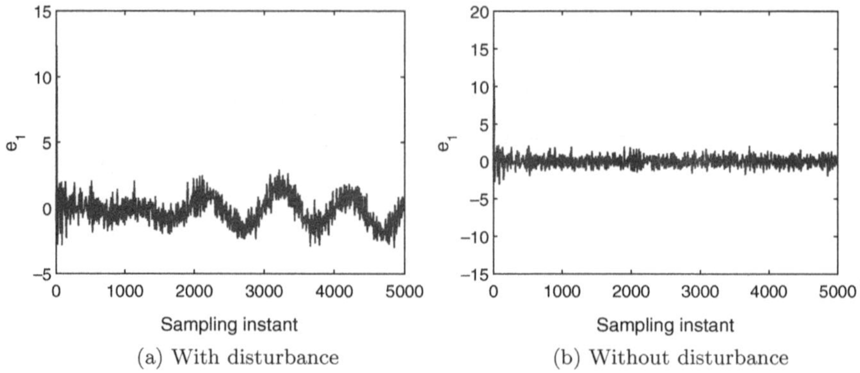

Figure 5.4 Comparison of error signals with and without disturbance (Example 5.1).

disturbance acting on $x_1(k)$ has an effect on both $y_1(k)$ and $y_2(k)$, also on $y_3(k)$ (not shown here).

By using the same observer with the identical initial conditions as in Example 4.5 with $\alpha = 0.96$, we obtain the estimated state variables $\hat{x}_1(k)$, $\hat{x}_2(k)$ and $\hat{x}_3(k)$. Figure 5.4(a) shows the error signal $\tilde{x}_1(k)$ in the presence of disturbance, and in comparison Figure 5.4(b) shows the error signal without the disturbance. It is seen from Figure 5.4(a) that there is a sinusoidal component in the estimation error, which means that without a compensation of the disturbance, the estimated state is no longer correct. $\tilde{x}_2(k)$ and $\tilde{x}_3(k)$ are similarly affected and are not shown here.

Example 5.1 motivates us to find a way to overcome the effect of disturbance by modelling it in the state estimation.

5.3.2 Input Disturbance Observer Design

We assume that system has an input disturbance $\mu(k)$ and state-space model

$$x_m(k+1) = A_m x_m(k) + B_m(u(k) + \mu(k)) \tag{5.9}$$

$$y(k) = C_m x_m(k), \tag{5.10}$$

where $x_m(k)$ is the $n_1 \times 1$ state vector, $u(k)$ and $\mu(k)$ are the $m \times 1$ input vectors, $y(k)$ is $m \times 1$ output vector.

As introduced in Section 5.2, the disturbance $\mu(k)$ is modelled as

$$\mu(k+1) = \frac{\epsilon(k)}{D(q^{-1})} \tag{5.11}$$

where $\epsilon(k)$ is an $m \times 1$ vector. With the disturbance model $D(q^{-1})$ defined by

$$D(q^{-1}) = 1 + d_1 q^{-1} + d_2 q^{-2} + d_3 q^{-3} + \ldots + d_\gamma q^{-\gamma} \tag{5.12}$$

we write the disturbance vector $\mu(k+1)$ in the following form of difference equation:

$$\mu(k+1) = -d_1\mu(k) - d_2\mu(k-1) - \ldots - d_\gamma\mu(k-\gamma+1) + \epsilon(k) \tag{5.13}$$

where the vector $\epsilon(k)$ has each entry being white noise with zero mean.

To convert the difference equation (5.13) into a state space representation, we will first define a vector $p(k)$ for the disturbance model as

$$p(k) = \left[\mu^T(k) \quad \mu^T(k-1) \quad \ldots \quad \mu^T(k-\gamma+1)\right]^T \tag{5.14}$$

where $p(k)$ has the dimension $m\gamma \times 1$. The disturbance component is captured by the following state space model:

$$p(k+1) = A_d p(k) + B_\epsilon \epsilon(k); \tag{5.15}$$

$$\mu(k) = C_\epsilon p(k) \tag{5.16}$$

where the A_d matrix has the following structure:

$$A_d = \begin{bmatrix} -d_1 I_m & -d_2 I_m & \ldots & -d_{\gamma-1} I_m & -d_\gamma I_m \\ I_m & 0_m & 0_m & \ldots & 0_m \\ 0_m & I_m & 0_m & \ldots & 0_m \\ \vdots & \ddots & \ddots & \vdots & \vdots \\ 0_m & 0_m & \ldots & I_m & 0_m \end{bmatrix} \tag{5.17}$$

B_ϵ ($m\gamma \times m$) and C_ϵ ($m \times m\gamma$) have the identity matrix for the first m columns and rows, respectively. The remaining columns and rows for both matrices are the zero matrices. The 0_m and I_m matrices denote the zero and identity matrices with dimensions $m \times m$, respectively.

It is important to verify that with this formulation we have the following relationship:

$$\det(qI_{m\gamma} - A_d) = D(q)^m \tag{5.18}$$

where $I_{m\gamma}$ is the identity matrix with dimension $m\gamma \times m\gamma$ and $D(q)$ is the disturbance model in forward shift form. This means that the A_d matrix contains m sets of eigenvalues that correspond to the zeros of the $D(q)$ polynomial.

To compensate the effect of disturbance on the estimated state vector $x_m(k)$, we will incorporate the disturbance vector $\mu(k)$ into the dynamic model so that they can be estimated together. For this purpose, we introduce the augmented state vector $x(k)$ as

$$x(k) = \left[x_m^T(k) \quad p(k)^T\right]^T.$$

Note that from (5.16) the original state space model (5.9) with the input disturbance can be re-written as

$$x_m(k+1) = A_m x_m(k) + B_m C_\epsilon p(k) + B_m u(k).$$

Together with (5.15), the augmented state space model that contains both the state variable $x_m(k)$ and $p(k)$ is expressed as

$$\begin{bmatrix} x_m(k+1) \\ p(k+1) \end{bmatrix} = \overbrace{\begin{bmatrix} A_m & B_m C_\epsilon \\ 0_{m\gamma \times n} & A_d \end{bmatrix}}^{A} \begin{bmatrix} x_m(k) \\ p(k) \end{bmatrix} + \overbrace{\begin{bmatrix} B_m \\ 0_{m\gamma \times m} \end{bmatrix}}^{B} u(k) + \begin{bmatrix} 0_{n \times m} \\ B_\epsilon \end{bmatrix} \epsilon(k), \tag{5.19}$$

where $0_{p\times q}$ denotes a zero matrix with dimension $p \times q$, A and B matrices, as illustrated, are defined as the augmented system matrices. It is worthwhile to note that the augmented state space model (5.19) has the input signal $u(k)$ and the white noise signal $\epsilon(k)$ as the process noise. The output equation for the state space model (5.19) becomes

$$y(k) = \overbrace{\begin{bmatrix} C_m & 0_{m\times m\gamma} \end{bmatrix}}^{C} \begin{bmatrix} x_m(k) \\ p(k) \end{bmatrix} \tag{5.20}$$

where $0_{m\times m\gamma}$ is a zero matrix with dimension $m \times m\gamma$ and C matrix is defined as the augmented output matrix.

Now, the observer design will be based on the state space model described by (5.19) and (5.20) using the pair of (A, C) matrices. To estimate the augmented state vector $x(k)$, we will choose an observer gain matrix K_{ob} either using pole-assignment design technique or discrete-time LQR technique such that the closed-loop observer error system is stable with a desired dynamic response speed.

With K_{ob} selected, the augmented state vector $x(k)$ is estimated using

$$\hat{x}(k+1) = A\hat{x}(k) + Bu(k) + K_{ob}(y(k) - C\hat{x}(k)) \tag{5.21}$$

where A, B, and C matrices are defined in (5.19) and (5.20).

The question remains under what conditions is the pair (A, C) observable? The answer to this question is that if observability holds for (A_m, C_m), and if the system does not have zeros corresponding to the disturbance modes contained in the polynomial $D(q)$, then the pair (A, C) is observable. The condition on the requirement of the system's zeros is illustrated with the following example.

Example 5.2

Consider that a discrete-time system has the following transfer function:

$$G(z) = \frac{z^2 - 2\cos\omega_d z + 1}{(z-0.9)^2(z-0.5)}$$

where $\omega_d = \frac{2\pi}{40}$. This transfer function corresponds to the following state space model:

$$A_m = \begin{bmatrix} 2.3 & -1.71 & 0.405 \\ 1 & 0 & 0 \\ 0 & 1 & 0 \end{bmatrix}; \; B_m = \begin{bmatrix} 1 \\ 0 \\ 0 \end{bmatrix}; C_m = \begin{bmatrix} 1 & -1.9754 & 1 \end{bmatrix}.$$

We will illustrate that the state space model for the pair (A_m, C_m) is observable, however, if the input disturbance $\mu(k)$ is a sinusoidal signal with discrete-time frequency $\omega_d = \frac{2\pi}{40}$, then the augmented system matrices (A, C) is not observable.

Solution

To check the observability of the pair (A_m, C_m), we can use MATLAB program **ctrb.m** *to form the observability matrix L_o with the following code:*

```
Lo=ctrb(Am',Cm')'
```

The rank of the observability matrix L_o is 3. Therefore, the pair A_m, C_m is observable.

Consider that the disturbance model for the sinusoidal disturbance $\mu(k)$ is described by

$$\mu(k+1) = \frac{\epsilon(k)}{1 - 2\cos\omega_d q^{-1} + q^{-2}}.$$

Therefore, we choose $D(q^{-1}) = 1 - 2\cos\omega_d q^{-1} + q^{-2}$. By following equation (5.19), we obtain the augmented A and B matrices:[1]

$$A = \begin{bmatrix} 2.3 & -1.71 & 0.405 & 1 & 0 \\ 1 & 0 & 0 & 0 & 0 \\ 0 & 1 & 0 & 0 & 0 \\ 0 & 0 & 0 & 1.9754 & -1 \\ 0 & 0 & 0 & 1 & 0 \end{bmatrix}; \; B = \begin{bmatrix} 1 \\ 0 \\ 0 \\ 0 \\ 0 \end{bmatrix}$$

Based on equation (5.20), we obtain the C matrix:

$$C = \begin{bmatrix} 1 & -1.9754 & 1 & 0 & 0 \end{bmatrix}$$

With the augmented system matrices, A and C, we again form the observability matrix and check its rank. The rank of the observability matrix is 3. Therefore the augmented system matrix is not observable because the augmented system has 5 states and the rank is less than 5.

For the more general case when the number of inputs is not necessarily equal to the number of outputs, then an additional requirement for observability is that the

1 One can follow Tutorial 5.1 presented in the next section to form the augmented system matrices.

number of outputs must be greater or equal to the number of inputs. If in doubt, we should always check the observability of the pair (A, C) matrices.

5.3.3 MATLAB Tutorial for Augmented State Space Model

The estimation of input disturbance will be used in the control system design in the later sections. It is important to correctly generate the augmented system matrices (A, B, C) for the observer design and for the compensation of disturbance in control design. The following tutorial presents how to generate these system matrices.

Tutorial 5.1 *Write a MATLAB function* **am4obs.m** *that will produce the augmented model for the observer that has the capacity to compensate input disturbance.*

Step by Step

1. *Create a new file called* **am4obs.m**.
2. *The input variables to the function are the plant system matrices: A_m, B_m, C_m, and the polynomial disturbance model $D(q^{-1})$ that is called Dmodel. The output variables from this function are the augmented system matrices A, B, and C. Enter the following program into the file:*

```
function [A,B,C]=am4obs(Am,Bm,Cm,Dmodel)
```

3. *Check the numbers of inputs and state variables, and the dimension of Dmodel. Continue entering the following program into the file:*

```
gamma=length(Dmodel)-1;
[n1,n_in]=size(Bm);
m1=n_in;
```

4. *Initialize the system matrices by using zero matrices. Continue entering the following program into the file:*

```
n=n1+gamma*n_in;
A=zeros(n,n);
B=zeros(n,n_in);
C=zeros(m1,n);
```

5. *Create the block matrices A_m and B_mC_ϵ in A. Continue entering the following program into the file:*

```
A(1:n1,1:n1)=Am;
A(1:n1,n1+1:n1+n_in)=Bm;
```

6. *Create the block matrix A_d in A. Continue entering the following program into the file:*

```
nk=1;
for kk=1:gamma
A(n1+1:n1+n_in,n1+nk:n1+kk*n_in)=-Dmodel(kk+1)*
eye(n_in);
nk=nk+n_in;
end
A(n1+n_in+1:n,n1+1:n-n_in)=eye((gamma-1)*n_in);
```

7. *Form the B and C matrices. Continue entering the following program into the file:*

```
B(1:n1,:)=Bm;
C(:,1:n1)=Cm;
```

8. *Test the function with the following matrices:*

```
Am=[1 -1; 2 1];
Bm=[1 2;2 1];
Cm=[1 0; 0 1];
Dmodel-[1 -2 -3];
```

9. *Check A matrix against the following answer:*

$$A = \left[\begin{array}{cc|cccc} 1 & -1 & 1 & 2 & 0 & 0 \\ 2 & 1 & 2 & 1 & 0 & 0 \\ \hline 0 & 0 & 2 & 0 & 3 & 0 \\ 0 & 0 & 0 & 2 & 0 & 3 \\ 0 & 0 & 1 & 0 & 0 & 0 \\ 0 & 0 & 0 & 1 & 0 & 0 \end{array}\right]$$

5.3.4 The Observer Error System

Define the error vector as

$$\tilde{x}(k) = x(k) - \hat{x}(k).$$

Then the observer error system is formed by subtracting (5.21) from (5.19):

$$\begin{bmatrix} \tilde{x}_m(k+1) \\ \tilde{p}(k+1) \end{bmatrix} = (A - K_{ob}C) \begin{bmatrix} \tilde{x}_m(k) \\ \tilde{p}(k) \end{bmatrix} + \begin{bmatrix} 0_{n \times m} \\ B_\epsilon \end{bmatrix} \epsilon(k). \tag{5.22}$$

In the deterministic case, the noise vector $\epsilon(k) = 0$. Then stability of the error system (5.22) leads to the convergence of the error states, which is $||\tilde{x}_m(k)|| \to 0$,

$||\tilde{p}(k)|| \to 0$, as $k \to \infty$. When $\epsilon(k) \neq 0$, we will examine the expectance of (5.22), which is

$$\begin{bmatrix} E[\tilde{x}_m(k+1)] \\ E[\tilde{p}(k+1)] \end{bmatrix} = (A - K_{ob}C) \begin{bmatrix} E[\tilde{x}_m(k)] \\ E[\tilde{p}(k)] \end{bmatrix} + \begin{bmatrix} 0_{n\times m} \\ B_\epsilon \end{bmatrix} E[\epsilon(k)]. \tag{5.23}$$

With the assumption of $\epsilon(k)$ that is a vector with each component being zero mean white noise, the expectance of $\epsilon(k)$ is a zero vector. Hence, the expectance of the observer error system becomes,

$$\begin{bmatrix} E[\tilde{x}_m(k+1)] \\ E[\tilde{p}(k+1)] \end{bmatrix} = (A - K_{ob}C) \begin{bmatrix} E[\tilde{x}_m(k)] \\ E[\tilde{p}(k)] \end{bmatrix}. \tag{5.24}$$

Because the observer gain K_{ob} is chosen such that the error system matrix $A - K_{ob}C$ is stable, the expectance of the error vectors $E[\tilde{x}_m(k)]$, $E[\tilde{p}(k)]$ will converge to zero as $k \to \infty$.

Example 5.3

In Example 5.1, we have demonstrated the effect of disturbance without compensation and the results show that the error signal exhibits a sinusoidal component. In this example, we will use the same set of output data as shown in Figure 5.3 for the estimation of state variables, however with the estimation of state vector $x_m(k)$ together with the disturbance vector $p(k)$.

Solution

For the sinusoidal disturbance, the discrete-time frequency $\omega_d = 2\pi f \Delta t$ is known, where $f = 0.1$ (Hz) and $\Delta t = 0.01$ (sec). Because there are three control signals for this example, we will use three input disturbances, but all of them share the same disturbance model with

$$D(q^{-1}) = 1 - 2\cos\omega_d q^{-1} + q^{-2} = 1 + d_1 q^{-1} + d_2 q^{-2}.$$

With the discrete state space system matrices A_m, B_m, C_m, and the information for the disturbance model $D(q^{-1})$, we form the following A and C matrices for the observer design (see (5.19) and (5.20)):

$$A = \begin{bmatrix} A_m & B_m C_\epsilon \\ 0_{m\gamma \times n} & A_d \end{bmatrix}; \; C = \begin{bmatrix} C_m & 0_{m \times m\gamma} \end{bmatrix}.$$

The dimensions of A and C are 12×12 and 3×12, respectively. Here, we used the program **am4obs.m** *produced in Tutorial 5.1 to find the A and C matrices.*

We will use the MATLAB program **dlqr.m** *to find the observer gain K_{ob}. The weighting matrix $R = 0.01I$ is selected to correspond to the variance of the measurement noise. For the Q matrix, we choose*

$$Q = \begin{bmatrix} 0_{n\times m} \\ B_\epsilon \end{bmatrix} \begin{bmatrix} 0^T_{n\times m} & B^T_\epsilon \end{bmatrix}$$

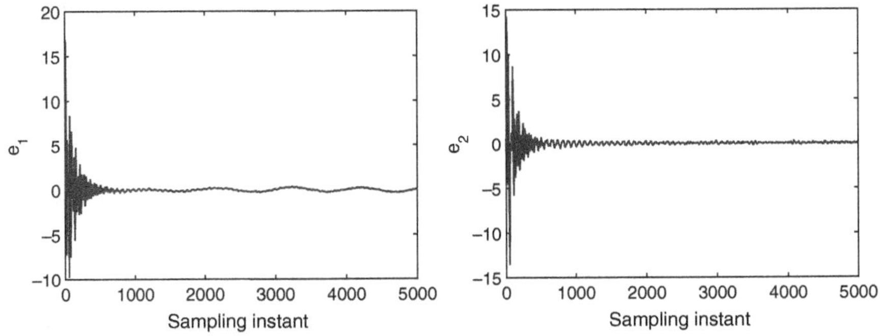

(a) $\tilde{x}_1(k)$　(b) $\tilde{x}_2(k)$

Figure 5.5 State error response in presence of disturbance (Example 5.3).

which basically has $Q(7,7) = Q(8,8) = Q(9,9) = 1$ and the remaining elements are zero. In the Kalman filter chapter, we will show that these choices correspond to the optimal selection of Q and R matrices. Figures 5.5(a)–(b) show the state error response for $\tilde{x}_1(k)$ and $\tilde{x}_2(k)$. The results show that the errors converge to zero with a small amount of noise.

5.3.5 Output Disturbance Observer Design

When there is an output disturbance, the state space model is described by

$$\begin{aligned} x_m(k+1) &= A_m x_m(k) + B_m u(k) \\ y(k) &= C_m x_m(k) + \beta(k), \end{aligned} \tag{5.25}$$

where without more detailed information, we have assumed that each output has an output disturbance acting on it. Similar to the case for input disturbance description, it is assumed that the ith component of the disturbance vector $\beta(k)$ is described by

$$\beta_i(k) = \frac{\epsilon_i(k)q^{-1}}{D(q^{-1})}, \tag{5.26}$$

where $\epsilon_i(k)$ is a white noise sequence with zero mean and standard deviation σ_i, $1 \leq i \leq m$. The same $D(q^{-1})$ is applied to each component in $\beta(k)$. With the disturbance model $D(q^{-1})$, we write the disturbance vector $\beta(k+1)$ in the following form

$$\beta(k+1) = -d_1\beta(k) - d_2\beta(k-1) - \ldots - d_\gamma\beta(k-\gamma+1) + \epsilon(k) \tag{5.27}$$

where the vector $\epsilon(k)$ has each entry being zero mean white noise.

Similar to the case with input disturbance, we will define a vector $p(k)$ for the disturbance model as

$$p(k) = \left[\beta^T(k) \quad \beta^T(k-1) \quad \ldots \quad \beta^T(k-\gamma+1)\right]^T. \tag{5.28}$$

The output disturbance component is captured by the following state space model:

$$p(k+1) = A_d p(k) + B_\epsilon \epsilon(k); \tag{5.29}$$

$$\beta(k) = C_\epsilon p(k) \tag{5.30}$$

where the A_d, B_ϵ, and C_ϵ matrices are identical to those in (5.15) and (5.16).

To compensate the output disturbance in the state estimation, the augmented state vector $x(k)$ is chosen as

$$x(k) = \begin{bmatrix} x_m^T(k) & p(k)^T \end{bmatrix}^T.$$

The augmented state space model that contains both the state variable $x_m(k)$ and $p(k)$ is given as

$$\begin{bmatrix} x_m(k+1) \\ p(k+1) \end{bmatrix} = \overbrace{\begin{bmatrix} A_m & 0_{n\times m\gamma} \\ 0_{m\gamma\times n} & A_d \end{bmatrix}}^{A} \begin{bmatrix} x_m(k) \\ p(k) \end{bmatrix} + \overbrace{\begin{bmatrix} B_m \\ 0_{m\gamma\times m} \end{bmatrix}}^{B} u(k) + \begin{bmatrix} 0_{n\times m} \\ B_\epsilon \end{bmatrix} \epsilon(k) \tag{5.31}$$

where $0_{p\times q}$ denotes a zero matrix with dimension $p \times q$, A and B matrices, as illustrated, are defined as the augmented system matrices. The output equation for the state space model (5.31) is

$$y(k) = \overbrace{\begin{bmatrix} C_m & C_\epsilon \end{bmatrix}}^{C} \begin{bmatrix} x_m(k) \\ p(k) \end{bmatrix}. \tag{5.32}$$

The pair A, C is observable if the original system (A_m, C_m) is observable. The observer gain K_{ob} is chosen based on the pair A and C matrices such that the observer error system matrix $A - K_{ob}C$ is stable with all eigenvalues strictly inside the unit circle of the complex plane.

The next step is to estimate the state vector $x_m(k)$ together with the output disturbance vector $p(k)$, where the estimated state is computed using

$$\hat{x}(k+1) = A\hat{x}(k) + Bu(k) + K_{ob}(y(k) - C\hat{x}(k)) \tag{5.33}$$

where A, B, and C matrices are defined in (5.31) and (5.32).

Defining the observer error vector as $\tilde{x}(k) = \begin{bmatrix} \tilde{x}_m(k)^T & \tilde{p}(k)^T \end{bmatrix}^T$, based on (5.31)–(5.33), we form the observer error system:

$$\begin{bmatrix} \tilde{x}_m(k+1) \\ \tilde{p}(k+1) \end{bmatrix} = (A - K_{ob}C) \begin{bmatrix} \tilde{x}_m(k) \\ \tilde{p}(k) \end{bmatrix} + \begin{bmatrix} 0_{n\times m} \\ B_\epsilon \end{bmatrix} \epsilon(k). \tag{5.34}$$

Its expectance is calculated as

$$\begin{bmatrix} E[\tilde{x}_m(k+1)] \\ E[\tilde{p}(k+1)] \end{bmatrix} = (A - K_{ob}C) \begin{bmatrix} E[\tilde{x}_m(k)] \\ E[\tilde{p}(k)] \end{bmatrix} + \begin{bmatrix} 0_{n\times m} \\ B_\epsilon \end{bmatrix} E[\epsilon(k)]$$

$$= (A - K_{ob}C) \begin{bmatrix} E[\tilde{x}_m(k)] \\ E[\tilde{p}(k)] \end{bmatrix} \tag{5.35}$$

where $E[\epsilon(k)] = 0$ from the assumption that $\epsilon(k)$ is a vector with each component to be zero mean white noise. Equation (5.35) says that the mean error $||E[\tilde{x}(k)]||$ will converge to zero as $k \to \infty$ if the error system matrix $A - K_{ob}C$ is stable.

Recall that when we assumed input disturbance $\mu(k)$ in Section 5.3.2, it is required that the zeros of the disturbance model do not correspond to the zeros of the plant transfer function in order to maintain the observability of the augmented system matrices. However, this requirement is not longer needed if we assume the disturbance occurred at system's output. This is verified through the following example.

Example 5.4

Consider that a discrete-time system has the following transfer function:

$$G(z) = \frac{z^2 - 2\cos\omega_d z + 1}{(z - 0.9)^2(z - 0.5)}$$

where $\omega_d = \frac{2\pi}{40}$. *We assume that the system has a sinusoidal output disturbance with frequency* ω_d *and illustrate the augmented system matrices A and C are observable.*

Solution

This system was used in Example 5.2 where A_m, B_m, *and* C_m *matrices can be found. From equation (5.31), we form the following A and B matrices:*

$$A = \begin{bmatrix} 2.3 & -1.71 & 0.405 & 0 & 0 \\ 1 & 0 & 0 & 0 & 0 \\ 0 & 1 & 0 & 0 & 0 \\ 0 & 0 & 0 & 1.9754 & -1 \\ 0 & 0 & 0 & 1 & 0 \end{bmatrix}; B = \begin{bmatrix} 1 \\ 0 \\ 0 \\ 0 \\ 0 \end{bmatrix}.$$

Based on equation (5.32), we form the C matrix:

$$C = \begin{bmatrix} 1 & -1.9754 & 1 & 1 & 0 \end{bmatrix}.$$

The observability matrix L_o *is calculated using MATLAB program* **ctrb.m**:

```
Lo=ctrb(A',C')';
```

The rank of L_o *is 5. Therefore, the augmented system is observable.*

In the observer applications, we are often given a set of output measurement data that contain low frequency disturbances. It is sometimes difficult to determine whether they are input disturbance or output disturbance. It is shown in the following example that for a linear time-invariant system, the effect of input disturbance can also be compensated through the modelling of output disturbance so that the state variables can be estimated correctly.

Example 5.5
In this example, we will use the set of output measurement data generated in Example 5.1, where the disturbance was generated by adding a sinusoidal signal to $x_1(k)$. *Demonstrate that correct state estimation can be achieved by capturing the effect of disturbance in the output.*

Solution
Since there are three output signals, we will use three output disturbances, each having the same disturbance model:

$$D(q^{-1}) = 1 - 2\cos\omega_d q^{-1} + q^{-2} = 1 + d_1 q^{-1} + d_2 q^{-2}$$

where $\omega_d = 2\pi f \Delta t, f = 0.1$ *(Hz) and* $\Delta t = 0.01$ *(sec).*
We form the augmented system matrices:

$$A = \begin{bmatrix} A_m & 0_{n\times m\gamma} \\ 0_{m\gamma\times n} & A_d \end{bmatrix}; C = \begin{bmatrix} C_m & C_\epsilon \end{bmatrix}$$

where A_m *and* C_m *are the original system matrices,* C_ϵ *has the identity matrix for the first 3 columns and rows while the remaining columns and rows are the zero matrices.*
In the design of observer gain K_{ob}, *the MATLAB program* **dlqr.m** *is used. The weighting matrix* $R = 0.01I$ *is selected to correspond to the variance of the measurement noise. For the Q matrix, we will use the structure of noise model on the process by choosing*

$$Q = \begin{bmatrix} 0_{n\times m} \\ B_\epsilon \end{bmatrix} \begin{bmatrix} 0_{n\times m}^T & B_\epsilon^T \end{bmatrix}$$

which basically has $Q(7,7) = Q(8,8) = Q(9,9) = 1$ *and the remaining elements are zero. With the choices of Q and R matrices, the eigenvalues of the error system consist of the three pairs:*

$$0.9446 \pm j0.2946,\ 0.9716 \pm j0.2123,\ 0.9960 \pm j0.0807$$

and the remaining three pairs at the same location of: $0.0187 \pm j0.0954$. *From the original example (see Example 4.4), the first three pairs of the eigenvalues correspond to the eigenvalues of* A_m, *which could not be changed with this particular combination of Q and R matrices. Figure 5.6(a) shows that the error* $\tilde{x}_2(k)$ *oscillates for a long time. At this point, we could try a different Q matrix in order to change the first three pairs of eigenvalues. But we need to choose it carefully because the noise effect could be amplified. We choose to have a small modification to the previous design by using*

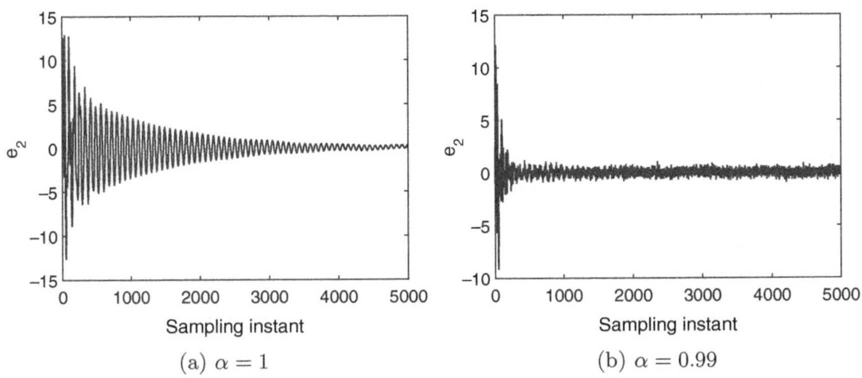

Figure 5.6 Comparison of state errors with and without prescribed degree of stability (Example 5.5).

the prescribed degree of stability. This is a convenient way to increase the convergence speed of the observer errors without drastically changing the effect of noise. Here, we will use $\alpha = 0.99$, which is to force all eigenvalues inside the circle of 0.99 radius. With the prescribed degree of stability, the eigenvalues of the observer error system are:

$$0.9628 \pm j0.2104, \;\; 0.9776 \pm j0.0792, \;\; 0.9446 \pm j0.2946$$

together with the three identical pairs at $0.0185 \pm j0.0944$. For comparison, Figure 5.6(b) shows that the convergence of the error $\tilde{x}_2(k)$ has been significantly improved, however, there is an amplification of noise.

5.3.6 Food for Thought

1. Assume that a first order system is given by

$$x_m(k+1) = 0.8x_m(k) + 0.1u(k) + 0.6\mu(k); \;\; y(k) = x_m(k),$$

   where $\mu(k) = \frac{\epsilon(k)}{1-0.9q^{-1}}$. If we neglected the disturbance $\mu(k)$ in the observer design and used the following equation for the estimation:

$$\hat{x}_m(k+1) = 0.8\hat{x}_m(k) + 0.1u(k) + K_{ob}(y(k) - \hat{x}_m(k)),$$

   where K_{ob} is chosen such that the pole for the closed-loop error system is at 0.3, do you think that the error $E[x_m(k) - \hat{x}_m(k)]$ would converge to zero as $k \to 0$? Why?
2. If you wish to capture the effect of disturbance $\mu(k)$ from Problem 1, what is the augmented model (A_d, C_d)?
3. In the formulation of disturbance estimation (see (5.13) and (5.14)), we have used the backward shift operator q^{-1}. If we wish to use the forward shift

operator q, what would be the state vector $p(k)$? Does this lead to the same A_d and C_d matrices?

4. Assume that a system has two inputs, two outputs and three state variables. In order to compensate a sinusoidal in the observer design, how many eigenvalues does the system matrix A_d have?
5. In the formulation of disturbance estimation (see Section 5.2.2), we simplified the problem by assuming that $B_d = B_m$. If $B_m \neq B_d$, will we obtain the correct estimate for $x_m(k)$? What about $\mu(k)$?
6. If the input disturbance to the system is

$$\mu(k) = \frac{q^{-1}}{1 - q^{-1}}\epsilon(k),$$

however, it was modelled as

$$\mu(k) = \frac{q^{-1}}{(1 - q^{-1})^2}\epsilon(k).$$

Do you think that the estimated state vector $\hat{x}_m(k)$ would converge to the true state vector $x_m(k)$? If so, what would be a downside of such an approach?

5.4 Disturbance-Observer-based State Feedback Control

In the previous section (see Section 5.3), we have discussed how to compensate the disturbances in the state estimation. From estimation point of view, it is shown that we can compensate them through a disturbance model either in the input variable or in the output variable in order to remove their effects from the estimated state variables. However, from control system design point of view, it is much more effective to derive a state feedback control system using the input disturbance estimation.

5.4.1 The Control Law

In the state feedback control design, we assume the disturbance occurring at the input variable with the following model:

$$x_m(k+1) = A_m x_m(k) + B_m(u(k) + \mu(k)) \tag{5.36}$$

$$y(k) = C_m x_m(k) \tag{5.37}$$

where $x_m(k)$ is the $n_1 \times 1$ state vector, $u(k)$ and $\mu(k)$ are the $m \times 1$ input vectors, $y(k)$ is $m \times 1$ output vector. In this section, we also assume that the pair A_m, B_m is controllable and the pair A_m, C_m is observable, and the system does not have zeros corresponding to the disturbance modes contained in the polynomial $D(q)$.

Introducing the intermediate control variable:

$$\tilde{u}(k) = u(k) + \mu(k)$$

we re-write (5.36) as

$$x_m(k+1) = A_m x_m(k) + B_m \tilde{u}(k). \tag{5.38}$$

The idea in disturbance-observer-based state feedback control is centered on designing the state feedback controller K for the intermediate control variable $\tilde{u}(k)$ and compensating the input disturbance $\mu(k)$ with the estimated variable $\hat{\mu}(k)$.

As in Section 5.3, we will form the augmented state space model using the disturbance model $D(q^{-1})$:

$$\begin{bmatrix} x_m(k+1) \\ p(k+1) \end{bmatrix} = \overbrace{\begin{bmatrix} A_m & B_m C_\epsilon \\ 0_{m\gamma \times n} & A_d \end{bmatrix}}^{A} \begin{bmatrix} x_m(k) \\ p(k) \end{bmatrix} + \overbrace{\begin{bmatrix} B_m \\ 0_{m\gamma \times m} \end{bmatrix}}^{B} u(k) + \begin{bmatrix} 0_{n\times m} \\ B_\epsilon \end{bmatrix} \epsilon(k)$$

$$y(k) = \overbrace{\begin{bmatrix} C_m & 0_{m\times m\gamma} \end{bmatrix}}^{C} \begin{bmatrix} x_m(k) \\ p(k) \end{bmatrix} \tag{5.39}$$

Here, we assume that we have learned the state estimation with input disturbance compensation from Section 5.3 to obtain both $\hat{x}_m(k)$ and $\hat{\mu}(k)$ based on the augmented system model (5.39).

The basic steps in designing disturbance-observe-based state feedback control are summarized as follows.

1. Find the controller K based on A_m and B_m so that the closed-loop control system matrix $A_m - B_m K$ is stable and the intermediate control signal $\tilde{u}(k)$ is computed using

$$\tilde{u}(k) = -K\hat{x}_m(k). \tag{5.40}$$

2. Find an observer gain K_{ob} based on A and C matrices in (5.39) and estimate the input disturbance $\mu(k)$ and the state variable $x_m(k)$ with the augmented state space model (5.39) as illustrated in Section 5.3, which will give $\hat{\mu}(k)$ and $\hat{x}_m(k)$.
3. Subtract $\hat{\mu}(k)$ from $\tilde{u}(k)$ to yield the control signal $u(k)$:

$$u(k) = \tilde{u}(k) - \hat{\mu}(k). \tag{5.41}$$

The remaining question for the disturbance-observer-based control is how the reference signal $r(k)$ enters the feedback control system. Given the reference vector $r(k)$, the observer equation can be written as

$$\hat{x}(k+1) = A\hat{x}(k) + Bu(k) + K_{ob}(y(k) - r(k) - C\hat{x}(k)) \tag{5.42}$$

where A, B, and C matrices are the augmented system matrices with input disturbance compensation, which was discussed in Section 5.3. The observer gain K_{ob} is chosen such that the observer error system matrix $A - K_{ob}C$ has all eigenvalues strictly inside the unit circle.

The next important step is to find the closed-loop eigenvalues of disturbance-observer-based state feedback control system, in order to establish its closed-loop stability. There are a few steps involved.

We will first establish the observer error system. Noting that the estimated $x_m(k)$ and $\hat{p}(k)$ directly follow the formulations proposed in Section 5.3, therefore, the following equation holds for the observer error system:

$$\begin{bmatrix} E[\tilde{x}_m(k+1)] \\ E[\tilde{p}(k+1)] \end{bmatrix} = (A - K_{ob}C) \begin{bmatrix} E[\tilde{x}_m(k)] \\ E[\tilde{p}(k)] \end{bmatrix}. \tag{5.43}$$

Second, we will establish the controlled dynamic system. Substituting the control signal: $u(k) = -K\hat{x}_m(k) - \hat{\mu}(k)$ into the original system equation (5.36) gives

$$\begin{aligned} x_m(k+1) &= A_m x_m(k) - B_m K\hat{x}_m(k) + B_m\mu(k) - B_m\hat{\mu}(k) \\ &= (A_m - B_m K)x_m(k) + B_m K\tilde{x}_m(k) + B_m\tilde{\mu}(k) \end{aligned} \tag{5.44}$$

where $\tilde{x}_m(k) = x_m(k) - \hat{x}_m(k)$ and $\tilde{\mu}(k) = \mu(k) - \hat{\mu}(k)$. Choosing C_ϵ as in (5.16), which has the identity matrix for the first m columns and rows, and the zero matrix for remaining columns and rows, we write $\tilde{\mu}(k) = C_\epsilon\tilde{p}(k)$, which leads to the expectance of (5.44) as

$$E[x_m(k+1)] = (A_m - B_m K)E[x_m(k)] + B_m KE[\tilde{x}_m(k)] + B_m C_\epsilon E[\tilde{p}(k)]. \tag{5.45}$$

The third step is to establish the closed-loop system. Combining (5.45) with the observer error system (5.43) gives the following description of the controlled dynamics:

$$\begin{bmatrix} E[x_m(k+1)] \\ E[\tilde{x}_m(k+1)] \\ E[\tilde{p}(k+1)] \end{bmatrix} = \begin{bmatrix} A_m - B_m K & B_m K & B_m C_\epsilon \\ 0_{n\times n} & A_m - K_{ob}^m C_m & B_m C_\epsilon \\ 0_{m\gamma\times n} & -K_{ob}^d C_m & A_d \end{bmatrix} \begin{bmatrix} E[x_m(k)] \\ E[\tilde{x}_m(k)] \\ E[\tilde{p}(k)] \end{bmatrix} \tag{5.46}$$

where the observer gain matrix is expressed as $K_{ob} = \begin{bmatrix} K_{ob}^m \\ K_{ob}^d \end{bmatrix}$, K_{ob}^m corresponding to the observer gain for the state vector $\hat{x}_m(k)$ and K_{ob}^d the observer gain for the disturbance vector $\hat{p}(k)$. $0_{p\times q}$ denotes a zero matrix with dimension $p \times q$.

As the system matrix in (5.46) is a block upper triangular matrix, its eigenvalues are the union of the solutions of the following two polynomial equations:

$$\det(\lambda I - (A_m - B_m K)) = 0 \tag{5.47}$$

$$\det\left(\lambda I - \begin{bmatrix} A_m - K_{ob}^m C_m & B_m C_\epsilon \\ -K_{ob}^d C_m & A_d \end{bmatrix}\right) = 0. \tag{5.48}$$

The solutions of (5.47) are the eigenvalues of the state matrix of the control system and those of (5.48) are the eigenvalues of the closed-loop observer error system. Both sets of eigenvalues lie inside the unit circle by design and hence the disturbance-observer-based control system is stable.

5.4.2 MATLAB Tutorial for Control Implementation

The implementation of the disturbance-observer-based control system is exceedingly simple. The following MATLAB program is created to demonstrate how the closed-loop system simulation is performed with respect to disturbance rejection and reference tracking.

Tutorial 5.2 *Write a MATLAB function that will perform closed-loop simulation of the disturbance-observer-based controller for a multi-input and multi-output system. The starting point is that we are given the discrete-time system matrices, A_m, B_m, and C_m, and have already obtained the augmented system matrices A, B, and C together with the state feedback controller K and the observer gain K_{ob}.*

Step by Step

1. *Create a new file called* **simu4disc.m**.
2. *The input variables to the function are the plant system matrices: A_m, B_m, C_m and the augmented system matrices: A, B, C, the number of samples for the simulation (Nsim), controller K, and observer gain K_{ob}, reference signal sp and disturbance signal dis. The output variables from this function are output y_1 and control signal u_1. The dimension of sp is $m_1 \times Nsim$ and the dimension of dis is $n_{in} \times Nsim$, where m_1 is the number of outputs and n_{in} is the number of control variables. Enter the following program into the file:*

```
function [y1,u1]=simu4disc(Am,Bm,Cm,A,B,C,Nsim,K,Kob,
sp,dis)
```

3. *Check numbers of states and augmented states, number of outputs and number of inputs. Continue entering the following program into the file:*

```
[n1,n_in]=size(Bm);
[m1,n]=size(C);
```

4. *Set initial conditions for the simulation. One can adjust the initial conditions to suit individual applications. Here, we choose them to be zero for simplicity. Continue entering the following program into the file:*

```
y=zeros(m1,1);
xm=zeros(n1,1);
xhat=zeros(n,1);
```

5. *Perform the simulation iteratively, which corresponds to a real-time control system. Continue entering the following program into the file:*

```
for kk=1:Nsim;
```

6. *Calculate the control signal with the knowledge that the first n_1 elements in the estimated state vector correspond to estimated state variables and the next n_{in} elements correspond to the estimated disturbance signal. Thus, the implementation of the control law defined by (5.40) and (5.41) becomes straightforward. Continue entering the following program into the file:*

```
u=-K*xhat(1:n1,1)-xhat(n1+1:n1+n_in,1);
```

7. *With the control signal calculated and the output signal measured at the current sampling time, the observer is used to calculate the next estimated state variables. At this stage the reference signal sp enters the computation for the calculation of estimated states. Continue entering the following program into the file:*

```
xhat=A*xhat+B*u+Kob*(y-sp(:,kk)-C*xhat);
```

8. *The above two steps capture the implementation of disturbance-observer-based control system. We save the control signal and output signal from this function. Continue entering the following program into the file:*

```
u1(1:n_in,kk)=u;
y1(1:m1,kk)=y;
```

9. *Because we are in a simulation environment, the control signal calculated is sent to the plant model for closed-loop simulation. In the closed-loop simulation, we will add the input disturbance to the system. For simplicity, we assume that the disturbance matrix is the same as the input matrix B_m. If they are not the same, one can adapt the computation to reflect the difference. This completes one-cycle of computation in the closed-loop simulation. Continue entering the following program into the file:*

```
xm=Am*xm+Bm*(u+dis(:,kk));
y=Cm*xm;
end
```

10. *Test this function with Example 5.6.*

Example 5.6

A discrete-time system is described by the following state space model:

$$x_m(k+1) = A_m x_m(k) + B_m u(k);\ y(k) = C_m x_m(k)$$

where the system matrices are

$$A_m = \begin{bmatrix} 1.8713 & -0.8752 \\ 1 & 0 \end{bmatrix}; \; B_m = \begin{bmatrix} 1 \\ 0 \end{bmatrix}; \; C_m = \begin{bmatrix} 0.0016 & 0.0015 \end{bmatrix}.$$

We assume that the state variables are not measured. The reference signal for the system comprises of ramp signals and constants as shown in Figure 5.7(a). Design a disturbance-observer-based state feedback control system to track the reference signal.

Solution

In order for the state feedback control system to track the ramp signal, the disturbance model $D(q^{-1})$ is chosen to have two zeros at 1, which effectively has embedded a double integrator into the controller. With this choice, we have the following $D(q^{-1})$:

$$D(q^{-1}) = 1 - 2q^{-1} + q^{-2}.$$

The augmented state space model becomes:

$$A = \begin{bmatrix} 1.8713 & -0.8752 & 1 & 0 \\ 1 & 0 & 0 & 0 \\ 0 & 0 & 2 & -1 \\ 0 & 0 & 1 & 0 \end{bmatrix}; \; B = \begin{bmatrix} 1 \\ 0 \\ 0 \\ 0 \end{bmatrix}; C = \begin{bmatrix} 0.0016 & 0.0015 & 0 & 0 \end{bmatrix}$$

where we have used the MATLAB function **am4obs.m** *created in Tutorial 5.1.*

We use MATLAB function **dlqr.m** *to design both controller and observer. For the observer design, we choose Q to have zero elements except $Q(3,3) = 1$ and $R = 1$, which leads to the observer gain:*

$$K_{ob} = \begin{bmatrix} 179.7962 & 143.2857 & 9.7114 & 8.9355 \end{bmatrix}^T.$$

For the controller design, we choose Q to be identity matrix and $R = 1$, which leads to the controller gain:

$$K = \begin{bmatrix} 1.3165 & -0.7146 \end{bmatrix}.$$

The reference signal is generated using the following MATLAB code:

```
r1=0:0.025:0.025*400;
r2=0.025*400:-0.025:0;
sp=[[zeros(1,300) r1 10*ones(1,400) r2 zeros(1,200)]];
```

With the disturbance signal equal to zero sequence of the same length, the closed-loop system for the disturbance-observer-based controller is simulated using the function **simu4disc.m** *produced in Tutorial 5.2. Figure 5.7(a) compares the closed-loop output response to the reference signal, which is seen to have an excellent tracking performance. The maximum tracking error is about 0.08. The control signal is shown in Figure 5.7(b), which is seen to have spikes when the reference signal changes.*

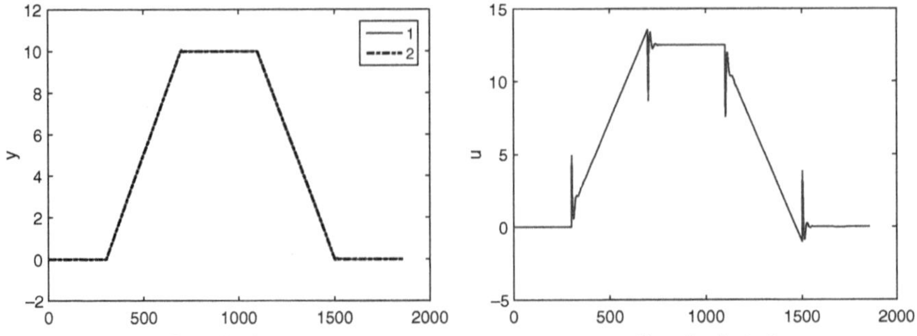

(a) Output (b) Control

Figure 5.7 Closed-loop control system response using disturbance observer (Example 5.6). Key: line (1) the output signal; line (2) the reference signal.

As an exercise, if we use the disturbance model $D(q^{-1}) = 1 - q^{-1}$ which is to have an integrator in the controller, then there will be steady-state errors for ramp reference signal and the maximum error becomes larger.

The next example will demonstrate disturbance-observer-based control for a multivariable system.

Example 5.7

A quadruple tank is illustrated in Figure 5.8, which has two input variables u_1 and u_2, and two output variables y_1 and y_2 (Johansson [2000], Goodwin et al. [2000]). The linearized transfer function model has the following form:

$$G(s) = \begin{bmatrix} \frac{3.7\gamma_1}{62s+1} & \frac{3.7(1-\gamma_2)}{(23s+1)(62s+1)} \\ \frac{4.7(1-\gamma_1)}{(30s+1)(90s+1)} & \frac{4.7\gamma_2}{90s+1} \end{bmatrix} \tag{5.49}$$

where $\gamma_1 = \gamma_2 = 0.8$. Design a disturbance-observer-based state feedback control system so that both outputs have the capability to track step reference changes and reject step input disturbances without steady-state errors. The system has sampling interval Δt as 1 second.

Solution

We will first enter the four transfer functions to form a transfer function model that has two inputs and two outputs, which is achieved using the MATLAB function **tf.m**. *Then, a state space model is obtained using the MATLAB function* **ss.m** *based on the two inputs and two outputs transfer function. With the sampling interval $\Delta t = 1$, we*

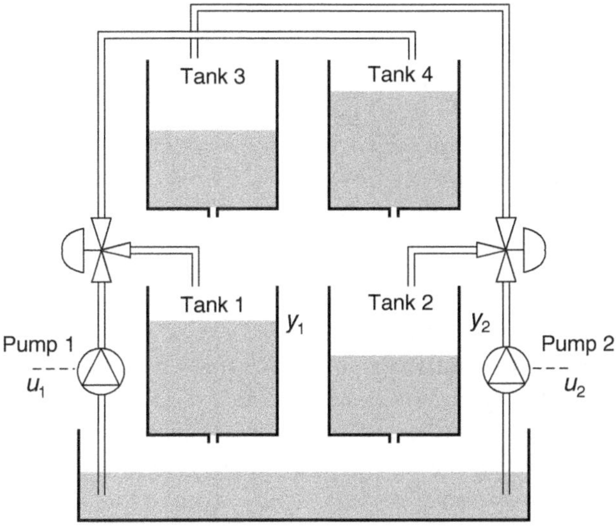

Figure 5.8 A quadruple tank system.

obtain the following system matrices A_m *and* B_m*:*

$$A_m = \begin{bmatrix} 0.9918 & 0.0090 & 0.0004 & -0.0005 \\ -0.0080 & 0.9524 & -0.0097 & 0.0027 \\ 0.0156 & 0.0097 & 0.9762 & -0.0053 \\ 0.0231 & -0.0083 & -0.0222 & 0.9772 \end{bmatrix}; \; B_m = \begin{bmatrix} 0.0993 & 0.0295 \\ -0.0304 & 0.0952 \\ 0.0590 & 0.0525 \\ -0.2097 & 0.0614 \end{bmatrix}$$

and the output matrix C_m*:*

$$C_m = \begin{bmatrix} 0.0674 & -0.0187 & 0.1691 & -0.1437 \\ 0.1783 & 0.0000 & -0.0000 & 0.0836 \end{bmatrix}.$$

In order for the closed-loop system to track step reference signals and reject step disturbances without steady-state errors, integrator will be added to the disturbance-observer-based control system, which is achieved by assuming constant input disturbance. From this regard, the disturbance model is chosen to be $D(q^{-1}) = 1 - q^{-1}$.

By using the MATLAB function **am4obs.m** *produced in Tutorial 5.1, the augmented system matrices* $A, \; B, \; C$ *are obtained for the observer design. Both controller and observer are designed using MATLAB program* **dlqr.m**. *To design the observer, we choose* $Q(5,5) = Q(6,6) = 1$ *and the rest of the elements to be zero; and the* R

matrix is an identity. For the controller design, we choose both Q and R matrices to be identity matrices with corresponding dimensions.

In the simulation studies, we assume at the steady-state operation, 4 units of the control signals correspond to 6 units of the output signals. Furthermore, two step reference signals are chosen and two input disturbance signals are added. In particular, the disturbance $\mu_1(k)$ acting on $u_1(k)$ is generated using the model:

$$\mu_1(k) = \frac{0.02\epsilon_1(k)}{1 - 0.98q^{-1}}$$

where $\epsilon_1(k)$ is a white noise with $\sigma^2 = 1$, and the disturbance $\mu_2(k)$ acting on u_2 is a constant disturbance with amplitude of -0.5.

By using the MATLAB function **simu4disc.m** *produced in Tutorial 5.2, we obtain the closed-loop simulation results. Figures 5.9(a)-(b) show the closed-loop responses to the reference signals and disturbance signals. From these two figures, we see that the outputs successfully track the step reference signals without steady-state errors while rejecting the step and low frequency disturbances. We also note that there are overshoots in the step responses for both outputs. The effect of noise is visible from y_1, but less so from y_2. Figures 5.10(a)–(b) show the control signals in response to the reference signals and disturbance signals. It is seen that the effect of disturbance is more on the control signal u_1.*

5.4.3 Food for Thought

1. A discrete-time system is described by the difference equations:

$$x_m(k+1) = 0.8x_m(k) + u(k); \; y(k) = 0.5x_m(k).$$

   If we wish the output to track a step reference signal, how do we choose the disturbance model $D(q^{-1})$?

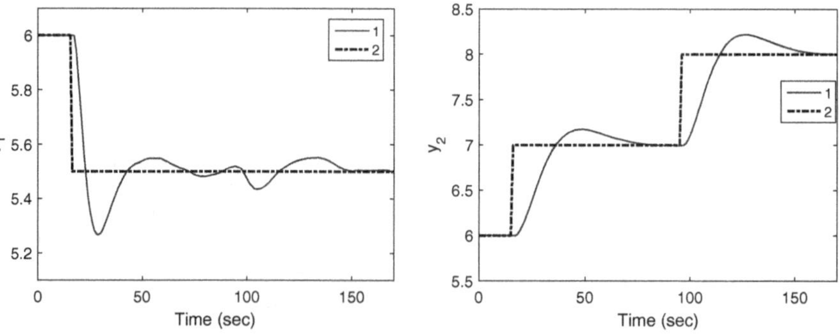

(a) Output y_1 (b) Output y_2

Figure 5.9 Closed-loop output response to reference and disturbance (Example 5.7). Key: line (1) output; line (2) reference.

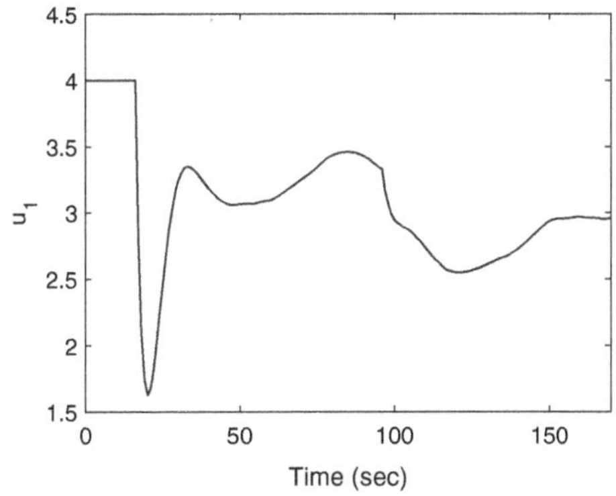

(a) Control u_1

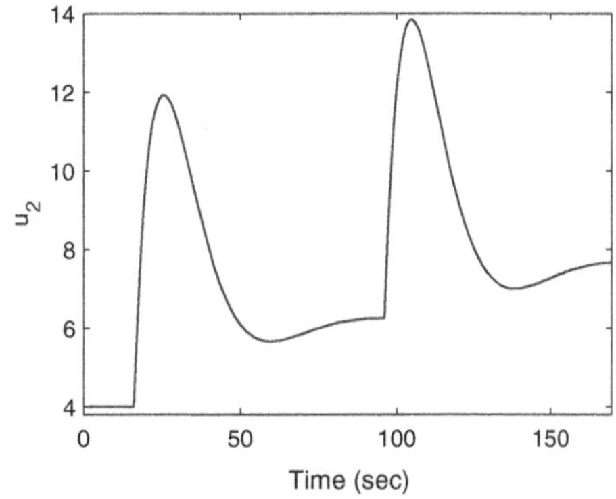

(b) Control u_2

Figure 5.10 Closed-loop control response to reference and disturbance (Example 5.7).

2. For the same system given in Problem 1, if we select the control system's pole at 0.7 and the observer poles at 0.6 and 0.5, what are the controller gain K and the observer gain K_{ob}? Where are the closed-loop poles for the disturbance-observer-based state feedback control system?
3. If the output $y(k)$ is severely corrupted by measurement noise and if we wish to reduce the effect of noise in the control system, would you increase the diagonal elements in R matrix when designing the observer with **dlqr.m**?
4. In the existing literature, sometimes, the reference signal $r(k)$ is proposed to enter the system using the following equation:

$$u(k) = -K\hat{x}_m(k) - \hat{\mu}(k) + K_r r(k)$$

   where $\hat{\mu}(k)$ is estimated using the disturbance observer. What problems do you think that we might have with this approach?
5. We have assumed that the input disturbance $\mu(k)$ shares the same input matrix as the control signal $u(k)$, which is $B_d = B_m$. What if in the applications, $B_d \neq B_m$, will the discrepancy affect the closed-loop control performance? Why?
6. If a closed-loop control system is designed to track a ramp signal and reject a sinusoidal disturbance with frequency 1 Hz where $\Delta t = 0.1$ sec, what disturbance model $D(q^{-1})$ do you propose to use?

5.5 Analysis of Disturbance-Observer-based Control System

5.5.1 Controller Transfer Function

The disturbance-observer-based control system is conceptually simple, which basically says that if there is a disturbance in the system, the disturbance should be compensated in the estimation as we did in Section 5.3, and their effect should be subtracted from the state feedback control signal. There are three questions to be answered in this section.

1. What is the role of the control system for disturbance rejection if the characteristics of the disturbance are known and considered in the observer?
2. Will the closed-loop control system track a reference signal with the same characteristics as $D(q^{-1})$ without steady-state error?
3. What happens if there is a mismatch between the frequency characteristic of the disturbance and the disturbance model $D(q^{-1})$?

To find the answers to the above questions, we will examine the frequency characteristics of the disturbance-observer- based control system. Thus, the first step is to derive the z-transfer function of the state feedback controller, which will establish that the common denominator of the z-transfer-function matrix of the feedback controller contains the disturbance model $\frac{1}{D(z)^m}$ as a factor where m is the number of inputs and outputs, and therefore it has infinite gain at the roots of $D(z)$.

Controller transfer function

By writing the observer gain as $K_{ob} = \begin{bmatrix} K_{ob}^m \\ K_{ob}^d \end{bmatrix}$, we obtain the estimated state $\hat{x}_m(k)$ and estimated disturbance vector $\hat{p}(k)$ in the following forms:

$$\hat{x}_m(k+1) = A_m\hat{x}_m(k) + B_m C_\epsilon \hat{p}(k) + B_m u(k) + K_{ob}^m y(k) - K_{ob}^m C_m \hat{x}_m(k) \quad (5.50)$$

$$\hat{p}(k+1) = A_d\hat{p}(k) + K_{ob}^d y(k) - K_{ob}^d C_m \hat{x}_m(k) \quad (5.51)$$

Now, substitute

$$u(k) = -K\hat{x}_m(k) - \hat{\mu}(k)$$

into equation (5.50) to obtain

$$\hat{x}_m(k+1) = (A_m - B_m K - K_{ob}^m C_m)\hat{x}_m(k) + K_{ob}^m y(k) \quad (5.52)$$

where the estimated disturbance $\hat{\mu}(k) = C_\epsilon \hat{p}(k)$ is used to cancel the same quantity in (5.50).

Defining the z-transforms of $\hat{x}_m(k)$ as $\hat{X}_m(z)$, $u(k)$ as $U(z)$, $\hat{p}(k)$ as $\hat{P}(z)$, the z -transform of the control signal is expressed as

$$\begin{aligned} U(z) &= -K\hat{X}_m(z) - C_\epsilon \hat{P}(z) \\ &= -K(zI - A_m + B_m K + K_{ob}^m C_m)^{-1} K_{ob}^m Y(z) \\ &\quad - C_\epsilon (zI - A_d)^{-1}[K_{ob}^d - K_{ob}^d C_m (zI - A_m + B_m K + K_{ob}^m C_m)^{-1} K_{ob}^m] Y(z) \\ &= -C(z)Y(z) \end{aligned} \quad (5.53)$$

where the z-transfer-function of the feedback controller $C(z)$ is written in terms of the output as

$$C(z) = K(zI - A_m + B_m K + K_{ob}^m C_m)^{-1} K_{ob}^m + C_\epsilon (zI - A_d)^{-1} [K_{ob}^d - K_{ob}^d C_m (zI - A_m + B_m K + K_{ob}^m C_m)^{-1} K_{ob}^m] \tag{5.54}$$

Note that A_d matrix is defined by (5.17), which is

$$A_d = \begin{bmatrix} -d_1 I_m & -d_2 I_m & \dots & -d_{\gamma-1} I_m & -d_\gamma I_m \\ I_m & 0_m & 0_m & \dots & 0_m \\ 0_m & I_m & 0_m & \dots & 0_m \\ \vdots & \ddots & \ddots & \vdots & \vdots \\ 0_m & 0_m & \dots & I_m & 0_m \end{bmatrix}.$$

Therefore, for a system with m inputs, it can be verified that the block matrix A_d has the following property:

$$\det(zI - A_d) = D(z)^m.$$

In (5.54), the first term on the right-hand side is the controller transfer-function of the state estimate feedback system without the disturbance observer, and the second term represents the controller transfer function from disturbance estimation that contains the disturbance model $D(z)^m$ in its denominator. Therefore, it has infinite gains at the solutions $D(z)^m = 0$.

5.5.2 Disturbance Rejection

To examine what happens to the control system for disturbance rejection, we consider the cases of input disturbance and output disturbance separately where the reference signal is assumed to be zero.

In the deterministic case, the input disturbance is modelled as

$$\mu(k) = \frac{q^{-1}}{D(q^{-1})} \delta(k)$$

where $\delta(k)$ is a vector that contains m unit pulses. Then, the z-transform of the disturbance signal becomes

$$Z[\mu(k)] = \frac{z^{-1}}{D(z^{-1})} V = \frac{z^{\gamma-1}}{D(z)} V \tag{5.55}$$

where $V = [1\ 1\ \dots\ 1]^T$ is a vector of m unit elements obtained from the z-transform of $\delta(k)$. The z-transform of the output $Y(z)$ is expressed in relation to the input disturbance as

$$Y(z) = G(z)(U(z) + Z[\mu(k)]) \tag{5.56}$$

where $G(z)$ is the z-transfer function of the system and $U(z)$ is the z-transform of the control signal. Substituting the control signal (see (5.53))

$$U(z) = -C(z)Y(z)$$

into (5.56) gives the closed-loop expression for the response of output to the input disturbance in z-transform:

$$\begin{aligned} Y(z) &= (I + G(z)C(z))^{-1}G(z)Z[\mu(k)] \\ &= (I + G(z)C(z))^{-1}G(z)\frac{z^{\gamma-1}}{D(z)}V \\ &= S_i(z)\frac{z^{\gamma-1}}{D(z)}V. \end{aligned} \tag{5.57}$$

Here, the transfer function

$$S_i(z) = (I + G(z)C(z))^{-1}G(z)$$

is the input disturbance sensitivity function, which governs the relationship between the input disturbance and the output. Noting that we have assumed that the system $G(z)$ does not contain any zeros that correspond to the zeros of the disturbance model $D(z)$, there is no pole-zero cancellation in the pair $G(z)C(z)$. The factor $D(z)$ contained in the denominator of the controller $C(z)$ (see (5.54)) appears in the numerator $(I + G(z)C(z))^{-1}G(z)$. This leads to the cancellation of $D(z)$ in (5.57).

With assumption that the controller $C(z)$ is designed to stabilize the system $G(z)$, the z-transform $(1 - z^{-1})Y(z)$ contains all poles strictly inside the unit circle, which is the condition for the final value theorem. Hence, with the application of final value theorem, we obtain the steady-state of the output $y(k)$ in response to the input disturbance $\mu(k)$ as

$$\lim_{k\to\infty} y(k) = \lim_{z\to 1} (1 - z^{-1})Y(z) = 0 \tag{5.58}$$

which means that the disturbance-observer-based control system will completely reject the input disturbance $\mu(k)$.

If the disturbance $\mu(k)$ occurs at the output, then the z-transform of the output is expressed as:

$$Y(z) = G(z)U(z) + Z[\mu(k)] \tag{5.59}$$

Substitution the controller $U(z) = -C(z)Y(z)$ gives the closed-loop z-transform of the output signal:

$$\begin{aligned} Y(z) &= (I + G(z)C(z))^{-1}\frac{z^{\gamma-1}}{D(z)}V \\ &= S(z)\frac{z^{\gamma-1}}{D(z)}V \end{aligned} \tag{5.60}$$

where the transfer function

$$S(z) = (I + G(z)C(z))^{-1} \tag{5.61}$$

is the sensitivity function that governs the relationship between the output disturbance and the output. When there is no pole-zero cancellation in the pair $G(z)C(z)$ as assumed, the disturbance model $D(z)$ contained in the denominator of the controller $C(z)$ will appear in the numerator of $(I + G(z)C(z))^{-1}$, leading to the cancellation of disturbance model $D(z)$. Thus, with the design that $C(z)$ stabilizes the system $G(z)$, the z-transform $(1 - z^{-1})Y(z)$ contains all poles strictly inside the unit circle. Then the application of final value theorem leads to the steady-state response of $y(k)$ to the output disturbance $\mu(k)$ as

$$\lim_{k\to\infty} y(k) = \lim_{z\to 1} (1 - z^{-1})Y(z) = 0 \tag{5.62}$$

where $Y(z)$ is described by (5.60), which is finite as $z \to 1$. This means that even though we have designed the disturbance-observer-based state feedback control system with the assumption of input disturbance, the same controller will effectively reject output disturbance.

5.5.3 Reference Tracking

To understand how the disturbance-observer-based control system tracks a reference signal $r(k)$, we will go back to the implementation equation (5.42), which is

$$\hat{x}(k+1) = A\hat{x}(k) + Bu(k) + K_{ob}(y(k) - r(k) - C\hat{x}(k)) \tag{5.63}$$

where we basically used the error signal $y(k) - r(k)$ as the measured signal for the observer. This is an important consideration for the disturbance-observer-based control system in order to eliminate steady-state errors for reference tracking.

When $r(k) \neq 0$, the z-transform of the control signal $U(z)$ in (5.53) is expressed as

$$\begin{aligned}
U(z) &= -K(zI - A_m + B_mK + K_{ob}^m C_m)^{-1}K_{ob}^m(Y(z) - R(z)) \\
&\quad - C_\epsilon(zI - A_d)^{-1}[K_{ob}^d - K_{ob}^d C_m(zI - A_m + B_mK + K_{ob}^m C_m)^{-1}K_{ob}^m](Y(z) - R(z)) \\
&= -C(z)(Y(z) - R(z))
\end{aligned}$$

where we replaced the original $Y(z)$ from the observer by $Y(z) - R(z)$.

For reference tracking, we consider the error signal $E(z) = R(z) - Y(z)$. Note with the control law, the output $Y(z)$ is

$$Y(z) = -G(z)C(z)(Y(z) - R(z))$$

and it is expressed in the closed-loop relationship to $R(z)$ as

$$Y(z) = (I + G(z)C(z))^{-1}G(z)C(z)R(z) \tag{5.64}$$

Therefore, the z-transform of the error signal $E(z)$ is

$$\begin{aligned} E(z) &= R(z) - Y(z) \\ &= (I + G(z)C(z))^{-1}R(z) = S(z)R(z) \end{aligned} \tag{5.65}$$

where (5.64) is used. Assuming that the reference signal $R(z)$ has the same characteristics as the disturbance signal so that the z-transform of the disturbance signal becomes

$$R(z) = \frac{z^{\gamma-1}}{D(z)}V. \tag{5.66}$$

Note that the sensitivity function $S(z)$ (see (5.61) and (5.65)) also represents the relationship between the reference signal $R(z)$ and the the error signal $E(z)$. With the same assumptions that there is no pole-zero cancellation in the pair $G(z)C(z)$ and the design that $C(z)$ stabilizes the system $G(z)$, the z-transform $(1 - z^{-1})E(z)$ contains all poles strictly inside the unit circle. Then the application of final value theorem leads to the steady-state response of the error $e(k)$ as

$$\lim_{k\to\infty} e(k) = \lim_{z\to 1} (1 - z^{-1})E(z) = 0. \tag{5.67}$$

5.5.4 A Case Study

The application of final value theorem only tells us the behavior of the disturbance-observer-based control system at the steady-state responses when the disturbance signal and the reference signal have the same characteristics as those contained in the $D(q)$ polynomial. It does not tell us what happens if there is a mismatch in the design, which commonly occurs in the disturbance rejection case. We will seek the answer to the question in the frequency response analysis.

Example 5.8

A continuous-time system with the transfer function is described by:

$$G(s) = \frac{1}{(s+2)^2}.$$

Design a disturbance-observer-based state feedback control system to track a reference signal

$$r(t) = 3\sin\left(\frac{2\pi}{3}t - \frac{\pi}{2}\right) + 3$$

and to reject a sinusoidal disturbance signal with period of 1.5 second that occurs at the input. The system has the sampling interval $\Delta t = 0.01$ second. Simulate the closed-loop response for reference tracking and disturbance rejection. Present the frequency response of the controller $C(z)$ and the sensitivity function $S(z)$.

Solution

The discrete-time transfer function for the second order system is

$$G(z) = \frac{0.4934z + 0.4868}{z^2 - 1.9604z + 0.9608}$$

which leads to the state space model

$$x_m(k+1) = A_m x_m(k) + B_m u(k); \ y(k) = C_m x_m(k)$$

where A_m, B_m, and C_m are

$$A_m = \begin{bmatrix} 1.9604 & -0.9608 \\ 1 & 0 \end{bmatrix}; B_m = \begin{bmatrix} 1 \\ 0 \end{bmatrix}; C_m = \begin{bmatrix} 0.4934 & 0.4868 \end{bmatrix}.$$

With the choice of $Q = I$ and $R = 1$, we design the state feedback controller K using the MATLAB **dlqr.m** *function, leading to*

$$K = \begin{bmatrix} 1.3840 & -0.7925 \end{bmatrix}$$

where the closed-loop eigenvalues are $0.2882 \pm j0.2920$. To design the observer, we consider the requirement of reference following and disturbance rejection. Since there is a constant term in the reference signal, the first factor to be included in the disturbance model is $1 - q^{-1}$, which results in integral action in the controller. Note that the sinusoidal signal for the reference has the frequency $\omega_1 = \frac{2\pi}{3}$ and the disturbance has the frequency $\omega_2 = \frac{2\pi}{1.5}$. Because these two frequencies are quite close to each other, we choose the frequency for the second factor to be included as ω_2. Therefore, the disturbance model is chosen to be

$$\begin{aligned} D(q^{-1}) &= (1 - q^{-1})(1 - 2\cos(\omega_2 \Delta t) q^{-1} + q^{-2}) \\ &= 1 - 2.9982 q^{-1} + 2.9982 q^{-2} - q^{-3} \end{aligned}$$

where the discrete frequency is $\omega_d = \omega_2 \Delta t$. The augmented system matrices for the observer design are

$$A = \begin{bmatrix} 1.9604 & -0.9608 & 1 & 0 & 0 \\ 1 & 0 & 0 & 0 & 0 \\ 0 & 0 & 2.9982 & -2.9982 & 1 \\ 0 & 0 & 1 & 0 & 0 \\ 0 & 0 & 0 & 1 & 0 \end{bmatrix};$$

$$C = \begin{bmatrix} 0.4934 & 0.4868 & 0 & 0 & 0 \end{bmatrix}.$$

In the observer design, we choose $Q(3,3) = 1$ and all other components to be zero, which is similar to the choice we made in the previous observer design (see Section 5.3). Together with the choice of $R = 1$ and constraint on the eigenvalues of the observer error system within the circle of radius 0.99, the observer gain K_{ob} is calculated using **dlqr.m**, *which is*

$$K_{ob} = 10^3 \begin{bmatrix} 5.7941 & 4.6241 & 0.2121 & 0.1942 & 0.1770 \end{bmatrix}^T.$$

The eigenvalues of the observer error system are

$$0.9322 \pm j0.1434, \ 0.8465, \ 0.8683 \pm j0.0820.$$

The estimated state vector $\hat{x}(k)$ *is computed using the following observer equation:*

$$\hat{x}(k+1) = A\hat{x}(k) + Bu(k) + K_{ob}(y(k) - C\hat{x}(k)),$$

and the state estimate feedback control law becomes

$$u(k) = -K\hat{x}_m(k) - \hat{\mu}(k)$$

where the $\hat{x}_m(k)$ *contains the first two elements of* $\hat{x}(k)$ *and* $\hat{\mu}(k)$ *is the third element of* $\hat{x}(k)$.

To evaluate the closed-loop control performance, a total simulation time is chosen to be 20 (sec) and a disturbance signal $10\sin\frac{2\pi}{1.5}t$ *is added at half of the simulation time. Figure 5.11(a) compares the reference signal with the output signal in the top diagram. The bottom diagram of Figure 5.11(a) shows that the error signal* $r(t) - y(t)$ *is within the limits of* ± 0.02 *and in the steady-state within the limits of* ± 0.0015. *Figure 5.11(b) shows the control signal generated from the disturbance-observer-based control system. It is seen that when the disturbance enters the system at* $t = 10$ *(sec), the control signal has responded to this disturbance signal.*

For the second part of this example, we will examine the frequency response of the controller in order to demonstrate why the disturbance-observer-based control system is capable of rejecting the disturbance and tracking the reference signal. For this purpose, we will construct the z-transfer function of the controller based on (5.54). The first part of the controller transfer function is:

$$K(zI - A_m + B_mK + K_{ob}^mC_m)^{-1}K_{ob}^m = \frac{4.3546z - 3.5567}{z^2 - 0.0654z + 0.2822}.$$

The second part of the controller transfer function is:

$$C_\epsilon(zI - A_d)^{-1}K_{ob}^d = \frac{212.0768z^2 - 405.3603z + 194.2310}{z^3 - 2.9982z^2 + 2.9982z - 1}$$

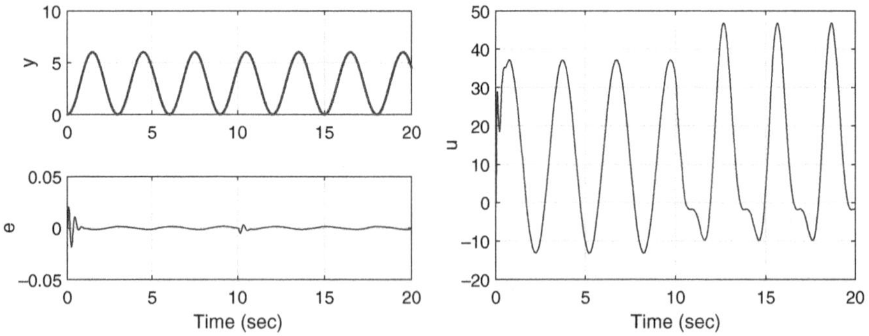

(a) Output and error (b) Control signal

Figure 5.11 Closed-loop response to reference and disturbance (Example 5.8).

which has D(z) as its denominator. The third part of the controller transfer function is:

$$C_m(zI - A_m + B_mK + K_{ob}^mC_m)^{-1}K_{ob}^m = \frac{0.5110z + 0.1139}{z^2 - 0.0654z + 0.2822}.$$

The controller transfer function becomes,

$$C(z) = \frac{4.3546z - 3.5567}{z^2 - 0.0654z + 0.2822} + \frac{212.0768z^2 - 405.3603z + 194.2310}{z^3 - 2.9982z^2 + 2.9982z - 1}\left(1 - \frac{0.5110z + 0.1139}{z^2 - 0.0654z + 0.2822}\right)$$

which is a 5th order controller with

$$C(z) = \frac{10^4(0.4567z^4 - 1.7140z^3 + 2.4184z^2 - 1.5199z + 0.3589)}{z^5 - 3.0636z^4 + 3.4766z^3 - 2.0423z^2 + 0.9116z - 0.2822}.$$

Indeed, it is seen that the denominator of the controller contains the disturbance model D(z). The frequency response of the controller is computed with $z = e^{j\omega}$, *where* $0 \le \omega \le \pi$. *For clarity, Figure 5.12(a) shows the magnitude of* $C(e^{j\omega})$ *in the lower frequency region. There are two magnitude spikes that correspond to the controller infinity gains at these two frequencies. The first spike occurs at the zero frequency and the second spike occurs at* $\omega_d = \omega_2\Delta t = 0.0419$ *(rad). Note that the controller gain at* $\omega_1\Delta t = 0.0209$ *remains to be very large although it is not infinite.*

In Figure 5.12(b), the magnitude of the sensitivity function

$$S(e^{j\omega}) = 1/(1 + G(e^{j\omega})C(e^{j\omega}))$$

is shown. At the points of spikes, the magnitude of the sensitivity function is zero (it is not seen as zero because of numerical computational limit). We note that the magnitude of sensitivity function at $\omega = 0.0209$ *(rad) is about* 4.7×10^{-4}. *This means that*

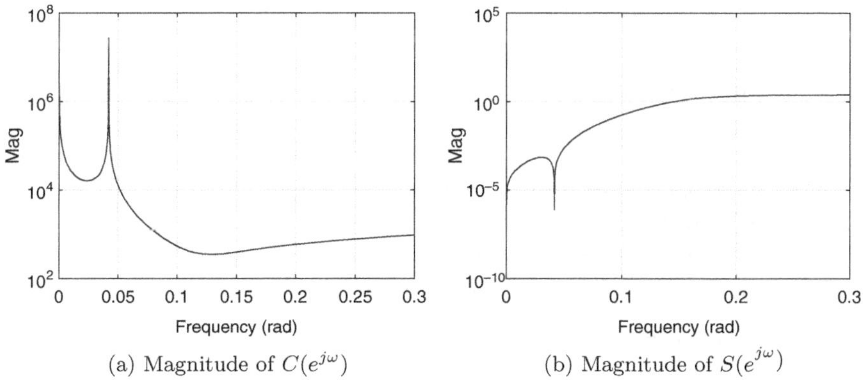

(a) Magnitude of $C(e^{j\omega})$ (b) Magnitude of $S(e^{j\omega})$

Figure 5.12 Frequency response of the controller and sensitivity (Example 5.8).

Table 5.1 Controller action corresponds to choice of disturbance model

Ref. or Dis	$D(q^{-1})$	Controller
Step or random walk	$1 - q^{-1}$	Integrator
Ramp signal	$(1 - q^{-1})^2$	Double integrator
Sinu. signal	$1 - 2\cos\omega_d q^{-1} + q^{-2}$	Harmonic component
Const. + sinu. signal	$(1 - q^{-1})(1 - 2\cos\omega_d q^{-1} + q^{-2})$	Integrator + harmonic component
Multiple sinu. signal	$\Pi_{k=1}^{n_s}(1 - 2\cos\omega_k q^{-1} + q^{-2})$	Multiple harmonic components

because of the very small magnitude of the sensitivity function, the tracking error for the reference tracking is very small although not zero.

This section is concluded with the following table, which illustrates the relationship between the types of controller and the choice of disturbance model $D(q^{-1})$. Once a reference signal or a disturbance signal is determined, a choice of $D(q^{-1})$ is made by following Table 5.1.

5.5.5 Food for Thought

1. Why is it important to find the controller transfer function for the disturbance-observer-based state feedback control system?
2. Is it correct to say that the disturbance-observer-based state feedback control system is relatively simple for the design and implementation, however it is less transparent for the controller transfer function representation?
3. Do you agree that the denominator of the controller transfer function containing the factor $D(z)^m$ is one of the key attributes for the disturbance-observer-based approach?
4. Assuming that the discrete-time system is described by

$$x_m(k+1) = A_m x_m(k) + B_m u(k) + B_d \mu(k);\; y(k) = C_m x_m(k),$$

   where $B_d \neq B_m$, would you say that the closed-loop system will completely reject the disturbance $\mu(k)$ as long as $\mu(k)$ has the same characteristics as $D(q)$ included in the augmented model? How would you reach this conclusion?
5. A discrete-time system is described by the difference equations:

$$x_m(k+1) = 0.6x_m(k) + u(k) + \mu(k);\; y(k) = 0.5x_m(k).$$

   The purpose of the control system is to maintain a constant reference signal without steady-state error. What is the order of controller transfer function?

6. Assume that a system is to track a step reference signal and reject a sinusoidal disturbance with frequency $f = 10$ Hz. What is your choice of the disturbance model $D(q)$? What is the controller gain at $\omega = 0$ and $\omega = 2\pi f \Delta t$?

5.6 Anti-windup Implementation of the Control Law

5.6.1 Algorithm for Anti-windup Implementation

Because the disturbance-observer-based control system has the capacity to have integrator, double integrator, and harmonic components in the control law, the inclusion of an anti-windup mechanism is paramount for its applications to handle the situation when the control signal reaches its maximum or minimum limits. The anti-windup implementation of the disturbance-observer-based control law is straightforward for the reason that the unstable modes in the controller such as integrator or harmonic components are implicitly realized through the estimator, which is stable by design. Therefore, the key steps for the anti-windup mechanism are first to enforce the control signal saturations and second to inform the observer that control signal saturation has been reached so that the corresponding states are correctly estimated.

Suppose that the control signal $u(k)$ is constrained between u^{min} and u^{max}, where u^{min} and u^{max} contain the saturation limits for all input signals. With the initial conditions specified for $\hat{x}_m(0)$ and $\hat{\mu}(0)$, and the controller gain K and the observer gain K_{ob} designed, we summarize the computational steps for the disturbance-observer-based control law with an anti-windup mechanism.

1. Calculate the intermediate control signal $\tilde{u}(k)$ using

$$\tilde{u}(k) = -K\hat{x}_m(k). \tag{5.68}$$

2. Subtract $\hat{\mu}(k)$ from $\tilde{u}(k)$ to yield the control signal $u(k)$:

$$u(k) = \tilde{u}(k) - \hat{\mu}(k). \tag{5.69}$$

3. Impose the saturation limits on the control signal with the following computations.
For the ith control signal

$$u_i(k)^{act} = \begin{cases} u_i^{min}, & \text{if } u_i(k) < u_i^{min}; \\ u_i(k), & \text{if } u_i^{min} \le u_i(k) \le u_i^{max}; \\ u_i^{max}, & \text{if } u_i(k) > u_i^{max}, \end{cases}$$

where $i = 1, 2, \ldots, m$.

4. Compute the state vector $\hat{x}(k+1)$ with the actual control signal $u(k)^{act}$:

$$\hat{x}(k+1) = A\hat{x}(k) + Bu(k)^{act} + K_{ob}(y(k) - r(k) - C\hat{x}(k)). \tag{5.70}$$

We can update the implementation program **simu4disc.m** produced in Tutorial 5.2 to include the anti-windup mechanism. For instance, after defining the maximum and minimum values of the control signal (umax,umin), we include the following program before the observer implementation:

```
for i=1:n_in
if (u(i,1)>umax(i,1)); u(i,1)=umax(i,1); end
if (u(i,1)<umin(i,1)); u(i,1)=umin(i,1); end
end
```

The following tutorial illustrates the implementation of the discrete-time disturbance-observer-based control algorithm in real-time simulation with respect to control signal constraints. It will produce a MATLAB embedded function that can be used in a Simulink simulation as well as in xPC Target implementation.

Tutorial 5.3 *Because the embedded functions require the dimensions of the variables to be specified, in the tutorial we will consider a second order system model with a second order disturbance model. The entire embedded MATLAB function completes one cycle of computation for the control signal. For every sampling period, it will repeat the same computation procedure.*

Step by Step

1. *Create a new Simulink file called* **dsfcD.slx**
2. *In Simulink's directory of User-Defined Functions, find the icon of the embedded MATLAB function and copy it to the dsfcD model.*
3. *Click on the icon of the embedded function, and define the input and output variables and the controller parameters to the model so that the embedded function has the following form:*

   ```
   function u=dsfcD(r,y,A,B,C,K,Kob,Umin,Umax)
   ```

   where u is the calculated control signal at the sampling time k, the first two elements (r and y) among the input variables are the reference signal and the output signal at sampling time k, A, B, and C are the augmented system matrices that are used in the state estimation, K is the controller gain and K_{ob} is the observer gain, and Umin and Umax are the upper and lower limits imposed on the control signal.
4. *We need to edit the input and output data ports in order to let the embedded function know which input ports are the real-time variables and which are the parameters. This editing task is performed using Model Explorer.*

- *click on 'r', on Scope, select 'input', assign port '1' and size '-1', complexity 'Inherited', type 'Inherit: Same as Simulink'. Repeat the same editing procedure for the output signal 'y'.*
- *The rest of the inputs to the embedded function are the parameters required in the computation. Click on 'A', on Scope, select 'Parameter' and click 'Tunable' and click 'Apply' to save the changes. Repeat the same editing procedure for the rest of the parameters.*
- *To edit the output port from the embedded function, click on 'u', on Scope, select 'Output', Port '1', Size '-1', Sampling Model 'Sample based', Type 'Inherit: Same as Simulink', and click on 'Apply' to save the changes.*

5. *In the following, the program will declare the variables stored in the embedded function during each iteration for their dimensions and initial values. For this control algorithm, the estimated state variables will be stored in the embedded function. We need to precisely define the dimension of $\hat{x}(k)$ for the embedded function. As an example, we consider a single input and single output system with two state variables, and a second order disturbance model, which gives the dimension of $\hat{x}$ to be 4×1. Continue entering the following program into the file:*

```
persistent xhat
if isempty(xhat)
xhat=zeros(4,1);
end
```

6. *After defining the variables to be stored, the calculation begins for the control signals at the sampling time k. The first two variables in $\hat{x}$ correspond to the estimated state variables and the third variable corresponds to the disturbance. Continue entering the following program into the file:*

```
u=-K*xhat(1:2,1)-xhat(3,1);
```

7. *If the control signal exceeds its maximum or minimum values, impose constraints on the amplitude of the control signal. Enter the following program into the file:*

```
if u<Umin;u=Umin;end
if u>Umax;u=Umax;end
```

8. *Calculate the estimated state variables for the next sampling period $k+1$ and complete the embedded function. Continue entering the following program into the file:*

```
xhat=A*xhat+Kob*(y-r-C*xhat)+B*u;
```

9. *Validate this embedded function by building a Simulink simulation program based on Example 5.6 and compare the simulation results. In the Simulink program, we may need to add a discrete-time delay to connect the plant output to y in this program.*

5.6.2 Heating Furnace Control

A fired heater is a heating furnace using gas fuel that is typically used for household heating in the winter times. For this type of system, the input to the heating furnace is the feed rate of the gas fuel and the output is the heater outlet or the room temperature in a house. Because the temperature sensors are located away from the heating source, there is a time delay in the measured temperature when the input feed rate changes. In this case study, we will use a transfer function model given in Ralhan and Badgwell [2000] to describe the operations of a gas fired heater at a high fuel operation, which will demonstrate the disturbance-observer-based control system in the presence of time delay with control signal constraints.

Example 5.9
In Ralhan and Badgwell [2000], the following transfer function model is obtained for a household gas fired heater:

$$G_H(s) = \frac{e^{-5s}}{(5s+1)^2} \frac{°\text{C}}{\text{m}^3\ \text{s}^{-1}} \tag{5.71}$$

where the time constant and delay are in minutes. Design a disturbance-observer-based control system for this gas fired heater to maintain a constant house temperature. The sampling interval Δt is one minute and the control signal is constrained above zero and 3 units, which is

$$0 \leq u(k) \leq 3.$$

The incremental reference signal is 1°C to start the simulation and increased to 2°C after 50 minutes.

Solution
Forming state space model with time delay. *The first step in the control system design is to find the discrete-time state space model for the time-delay system (5.71). The transfer function without time-delay is converted into a state space model with MATLAB function* **tf2ss.m**, *followed by discretizaton to obtain the state space model:*

$$x_0(k+1) = A_0 x_0(k) + B_0 u(k);\ y_0(k) = C_0 x_0(k),$$

where $x_0(k)$ contains two unknown state variables and $y_0(k)$ is temperature inside the fired furnace. The system matrices are

$$A_0 = \begin{bmatrix} 0.6550 & -0.0327 \\ 0.8187 & 0.9825 \end{bmatrix};\ B_0 = \begin{bmatrix} 0.8187 \\ 0.4381 \end{bmatrix};\ C_0 = \begin{bmatrix} 0 & 0.0400 \end{bmatrix}.$$

To include the measurement delay of 5 minutes in the state space model, we consider that the number of samples for the time delay is 5 because the sampling interval Δt is 1 minute. Therefore, the measured output $y(k) = y_0(k-5)$. In order to obtain *the state space model for the disturbance-observer-based control system, we define the state variable vector $x_m(k)$ as*

$$x_m(k) = \begin{bmatrix} x_0(k)^T & y_0(k-1) & y_0(k-2) & y_0(k-3) & y_0(k-4) & y_0(k-5) \end{bmatrix}^T,$$

leading to the following state space model:

$$\begin{bmatrix} x_0(k+1) \\ y_0(k) \\ y_0(k-1) \\ y_0(k-2) \\ y_0(k-3) \\ y_0(k-4) \end{bmatrix} = \overbrace{\begin{bmatrix} A_0 & 0 & 0 & 0 & 0 & 0 \\ C_0 & 0 & 0 & 0 & 0 & 0 \\ 0 & 1 & 0 & 0 & 0 & 0 \\ 0 & 0 & 1 & 0 & 0 & 0 \\ 0 & 0 & 0 & 1 & 0 & 0 \\ 0 & 0 & 0 & 0 & 1 & 0 \end{bmatrix}}^{A_m} \begin{bmatrix} x_0(k) \\ y_0(k-1) \\ y_0(k-2) \\ y_0(k-3) \\ y_0(k-4) \\ y_0(k-5) \end{bmatrix} + \overbrace{\begin{bmatrix} B_0 \\ 0 \\ 0 \\ 0 \\ 0 \\ 0 \end{bmatrix}}^{B_m} u(k). \tag{5.72}$$

with the measured output $y(k)$ being

$$y(k) = \overbrace{\begin{bmatrix} 0 & 0 & 0 & 0 & 0 & 1 \end{bmatrix}}^{C_m} \begin{bmatrix} x_0(k) \\ y_0(k-1) \\ y_0(k-2) \\ y_0(k-3) \\ y_0(k-4) \\ y_0(k-5) \end{bmatrix}. \tag{5.73}$$

The state space model given by (5.72) and (5.73) is used in the control system design. The system matrix A_m has 7 eigenvalues with two from the system without delay and the remainder being on the origin of the complex plane. It can be verified that the pair of matrices A_m, B_m is controllable and A_m, C_m is observable.

Controller and observer design. *Because the control objective is to maintain a constant house temperature, the controller needs to have integral action. For this reason, we choose the disturbance model $D(q^{-1}) = 1 - q^{-1}$. Based on this disturbance model, augmented system matrices (A, B, C) are obtained by using MATLAB function* **am4obs.m** *written in Tutorial 5.1.*

The controller design is based on (A_m, B_m) matrices. We choose $Q = I$ and $R = 1$, the application of MATLAB function **dlqr.m** *leads to the controller:*

$$K = \begin{bmatrix} 0.8216 & 0.4869 & -2 \times 10^{-17} & -6 \times 10^{-17} & 1 \times 10^{-16} & -5 \times 10^{-17} & 0 \end{bmatrix}.$$

The closed-loop eigenvalues are at $0.3757 \pm j0.2078$, and the remainder at 0.

The observer design is based on the augmented system matrices (A, C), where Q has all zero components except $Q(8, 8) = 1$ and $R = 1$. The application of MATLAB *function* **dlqr.m** *gives the following observer gain:*

$$K_{ob} = \begin{bmatrix} 0.8776 & 14.5318 & 0.5442 & 0.5031 & 0.4581 & 0.4093 & 0.3575 & 0.8346 \end{bmatrix}^T$$

and the following closed-loop eigenvalues for the observer error system: $0.7983 \pm j0.2143$, 0.6833, *and the rest being 0. With the optimal controller and observer design using linear quadratic regulator, the eigenvalues in the state space model caused by the time-delay are unchanged, which is an advantage from using the optimal controller design method. For closed-loop simulation of the control system with reference changes, we can either build a Simulink simulator based on Tutorial 5.3 or use the MATLAB function* **simu4disc.m** *written in Tutorial 5.2. The simulation results are discussed as follows.*

1. *Without operational constraints on the control signal, Figure 5.13(a) shows the temperature response with the reference increasing of 1°C while Figure 5.13(b) shows the feed rate of gas fuel (see the solid line in both figures). Clearly, the operational constraints that require* $0 \leq u(k) \leq 3$ *are violated.*
2. *With the anti-windup implementation of the disturbance-observer-based control signal, the control signal constraints are imposed in the simulation. In the presence of operational constraints, Figure 5.13(a) compares the closed-loop responses with and without constraints while Figure 5.13(b) compares the control signals. It is seen that the control signal constraints are satisfied and in the presence of constraints the closed-loop response has a good performance.*
3. *It is worthwhile to point out that this gas fired heater example was used in Wang [2020] for a PID control system, where the closed-loop response was quite oscillatory because of the large time delay. When using the discrete-time state feedback control, the time delay can be accurately taken into account in the control system*

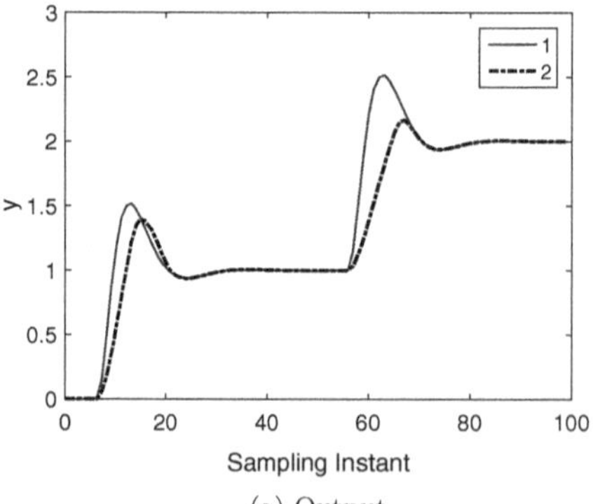

(a) Output

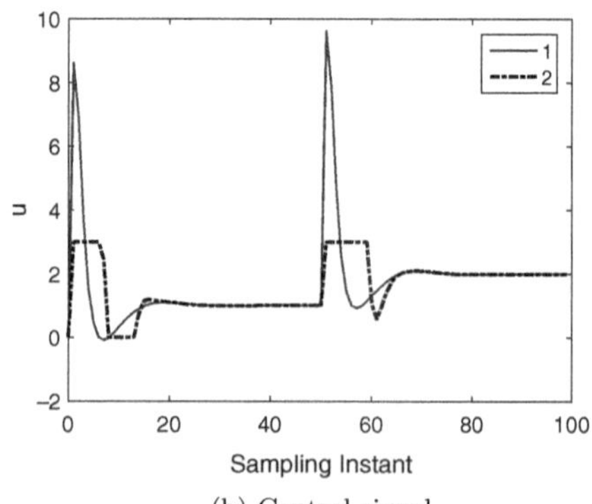

(b) Control signal

Figure 5.13 Closed-loop response to reference change with constraints (Example 5.9). Key: line (1) response without constraints; line (2) response with constraints.

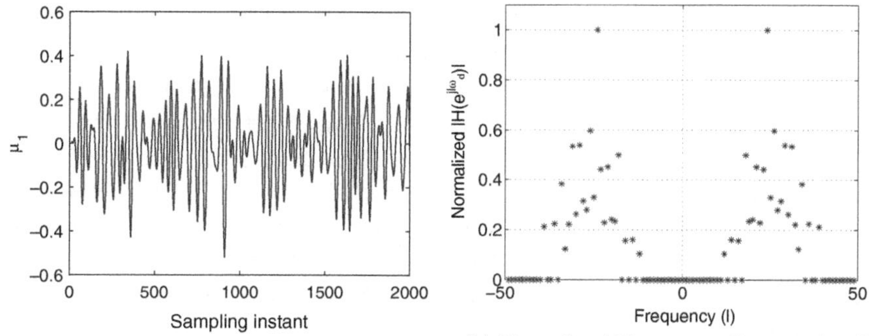

(a) Disturbance signal $\mu_1(k)$ (b) Normalized Fourier coefficients of $\mu_1(k)$

Figure 5.14 Disturbance signal and normalized Fourier coefficients (Example 5.10).

design and implementation. Therefore, the existence of time delay is no longer a challenge in control system design.

5.6.3 Example for Bandlimited Disturbance

The following example is used to demonstrate the application of disturbance-observer-based control system with anti-windup mechanism for a multivariable system in the presence of bandlimited disturbance.

Example 5.10

A discrete-time two input and two output system is described by the following difference equations:

$$\begin{bmatrix} x_1(k+1) \\ x_2(k+1) \end{bmatrix} = \begin{bmatrix} 0.9 & -0.1 \\ 0.2 & 0.8 \end{bmatrix} \begin{bmatrix} x_1(k) \\ x_2(k) \end{bmatrix} + \begin{bmatrix} 1 & 0 \\ 0 & 1 \end{bmatrix} \begin{bmatrix} u_1(k) + \mu_1(k) \\ u_2(k) + \mu_2(k) \end{bmatrix} \tag{5.74}$$

$$\begin{bmatrix} y_1(k) \\ y_2(k) \end{bmatrix} = \begin{bmatrix} 0.5 & 0.1 \\ 0.3 & 0.8 \end{bmatrix} \begin{bmatrix} x_1(k) \\ x_2(k) \end{bmatrix}, \tag{5.75}$$

where the input disturbance $\mu_1(k)$ is a bandlimited signal shown in Figure 5.14(a), and $\mu_2(k)$ is a step disturbance with amplitude -0.5. The state variables $x_1(k)$ and $x_2(k)$ are not measurable so an observer is needed for their estimations. Design a disturbance-observer-based control system with anti-windup mechanism to reject the disturbances with the following control signal amplitude constraints:

$$-0.4 \le u_1(k) \le 0.4; \; 0.3 \le u_2(k) \le 0.6.$$

Solution

Determining the disturbance model. *We perform Fourier analysis of the disturbance signal $\mu_1(k)$ using MATLAB* **fft.m** *function to determine its frequency band.*

From Figure 5.14(b), it is seen that the disturbance is bandlimited with the center frequency at $l = \pm 25$. Because the data length for disturbance is $L = 999$, the unit frequency from the Fourier analysis is $\frac{2\pi}{999}$. Thus, the discrete frequency corresponding to the peak is $\omega_d = 25 \times \frac{2\pi}{999}$. This leads to the following factor that will be included in the disturbance model: $1 - 2\cos\omega_d q^{-1} + q^{-2}$. By including this factor in the disturbance model, the bandlimited disturbance should be suppressed. Considering that one of the objectives for disturbance rejection is to maintain steady-state operation, including the $1 - q^{-1}$ factor in the disturbance model will serve this purpose. By combining these two factors together, an effective choice for disturbance rejection of this bandlimited signal is

$$D_1(q^{-1}) = (1 - q^{-1})(1 - 2\cos\omega_d q^{-1} + q^{-2}). \tag{5.76}$$

Since the disturbance $\mu_2(k)$ is a constant, $D_2(q^{-1}) = (1 - q^{-1})$ is selected. This factor is also contained in $D_1(q^{-1})$. Thus, $D(q^{-1}) = D_1(q^{-1})$ containing the factors required for disturbance rejections in this two input and two output system.

Controller and observer design. *By using the MATLAB function* **am4obs.m** *produced in Tutorial 5.1, we obtain the augmented system matrices A, B, C. To design the state feedback controller K based on the state space model (A_m, B_m, C_m), we use the MATLAB function* **dlqr.m** *where the weighting matrices Q and R are chosen to be identity matrices, leading to the controller gain K and the following closed-loop eigenvalues: $0.3472 \pm j0.0588$. To design the observer, we choose $Q(3,3) = 1$, and $Q(4,4) = 1$ while the rest of the elements in Q are zero, R is identity matrix. This combination of Q and R matrices gives the observer gain K_{ob} and the corresponding closed-loop eigenvalues:*

$$0.3845 \pm j0.5367,\ 0.5201 \pm j0.4882,\ 0.4498 \pm j0.1321,\ 0.3892 \pm j0.1453.$$

Clearly, the observer dynamic response speed is slower than the dynamic response speed by the controller. Therefore, we will increase the response speed of the observer. To do so, we use the prescribed degree of stability to re-design the observer where the parameter α is chosen to be 0.5, which gives the observer gain K_{ob} and the corresponding closed-loop eigenvalues:

$$0.2256 \pm j0.1963,\ 0.2685 \pm j0.1398,\ 0.1866 \pm j0.0414,\ 0.1682 \pm j0.0496.$$

The closed-loop simulation results are discussed as follows.

1. *In order to establish the closed-loop performance achieved by the controller, we will first evaluate the open-loop responses to the disturbance signals $\mu_1(k)$ and $\mu_2(k)$ where measurement noises with standard deviation of 0.0001 were added to the simulations. As a benchmark, Figure 5.15 shows the open-loop output responses to the disturbance signals $\mu_1(k)$ and $\mu_2(k)$. Because of the interactions*

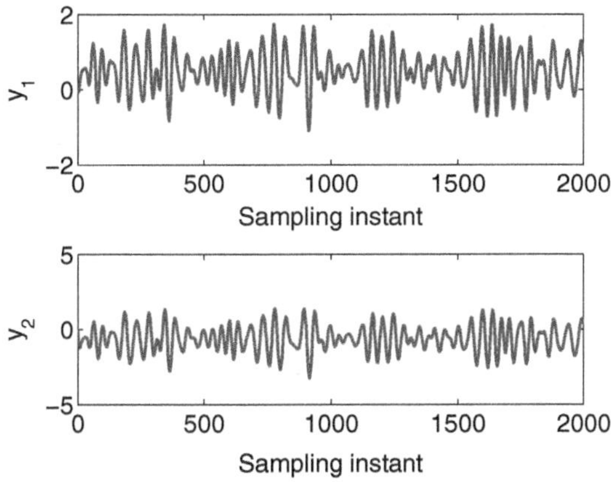

Figure 5.15 Open-loop output responses to input disturbances (Example 5.10).

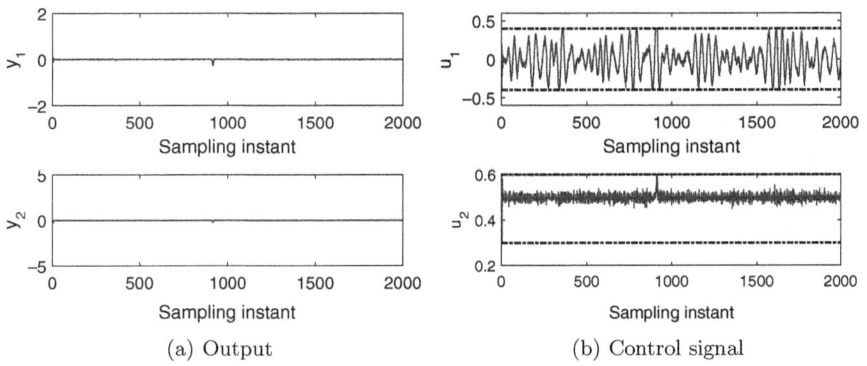

Figure 5.16 Closed-loop response to disturbances with constraints (Example 5.10).

in the multi-input and multi-output system, we see that the output $y_1(k)$ is affected by $\mu_2(k)$ and $y_2(k)$ by $\mu_1(k)$.

2. *Figure 5.16(a) shows the outputs y_1 and y_2 in response to the bandlimited disturbance and the step disturbance while maintaining steady-state condition at 0.*
3. *Figure 5.16(b) shows the control signals generated by the disturbance-observer-based control system with constraints. We can see that all constraints are satisfied. We can verify that without the constraints, the maximum control amplitude for u_1 is 0.5186 and u_2 is 2.541.*

5.6.4 Food for Thought

1. A discrete-time system designed to track a step reference signal has the state feedback controller K and the observer gain K_{ob}. Assuming the augmented system matrices are A, B, and C, what are the key steps for the implementation of the controller?

2. Continue from Problem 1. If the control signal is constrained between u^{min} and u^{max}, how do we impose the constraints?
3. Continue from Problem 1. If we wish to impose constraints on $\Delta u(k)$ ($\Delta u(k) = u(k) - u(k-1)$), where

$$\Delta u^{min} \leq \Delta u(k) \leq \Delta u^{max},$$

how do we impose the constraints?
4. In the observer implementation, without modification to the computational steps, could we use the following equation:

$$\hat{x}(k) = A\hat{x}(k-1) + Bu(k-1) + K_{ob}(y(k-1) - C\hat{x}(k-1))?$$

If no, what modifications would you propose so that this observer implementation could be used?
5. What are the key mechanisms that make the anti-windup implementation work?
6. Do you think that implementation of an anti-windup mechanism is necessary for control systems as long as they have integral control or controllers with harmonic components?

5.7 Summary

This chapter has discussed design, simulation and implementation of disturbance-observer-based control systems. By assuming the existence of an input disturbance, the essential steps in the design are to: (i) capture the characteristics of the disturbance signal with a disturbance model $D(q^{-1})$; (ii) augment the plant model (A_m, B_m, C_m) with the disturbance model to obtain the augmented state space model (A, B, C); (iii) design the observer using the A and C matrices and the controller using the A_m and B_m matrices.

The other important aspects of this chapter are summarized as follows.

1. The choice of a disturbance model is dependent on control system specifications. For instance, if a closed-loop system is required to track a step reference signal, then the disturbance model is selected to be $D(q^{-1}) = 1 - q^{-1}$ to ensure that the disturbance-observer-based control system has integral action. The same control system will reject a step disturbance or a low frequency disturbance.
2. The frequency analysis of the disturbance- observer-based control system leads to the conclusion that the proposed control law essentially has an infinite gain at the frequencies corresponding to the solutions $D(e^{-j\omega}) = 0$ where $0 \leq \omega < \pi$. For example, in the case when $D(z^{-1}) = 1 - z^{-1}$, this means that an integrator implicitly is used in the controller because an infinite controller gain is obtained at $\omega = 0$.

3. The implementation of the control system is based on the combination of the observer that estimates the state and the disturbance with the state feedback control that subtracts the estimated disturbance. The implementation of anti-windup mechanism for the disturbance- observer-based control system is simple and straightforward because the observer provides a stable realization of the control law.

5.8 Further Reading

1. A book for disturbance-observer-based linear and nonlinear control systems is given by (Li et al. [2014]). The paper (Chen et al. [2016]) presents the survey of disturbance-observer-based control. The same research group has also worked on various topics of disturbance-observer-based control systems (Li et al. [2012], Yang et al. [2012]).
2. The early work on motion control using disturbance-observer-based control includes Komada et al. [1991]. Disturbance-observer-based control is used for rigid mechanical systems in Schrijver and van Dijk [2002] and for robotic manipulator in Chen et al. [2000]. Controller design for disturbance rejection is proposed using disturbance observer with estimation of frequency in Jia [2009]. Stability and robust performance of motion control using disturbance observer is analyzed in Sariyildiz and Ohnishi [2015].
3. PID control and active disturbance rejection are introduced in Han [2009].
4. H_∞ control was compared with disturbance-observer-based control methods in Mita et al. [1998].
5. Disturbance rejection in dual-stage feed drive control system was discussed in She et al. [2011].
6. A recent survey is presented for disturbance estimation and attenuation in PMSM drives (Yang et al. [2017]) using the framework of disturbance-observer-based control.
7. Disturbance-observer-based predictive repetitive control is presented in Wang et al. [2020], Wang et al. [2017].
8. The disturbance-observer-based controller is used to control a single phase voltage converter with experimental validation in McNabb et al. [2017].

Problems

Problem 5.1 A continuous-time system is described by the transfer function model:

$$G(s) = \frac{1}{s^2 + 1}$$

With the sampling interval $\Delta t = 0.1$ (sec), a discrete-time state space model is obtained. Design a disturbance-observer-based state feedback control system that will track a step reference signal and reject a step input disturbance.

1. Determine the disturbance model $D(q^{-1})$.
2. Calculate the state feedback controller gain K by positioning the closed-loop poles around $e^{-\Delta t}$.
3. Calculate the observer gain K_{ob} by positioning the closed-loop poles around $e^{-2\Delta t}$.

Problem 5.2 A discrete-time system is described by the following state space model:

$$x_m(k+1) = A_m x_m(k) + B_m u(k);\ y(k) = C_m x_m(k)$$

where the system matrices are

$$A_m = \begin{bmatrix} 0.1 & 2 & 0 \\ -0.1 & 0.9 & 1 \\ 0 & 0.4 & 0.2 \end{bmatrix};\ B_m = \begin{bmatrix} 1 & 0 \\ 2 & 1 \\ 1 & 1 \end{bmatrix};\ C_m = \begin{bmatrix} 1 & 1 & 2 \\ 0 & 1 & 1 \end{bmatrix}.$$

Design a disturbance-observer-based state feedback control system that will track a step reference signal for output $y_1(k)$ and a sinusoidal signal with frequency $\omega_d = \frac{2\pi}{100}$ for $y_2(k)$.

1. Determine the disturbance model $D(q^{-1})$.
2. Calculate the state feedback controller gain K using DLQR design where the weighting matrices Q and R are identity matrices.
3. Assuming that the state vector $x_m(k)$ is not measured, calculate the observer gain K_{ob} using DLQR design where $Q_{ob} = I$, $R_{ob} = 0.1I$ and I is an identity matrix with a compatible dimension.

Problem 5.3 In Kailath [1980], the state space model to describe the dynamics of a hot air balloon (see Figure 5.17) was given as:

$$\begin{bmatrix} \dot{T}(t) \\ \dot{v}(t) \\ \dot{h}(t) \end{bmatrix} = \begin{bmatrix} -\frac{1}{\tau_1} & 0 & 0 \\ \beta & -\frac{1}{\tau_2} & 0 \\ 0 & 1 & 0 \end{bmatrix} \begin{bmatrix} T(t) \\ v(t) \\ h(t) \end{bmatrix} + \begin{bmatrix} \alpha \\ 0 \\ 0 \end{bmatrix} u(t) + \begin{bmatrix} 0 \\ \frac{1}{\tau_2} \\ 0 \end{bmatrix} w(t) \tag{5.77}$$

where the state variable $T(t)$ is the temperature change of air in the balloon, $u(t)$ is the manipulated variable and is proportional to change in heat added to air balloon, $v(t)$ is the vertical velocity of the balloon, $h(t)$ is the change in altitude from steady-state altitude, $w(t)$ is the vertical wind velocity, which is the disturbance for the control problem. The normalized constants in the model are $\tau_1 = 1$, $\tau_2 = 10$,

Figure 5.17 Hot air balloon system.

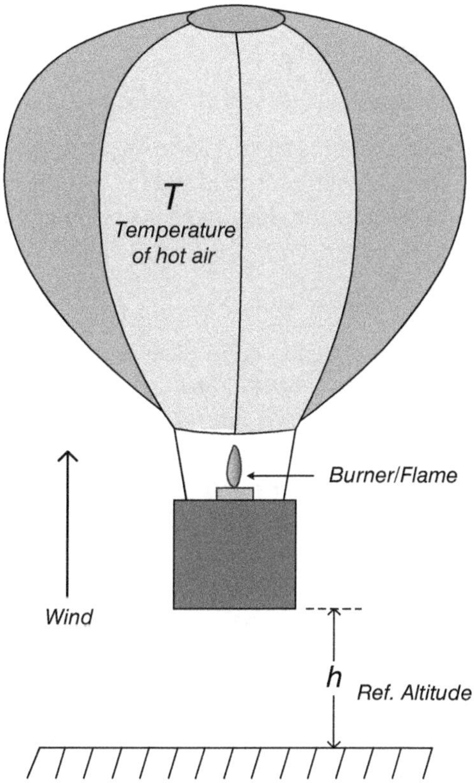

$\alpha = 1$ and $\beta = 0.01$. A discrete-time state space model is obtained with sampling interval $\Delta t = 0.1$ (sec). In order to save the cost for the air balloon controller, we only measure the altitude $h(k)$. Design a disturbance-observer-based control system that will reject the wind disturbance and track a step reference signal for the altitude, where the wind disturbance is considered to be a signal that has a low frequency content.

1. Determine the disturbance model $D(q^{-1})$.
2. Calculate the state feedback controller gain K with closed-loop poles positioned around $e^{-0.1\Delta t}$.
3. Calculate the observer gain K_{ob} with closed-loop poles positioned around $e^{-\Delta t}$.
4. Generate a wind disturbance using the following code where *Nsim* is the number of samples for the simulation.

```
Nsim=4000;
dis=randn(1,Nsim);
dis=filter(0.01,[1 -0.99],dis);
dis=dis+2*[zeros(1,Nsim/2) ones(1,Nsim/2)];
```

5. Simulate the closed-loop response with the reference signal generated by the following code:

```
r=[ones(1,Nsim/2) zeros(1, Nsim/2)];
```

   Add the wind disturbance signal to the simulation. Present the closed-loop response of the altitude and the control signal. What are your observations?
6. Investigate the effect of compensating the disturbance $\hat{\mu}(k)$ in the control law. For comparison, calculate the control signal using

$$u(k) = -K\hat{x}_m(k).$$

   Simulate the closed-loop response with the same reference and disturbance signal. Present the closed-loop response of the altitude and the control signal. Compare the two control systems' responses and discuss the effect of disturbance compensation.
7. Can we measure velocity $v(k)$ and temperature $T(k)$ and construct an observer to estimate the altitude $h(k)$? why?

Problem 5.4 Continue from Problem 5.3. We will investigate the implementation of anti-windup mechanism with respect to constraints on the control signal. The core strategy in avoiding integrator windup is to correctly inform the disturbance observer that control signal saturation has been reached. If the disturbance observer were not informed, then the closed-loop performance would deteriorate.

1. Determine the minimum and maximum control amplitude from Problem 5.3, which are denoted as u^{min} and u^{max}. Assume that the control signal is constrained between $0.8u^{min}$ and $0.8u^{max}$. Explore the integrator windup scenario by using the constrained control signal for the plant simulation, however using the unconstrained control signal in the observer implementation. Present the closed-loop response of the altitude and the control signal. What are your observations?
2. Implement the control signal constraints with anti-windup mechanism by using the constrained control signal for the plant simulation and for the observer implementation. Present the closed-loop response of the altitude and the control signal.
3. Discuss the closed-loop performance difference and the role of anti-windup mechanism when the control signal reaches its saturation limits.

Problem 5.5 A model for the drive train of a wind turbine is described by a two-mass system (Perdana [2008]) where two inertias are connected together through a spring that represents the low stiffness of the drive-train shaft and damping is also used in the two-mass model. We have used this model in the continuous-time control system design in Section 2.4 of Chapter 2 (see Figure 2.11).

For completeness of this exercise, defining the angular velocities of the turbine and generator as ω_t and ω_r, and the angular positions as θ_t and θ_r, the drive train of the wind turbine is described by the following differential equations:

$$2H_t \frac{d\omega_t}{dt} = T_t - T_w - K_s(\theta_r - \theta_t) - D_m(\omega_r - \omega_t) \tag{5.78}$$

$$2H_g \frac{d\omega_r}{dt} = -T_e + T_L + K_s(\theta_r - \theta_t) + D_m(\omega_r - \omega_t) \tag{5.79}$$

$$\frac{d\theta_t}{dt} = \omega_t \tag{5.80}$$

$$\frac{d\theta_r}{dt} = \omega_r \tag{5.81}$$

where H_t and H_g are the inertial time constants for the turbine and generator, K_s is the shaft stiffness constant and D_m is the mutual damping coefficient. T_w is the input disturbance generated by wind changes and T_L represents the load disturbance reflecting the demand of power output. For the current study, as in Perdana [2008], the turbine inertial constant $H_t = 2.6$ second, and the generator inertial constant $H_g = 0.22$ second, the stiffness constant $K_s = 141.0$ pu, and mutual damping coefficient $D_m = 3.0$ pu.

The control system for the wind turbine drive train is part of a larger system. The manipulated variables are the turbine torque T_t and the motor torque T_e. In order to optimize the wind energy generation, the angular velocity of the generator is required to be under control. Thus, the control objective is to achieve reference tracking of ω_r while rejecting wind disturbance T_w and load disturbance T_L. To avoid causing mechanical stress to the wind turbine generator, the angular velocity of the turbine ω_t needs to be synchronized with that of the generator. Namely, in the steady-state operation, the control objective is $\omega_t = \omega_r$.

1. Using sampling interval $\Delta t = 0.01$ (sec), find the discrete-time state space model:

$$x_m(k+1) = A_m x_m(k) + B_m u(k) + B_d \mu(k)$$

2. Assuming that θ_t and θ_r are measured, design a disturbance-observer-based state feedback control system for the wind turbine drive train, which will have integral control for reference signal tracking and disturbance rejection. For both controller design and observer design, MATLAB function **dlqr.m** is used with both Q and R being identity matrices that have compatible dimensions. What is K? What is K_{ob}?
3. Simulate the closed-loop responses for reference tracking and disturbance rejection with your choice of N_{sim}. For simplicity, all initial conditions are

chosen to be zero. The reference signals to θ_t and θ_r are unit ramp signals in discrete-time. The wind disturbance is a random walk disturbance generated using

$$T_w = \frac{0.001}{1 - q^{-1}} \epsilon(k)$$

where $\epsilon(k)$ is a zero mean white noise with variance $\sigma^2 = 0.36$ in the simulation studies. The load disturbance T_L is a negative step disturbance entering the system at half of the simulation time. Present the control signals, estimated state variables $\hat{\omega}_r$ and $\hat{\omega}_t$, and the output signals θ_r and θ_t.

Problem 5.6 Continue from Problem 5.5. Because the system is unstable, it requires a minimum control amplitude to produce a stable closed-loop system.

1. Determine the minimum and maximum control amplitude from Problem 5.5, which are denoted as u^{min} and u^{max}.
2. Implement the control signal constraints with anti-windup mechanism where the constrained control amplitudes are $0.98u^{min}$ and $0.98u^{max}$. Is the closed-loop control system stable? What are the minimum control signal amplitudes for the turbine torque T_t and the motor torque T_e that are required to stabilize the closed-loop system?
3. Propose an alternative approach that could be used to reduce the control signal amplitude for the unstable system in order to avoid its saturation. Validate the approach through simulation studies.

Problem 5.7 Continue from Problem 5.5. Assume that there is a periodic disturbance in the load torque T_L, which means that T_L is a combination of a constant and a sinusoidal signal. It is known that the periodic disturbance has frequency of 50 Hz.

1. Determine the disturbance model $D(q^{-1})$ for the disturbance-observer-based control system, where the closed-loop control system will reject the additional periodic disturbance while maintaining tracking the ramp reference signals and rejecting the wind and load disturbance as in Problem 5.5.
2. Design the disturbance-observer-based control system using MATLAB program **dlqr.m**. For both controller design and observer design, MATLAB function **dlqr.m** is used with both Q and R being identity matrices that have compatible dimensions. What is K? What is K_{ob}?
3. Simulate the closed-loop responses for reference tracking and disturbance rejection with your choice of N_{sim}. For simplicity, all initial conditions are chosen to be zero. The reference signals to θ_t and θ_r are unit ramp signals in discrete-time. The wind disturbance remains the same, however, the

load disturbance entering at half of the simulation time, has the additional sinusoidal signal with amplitude 0.2. Present the control signals, estimated state variables $\hat{\omega}_r$ and $\hat{\omega}_t$ and the output signals θ_r and θ_t.

Problem 5.8 Drum boiler is one of the processes in an electrical power generation plant. We have used this system in the continuous-time control system design in Section 2.2.6 of Chapter 2 (see Figure 2.3). For a drum boiler control system, it has three inputs and three outputs. The three input signals are feedwater flow rate (u_1 (kg/s)), fuel flow rate (u_2 (kg/s)), and attemperator spray flow rate (u_3 (kg/s)). Here attemperators are used to control steam temperature from the boiler. The three outputs are drum water level (y_1 (m)), drum pressure (y_2 (MPa)), and steam temperature (y_3 (^oC)). For convenience, the same transfer function model given in Tan et al. [2002] is presented as follows:

$$G(s) = \begin{bmatrix} G_{11}(s) & G_{12}(s) & G_{13}(s) \\ G_{21}(s) & G_{22}(s) & G_{23}(s) \\ G_{31}(s) & G_{32}(s) & G_{33}(s) \end{bmatrix} \tag{5.82}$$

where

$$G_{11}(s) = \frac{10^{-3}(-0.16s^2 + 0.052s + 0.0014)}{(s^2 + 0.0168s)(0.1s + 1)}, \; G_{12}(s) = \frac{10^{-3}(3.1s - 0.032)}{s^2 + 0.0215s},$$

$$G_{13}(s) = 0, \; G_{21}(s) = \frac{-10^{-3} \times 0.0395}{s + 0.018}, \; G_{22}(s) = \frac{10^{-3} \times 2.51}{s + 0.0157},$$

$$G_{23}(s) = \frac{10^{-3}(0.588s^2 + 0.2015s + 0.0009)}{(s^2 + 0.0352s + 0.000142)(0.1s + 1)},$$

$$G_{31}(s) = \frac{-0.00118s + 0.000139}{s^2 + 0.01852s + 0.000091},$$

$$G_{32}(s) = \frac{0.448s + 0.0011}{s^2 + 0.0127s + 0.000095}, \; G_{33}(s) = \frac{0.582s - 0.0243}{s^2 + 0.1076s + 0.00104}$$

1. To convert the transfer function model into a discrete-time state space model, we will first use the MATLAB function **tf.m** to produce a three input and three output transfer function and obtain a minimal state space realization using MATLAB function **ss.m**. The continuous-time state space model (A_p, B_p, C_p, D_p) is produced with the MATLAB function **ssdata.m**. With the sampling interval $\Delta t = 0.3$ (sec), convert the continuous-time state space model to a discrete-time state space model.
2. The overall control objective is to maintain the steady-state operations of the system for plant safety and achieving desired efficiency. Therefore, integral action is required in the boiler control system. Design a disturbance-observer-based state feedback control system using MATLAB function **dlqr.m**,

where both Q and R are identity matrices with compatible dimensions. What is the state feedback controller gain K and what is the observer gain K_{ob}? Where are the closed-loop eigenvalues for the control system and the observer error system?

3. Demonstrate that the control system designed can track step reference signals without steady-state errors. Simulate the disturbance-observer-based control system with all zero initial conditions, where for simplicity we assume that all the steady-state operating conditions are zero. Choosing the number of samples in simulation $N_{sim} = 18000$, the reference signal r_1 to y_1 is a unit step signal; the reference signal r_2 to y_2 is a unit step signal entering the simulation at $N_{sim}/3$, and the reference signal to y_3 is a unit step signal with negative amplitude entering the simulation at $2 \times N_{sim}/3$. Present the control signals and output signals from the closed-loop simulation studies.
4. Demonstrate that the control system designed can reject step disturbances without steady-state errors. Choosing all three reference signals to be zero, add input step disturbances to the system simulation. The disturbance signal added to u_1 is a unit step signal; the disturbance signal added to u_2 is a unit step signal entering the simulation at $N_{sim}/3$ and the disturbance signal added to u_3 is a unit step signal with negative amplitude entering the simulation at $2 \times N_{sim}/3$. Present the control signals and output signals.
5. Evaluate the performance of the control system with control signal constraints. From the disturbance rejection case, determine the maximum and minimum amplitudes of the control signals and choose the amplitudes of constrained control signals to be 80 percent of the maximum and minimum values. Simulate the closed-loop response in the presence of control signal constraints with the same disturbances and references as the previous case. Present the control signals and output signals.

Problem 5.9 Continue from Problem 5.8. In this exercise, we will design the controller and observer with constraints on the locations of the closed-loop eigenvalues based on the technique of prescribed degree of stability as introduced in Section 4.5 of Chapter 4.

1. In controller design, assuming that the largest time constant for the closed-loop system is constrained to be less than 3000 seconds, determine the corresponding closed-loop eigenvalue in the discrete-time where the sampling interval $\Delta t = 0.3$ (sec). Using the same Q and R matrices from Problem 5.8, compute the state feedback controller K using MATLAB program **dlqr.m** with prescribed degree of stability. What is K? Where are the closed-loop eigenvalues?
2. In the observer design, assuming that the largest time constant for the observer error system is constrained to be less than 2000 seconds, determine the corresponding closed-loop eigenvalue in the discrete-time. Compute the

observer gain K_{ob} using MATLAB program **dlqr.m** with prescribed degree of stability. What is K_{ob}? Where are the closed-loop eigenvalues of the observer error system?

3. Repeat the simulation studies with the same conditions as outlined in Problem 5.8, where the closed-loop control system is validated with respect to reference tracking, disturbance rejection, and control signal amplitude constraints. Compare the simulation results with those obtained from Problem 5.8. What are your observations?

Problem 5.10 Continue from Problem 5.9. We will explore an alternative strategy for the implementation of disturbance- observer-based control system. For reference tracking, in order to reduce the control signal amplitude at the initial period, we will calculate the control signal using

$$u(k) = -K\hat{x}_m(k)$$

which is the state feedback control without the integral control. After the initial period, we will calculate the control signal using

$$u(k) = -K\hat{x}_m(k) - \hat{\mu}(k)$$

which is the state feedback control with the integral action. Explore this strategy in the implementation of disturbance-observer-based control system designed in Problem 5.9. What are the advantages and disadvantages in comparison to the original implementation method?

6

Disturbance Rejection and Reference Tracking via Control Design

6.1 Introduction

The disturbance-observer-based approach (see Chapter 5) is simple in both design and implementation in addition to the inclusion of anti-windup mechanism when the control signal reaches its limitations. However, one of the main drawbacks for this type of methods is that it lacks a second degree of freedom that will separate the performance achieved in reference tracking from that in disturbance rejection.

The objective of this chapter is to introduce state feedback control for disturbance rejection and reference tracking through controller design. This is a complementary approach to that introduced in Chapter 5, and it offers a control system realization that separates the performance achieved in reference tracking from that in disturbance rejection. The chapter begins by embedding the disturbance model into the plant model to form an augmented state space model for the controller design (see Section 6.2). Based on this augmented model, state feedback controller is designed (see Section 6.3). If the state vector is not measured, then an observer is designed based for the implementation of the control law. In Section 6.4, three practical aspects of the control system are discussed. The first aspect is to introduce the two-degree-of-freedom control configuration for reducing overshoot in reference tracking. The second aspect is to present an anti-windup mechanism for the controller implementation, and the third aspect is to propose the control system without utilization of an observer when a transfer function model is given. In Section 6.5, the repetitive control systems are introduced as an extension of the state feedback control systems designed. The repetitive control systems have the capability for reference tracking and disturbance rejection of complex periodic exogenous signals.

Both MATLAB and Simulink tutorials are given in this chapter for the design and implementation of this state feedback controller.

State Feedback Control and Kalman Filtering with MATLAB/Simulink Tutorials, First Edition.
Liuping Wang and Robin Ping Guan.

Companion website: www.wiley.com/go/feedback19

6.2 Embedding Disturbance Model into Controller Design

For simplicity, in the first stage of derivation of the control law, the reference signal $r(k)$ is assumed to be zero and the control system is designed for disturbance rejection.

6.2.1 Formulation of Augmented State Space Model

The starting point for the control system design is based on the same assumption of the state space model as in Chapter 5 with input disturbance where it is assumed that the disturbance model $D(q^{-1})$ is available for the control system design.

Suppose that the plant to be controlled has m inputs and m outputs and is represented by the state space model

$$x_m(k+1) = A_m x_m(k) + B_m u(k) + B_m \mu(k), \tag{6.1}$$

$$y(k) = C_m x_m(k), \tag{6.2}$$

where $x_m(k)$ is the $n \times 1$ state vector, $u(k)$ and $y(k)$ are the $m \times 1$ input and output vectors. $\mu(k)$ represents a vector of disturbance signals. The disturbance signal $\mu(k)$ is

$$\mu(k) = \frac{\epsilon(k)}{D(q^{-1})}$$

where $\epsilon(k)$ contains m components with each being zero-mean white noise, and $D(q^{-1})$ is expressed as

$$D(q^{-1}) = 1 + d_1 q^{-1} + d_2 q^{-2} + d_3 q^{-3} + \ldots + d_\gamma q^{-\gamma}. \tag{6.3}$$

The formulation of the control law here is different from the disturbance observer based approach. Instead of trying to estimate the disturbance $\mu(k)$ and compensate it in the control signal calculation, the proposed approach is to filter the state variables with the disturbance model $D(q^{-1})$.

Applying the operator $D(q^{-1})$ to the state equation (6.1) gives

$$x_s(k+1) = A_m x_s(k) + B_m u_s(k) + B_m \epsilon(k) \tag{6.4}$$

where the filtered state $x_s(k)$ and control $u_s(k)$ are defined as

$$x_s(k) = D(q^{-1}) x_m(k), \; u_s(k) = D(q^{-1}) u(k).$$

Note that the input disturbance to the filtered state vector $x_s(k)$ is $\epsilon(k)$, which consists of m white noise components.

Applying $D(q^{-1})$ to the output equation in (6.2) gives

$$y_s(k) = C_m x_s(k), \tag{6.5}$$

where $y_s(k)$ is the filtered output signal defined as

$$y_s(k) = D(q^{-1})y(k).$$

To continue the formulation of control law, the output equation (6.5) is written as

$$\begin{aligned} y_s(k+1) &= C_m x_s(k+1) \\ &= C_m A_m x_s(k) + C_m B_m u_s(k) + C_m B_m \epsilon(k). \end{aligned} \tag{6.6}$$

Because $y_s(k+1)$ is the filtered output signal at the sampling instant $k+1$ (see (6.5)), equation (6.6) becomes

$$\begin{aligned} y(k+1) &= -d_1 y(k) - d_2 y(k-1) - \dots - d_{\gamma-1} y(k-\gamma+2) \\ &-d_\gamma y(k-\gamma+1) + C_m A_m x_s(k) + C_m B_m u_s(k) + C_m B_m \epsilon(k). \end{aligned} \tag{6.7}$$

Introduce the augmented state vector:

$$x(k) = \begin{bmatrix} x_s(k)^T & \overline{y(k)}^T \end{bmatrix}^T,$$

where $\overline{y(k)}$ is the output vector with dimension $\gamma m \times 1$ defined by

$$\overline{y(k)} = \begin{bmatrix} y(k)^T & y(k-1)^T & \dots & y(k-\gamma+1)^T \end{bmatrix}^T.$$

Based on (6.4) and (6.7), with the A, B, and C matrices as illustrated, the following augmented model is formed:

$$\begin{bmatrix} x_s(k+1) \\ \hline \overline{y(k+1)} \end{bmatrix} = \overbrace{\begin{bmatrix} A_m & 0_{n\times\gamma m} \\ \hline \overline{C_m A_m} & A_d \end{bmatrix}}^{A} \begin{bmatrix} x_s(k) \\ \hline \overline{y(k)} \end{bmatrix} + \overbrace{\begin{bmatrix} B_m \\ \hline \overline{C_m B_m} \end{bmatrix}}^{B} u_s(k) + \overbrace{\begin{bmatrix} B_m \\ \hline \overline{C_m B_m} \end{bmatrix}}^{B} \epsilon(k) \tag{6.8}$$

$$\overline{y(k)} = \overbrace{\begin{bmatrix} 0_{\gamma m\times n} & I_{\gamma m\times\gamma m} \end{bmatrix}}^{C} \begin{bmatrix} x_s(k) \\ \hline \overline{y(k)} \end{bmatrix} \tag{6.9}$$

where $0_{p\times q}$ denotes the zero matrix with dimension $p \times q$, $I_{\gamma m\times\gamma m}$ denotes the identity matrix with dimension $\gamma m \times \gamma m$, $\overline{C_m A_m}$ denotes a matrix with dimension $\gamma m \times n$ that has its first m rows and n columns equal to the matrix $C_m A_m$ and the remaining rows and columns are zero, $\overline{C_m B_m}$ denotes a matrix with dimension

$\gamma m \times m$ that has its first m rows and m columns equal to the matrix $C_m B_m$ and the remaining rows and columns are zero.

The A_d matrix with dimension $\gamma m \times \gamma m$ is the same as before in Section 5.3.2, defined by

$$A_d = \begin{bmatrix} -d_1 I_m & -d_2 I_m & \dots & -d_{\gamma-1} I_m & -d_\gamma I_m \\ I_m & 0_m & 0_m & \dots & 0_m \\ 0_m & I_m & 0_m & \dots & 0_m \\ \vdots & \ddots & \ddots & \vdots & \vdots \\ 0_m & 0_m & \dots & I_m & 0_m \end{bmatrix}.$$

It is emphasized that with this formulation,

$$\det(qI - A_d) = D(q)^m \tag{6.10}$$

where $D(q)$ is the disturbance model in forward shift form. This means that the A_d matrix contains m sets of eigenvalues that correspond to the zeros of the $D(q)$ polynomial.

Noting that the A matrix is a block lower triangular matrix, the characteristic equation for system matrix A in (6.8) is given by

$$\det(zI - A_m)\det(zI - A_d) = 0. \tag{6.11}$$

Hence, the eigenvalues of the matrix A are the union of the eigenvalues of the original plant model and the zeros of the disturbance model $D(z)$, because $\det(zI - A_d) = D(z)^m$, where m is the number of outputs.

6.2.2 MATLAB Tutorial

The following MATLAB tutorial will produce the augmented state space model given by (6.8).

Tutorial 6.1 *Write a MATLAB function am4csd.m that will produce the augmented model for the controller design where the disturbance model is directly embedded in the augmented state space model.*

Step by Step

1. *Create a new file called* **am4csd.m**.
2. *The input variables to the function are the plant system matrices: A_m, B_m, C_m, and the polynomial disturbance model $D(q^{-1})$ that is called Dmodel. The output variables from this function are the augmented system matrices A, B, and C. Enter the following program into the file:*

```
function [A,B,C]=am4csd(Am,Bm,Cm,Dmodel)
```

3. *Check the numbers of inputs,outputs and state variables, and the dimension of Dmodel. Continue entering the following program into the file:*

```
gamma=length(Dmodel)-1;
[n1,n_in]=size(Bm);
[m1,n1]=size(Cm);
```

4. *Initialize the system matrices by using zero matrices. Continue entering the following program into the file:*

```
n=n1+gamma*m1;
A=zeros(n,n);
B=zeros(n,n_in);
C=zeros(m1*gamma,n);
```

5. *Create the block matrices A_m and $\overline{C_m A_m}$ in A. Continue entering the following program into the file:*

```
A(1:n1,1:n1)=Am;
A(n1+1:n1+m1,1:n1)=Cm*Am;
```

6. *Create the block matrix A_d in A. Continue entering the following program into the file:*

```
nk=1;
for kk=1:gamma
    A(n1+1:n1+m1,n1+nk:n1+kk*m1)=-Dmodel(kk+1)*eye(m1);
    nk=nk+m1;
end
A(n1+m1+1:n,n1+1:n-m1)=eye((gamma-1)*m1);
```

7. *Form the B matrix. Continue entering the following program into the file:*

```
B(1:n1,:)=Bm;
B(n1+1:n1+m1,:)=Cm*Bm;
```

8. *Form the C matrix that corresponds to the output vector $\overline{y(k)}$. Continue entering the following program into the file:*

```
C(:,n1+1:n)=eye(m1*gamma,m1*gamma);
```

9. *Test the function with the following matrices:*

```
Am=[1 -1; 2 1];
Bm=[1 2;2 1];
Cm=[1 0; 0 1];
Dmodel=[1 -2 -3];
```

10. *Check A and B matrices against the following answers:*

$$A = \begin{bmatrix} 1 & -1 & 0 & 0 & 0 & 0 \\ 2 & 1 & 0 & 0 & 0 & 0 \\ 1 & -1 & 2 & 0 & 3 & 0 \\ 2 & 1 & 0 & 2 & 0 & 3 \\ 0 & 0 & 1 & 0 & 0 & 0 \\ 0 & 0 & 0 & 1 & 0 & 0 \end{bmatrix}; \ B = \begin{bmatrix} 1 & 2 \\ 2 & 1 \\ 1 & 2 \\ 2 & 1 \\ 0 & 0 \\ 0 & 0 \end{bmatrix}.$$

11. *The eigenvalues of the system matrix are* $1 \pm j1.4142$ *and the roots of* $D(q)$ *are* $3, -1$. *There are six eigenvalues of the augmented system matrix* A, *which are* $1 \pm j1.4142,\ 3,\ 3$ *and* $-1, -1$.

6.2.3 Controllability and Observability

Before performing the control system design, one needs to ask under what conditions the pair of augmented system matrices $A,\ B$ is controllable, if the original system is both controllable and observable. The answer to the question of controllability of $A,\ B$ matrices is that we need an additional assumption imposed on the zeros of the plant transfer function $G(z) = C_m(zI_m - A_m)^{-1}B_m$ so that there is no pole-zero cancellation between $G(z)$ and the disturbance model $D(z)$.

We summarize the conditions for controllability and observability of the augmented system. Assuming that the original system transfer function is expressed as $G(z) = C_m(zI_m - A_m)^{-1}B_m$ and is both controllable and observable, if there are no zeros in the plant transfer function $G(z)$ corresponding to the zeros of the disturbance model $D(z)$, then the augmented system matrices $A,\ B$ are controllable and $A,\ C$ are observable, where

$$A = \left[\begin{array}{cc} A_m & 0_{n\times\gamma m} \\ \hline C_m A_m & A_d \end{array}\right]; \ B = \left[\begin{array}{c} B_m \\ \hline C_m B_m \end{array}\right]; \ C = \begin{bmatrix} 0_{\gamma m\times n} & I_{\gamma m\times\gamma m} \end{bmatrix}.$$

A way to show that the above statement is true is to find the minimal realization of the state space model $A,\ B,\ C$ to the augmented transfer function $C(zI - A)^{-1}B$. When the triplet $(A,\ B,\ C)$ is a minimal realization, the pair $A,\ B$ is controllable and the pair $A,\ C$ is observable (Chen [1998], Bay [1999], Kailath [1980]). In Wang [2009], a similar result was presented for model predictive control using an augmented model.

The following example demonstrates loss of controllability when there is a zero in the transfer function corresponding to the zero of the disturbance model.

Example 6.1

Consider the discrete-time system with the following z-transfer function:

$$G(z) = \frac{z^2 - 2\cos\omega_d z + 1}{(z - 0.9)^2(z - 0.5)}. \tag{6.12}$$

Show that if the disturbance model $D(z) = z^2 - 2\cos\omega_d z + 1$*, where* $\omega_d = \frac{2\pi}{40} = 0.1571$*, the augmented state space model* A*,* B *is not controllable.*

Solution

By using the MATLAB function **tf2ss.m***, we convert the transfer function model (6.12) into the state space model:*

$$x_m(k+1) = A_m x_m(k) + B_m u(k);\ y(k) = C_m x_m(k),$$

where the system matrices are:

$$A_m = \begin{bmatrix} 2.3 & -1.71 & 0.4050 \\ 1 & 0 & 0 \\ 0 & 1 & 0 \end{bmatrix};\ B_m = \begin{bmatrix} 1 \\ 0 \\ 0 \end{bmatrix};$$

$$C_m = \begin{bmatrix} 1 & -1.9754 & 1 \end{bmatrix}.$$

With the MATLAB function **am4csd.m** *created in Tutorial 6.1, we obtain the pair of* A*,* B *matrices for the augmented state space model, where*

$$A = \begin{bmatrix} 2.3 & -1.71 & 0.4050 & 0 & 0 \\ 1 & 0 & 0 & 0 & 0 \\ 0 & 1 & 0 & 0 & 0 \\ 0.3246 & -0.7100 & 0.4050 & 1.9754 & -1 \\ 0 & 0 & 0 & 1 & 0 \end{bmatrix};B = \begin{bmatrix} 1 \\ 0 \\ 0 \\ 1 \\ 0 \end{bmatrix}.$$

The controllability matrix L_c*, where*

$$L_c = \begin{bmatrix} B & AB & A^2B & A^3B & A^4B \end{bmatrix},$$

is calculated using the MATLAB function **ctrb.m***. Its condition number is calculated using MATLAB function* **cond.m***, which is* 8.0246×10^{16}*, indicating that the rank of* L_c *is less than 5 (equal to 3) and the state space model is uncontrollable.*

If one uses the pole-assignment controller design technique, the MATLAB function **place.m** *fails to produce any results with an error message saying that the* (A, B) *is nearly uncontrollable. With the MATLAB function* **dlqr.m***, a state feedback controller* K *can be calculated, however there is a pair of closed-loop poles* $0.9877 \pm j0.1564$ *that can not be changed by the state feedback controller* K*. Moreover, this pair corresponds to the zeros of the disturbance model* $D(z)$ *on the unit circle, meaning that the closed-loop system will not be asymptotically stable.*

6.2.4 Food for Thought

1. Assume that a first order system is given by

$$x_m(k+1) = 0.8x_m(k) + 0.1u(k) + 0.6\mu(k);\ y(k) = x_m(k),$$

where $\mu(k) = \frac{\epsilon(k)}{1-q^{-1}}$. What are the augmented triplet $(A,\ B,\ C)$ matrices?

2. For the same state space model as in Problem 1, however with $\mu(k) = \frac{\epsilon(k)}{(1-q^{-1})^2}$, what are the augmented triplet (A, B, C) matrices?
3. For the same state space model as in Problem 1, however, it is known that a sinusoidal disturbance with frequency $\omega_0 = 2$ (rad/s) in the continuous-time occurred. Supposing that the sampling interval $\Delta t = 0.01$ (sec) is used to obtain the discrete system, what are the augmented triplet (A, B, C) matrices?
4. Assume that a system has two inputs, two outputs and three state variables. In order to overcome a sinusoidal disturbance in the system, what are the dimensions of the augmented system matrices (A, B, C)?
5. Suppose that the transfer function of a discrete-time system is given as

$$G(z) = \frac{(z-1)}{(z-0.6)(z-0.8)}.$$

Can we design the state feedback controller with an integrator by using the disturbance model $D(q^{-1}) = 1 - q^{-1}$? Why?
6. When a system is complex, it is not straightforward to reveal its zeros. What would you propose for a simple way to check if there is any pole-zero cancelation caused by embedding the disturbance model into the controller design?

6.3 Controller and Observer Design

The control system design will be based on the pair of augmented system matrices (A, B) which are illustrated in the state space model (6.8) and created in Tutorial 6.1.

6.3.1 Controller Design and Control Signal Calculation

Assuming controllability for following augmented model

$$\overbrace{\left[\begin{array}{c} x_s(k+1) \\ \hline y(k+1) \end{array}\right]}^{x(k+1)} = \overbrace{\left[\begin{array}{cc} A_m & 0_{n\times\gamma m} \\ \hline C_m A_m & A_d \end{array}\right]}^{A} \overbrace{\left[\begin{array}{c} x_s(k) \\ \hline y(k) \end{array}\right]}^{x(k)} + \overbrace{\left[\begin{array}{c} B_m \\ \hline C_m B_m \end{array}\right]}^{B} u_s(k) + \overbrace{\left[\begin{array}{c} B_m \\ \hline C_m B_m \end{array}\right]}^{B} \epsilon(k), \tag{6.13}$$

the control system design follows from the discrete-time linear quadratic regulator (LQR) design. For this purpose, the following objective function is chosen

$$J = \sum_{k=0}^{\infty} (x(k)^T Q x(k) + u_s(k)^T R u_s(k)) \tag{6.14}$$

where Q is semi positive definite and R is positive definite. Here, we assume that the pair of A, B matrices is controllable and the pair of A, Γ matrices is observable, where $Q = \Gamma^T\Gamma$. Then, the state feedback controller K is found as

$$K = (R + B^T PB)^{-1} B^T PA \tag{6.15}$$

where matrix P is calculated using the algebraic Riccati equation:

$$A^T[P - PB(R + B^T PB)^{-1} B^T P]A + Q - P = 0$$

Upon finding the controller K, the filtered control signal $u_s(k)$ is calculated as

$$u_s(k) = -Kx(k).$$

Since the filtered control signal $u_s(k)$ is defined as

$$u_s(k) = D(q^{-1})u(k),$$

where the disturbance model $D(q^{-1})$ is defined as

$$D(q^{-1}) = 1 + d_1 q^{-1} + d_2 q^{-2} + d_3 q^{-3} + \ldots + d_\gamma q^{-\gamma}, \tag{6.16}$$

the control signal $u(k)$ is constructed using the disturbance model $D(q^{-1})$, past control signal states and $u_s(k)$:

$$u(k) = u_s(k) - d_1 u(k-1) - \ldots - d_\gamma u(k-\gamma) \tag{6.17}$$

$$= -Kx(k) - d_1 u(k-1) - \ldots - d_\gamma u(k-\gamma). \tag{6.18}$$

The control signal calculation presented in (6.18) is for disturbance rejection where the reference signal is assumed to be zero.

Noting that the output equation for the augmented state space model (6.13) is

$$\overline{y(k)} = \overbrace{\begin{bmatrix} 0_{\gamma m \times n} & I_{\gamma m \times \gamma m} \end{bmatrix}}^{C} \begin{bmatrix} x_s(k) \\ \hline \overline{y(k)} \end{bmatrix} \tag{6.19}$$

one natural selection for Q is to choose $Q = C^T C$, where C is the output matrix for the augmented state space model. With this particular choice, the objective function J in (6.19) becomes

$$J = \sum_{k=0}^{\infty} (\overline{y(k)}^T \overline{y(k)} + u_s(k)^T R u_s(k)) \tag{6.20}$$

meaning that the sum of squares of the output signals is minimized for the optimal solution of the controller K (see the first term) whilst considering the weighting on the filtered control signal (see the second term). The output matrix C is computed in the program **am4csd.m** produced Tutorial 6.1.

6.3.2 Adding Reference Signal

The next question is how the reference signals should enter the control signal computation. When the controller is used for reference tracking, the reference signal will enter the computation through the augmented output variables.

Assuming that the reference signal is a vector $r(k)$ that has the same dimension as that of the output signal $y(k)$, we form the feedback error $e(k)$ as

$$e(k) = y(k) - r(k). \tag{6.21}$$

Note that we have defined the feedback error differently from the conventional way, which would be $r(k) - y(k)$, to give a consistent expression in the state feedback control. With the definition of $e(k)$, the state vector $x(k)$ is chosen to be

$$\begin{aligned} x(k) &= \left[x_s(k)^T \quad e(k)^T \quad e(k-1)^T \quad \dots \quad e(k-\gamma+1)^T\right]^T \\ &= \left[x_s(k)^T \quad (y(k)-r(k))^T \quad \dots \quad (y(k-\gamma+1)-r(k-\gamma+1))^T\right]^T, \end{aligned} \tag{6.22}$$

which contains the filtered state vector $x_s(k)$, and the feedback errors $y(k) - r(k)$, $y(k-1) - r(k-1), \dots, y(k-\gamma+1) - r(k-\gamma+1)$. These error signals will form the state vector in the computation of the filtered control signal $u_s(k)$. Therefore, with the reference signal $r(k)$, the control signal calculation becomes:

$$u(k) = -Kx(k) - d_1 u(k-1) - \dots - d_\gamma u(k-\gamma) \tag{6.23}$$

where the state vector $x(k)$ is defined by (6.22).

The proposed control system design is applicable to systems with unstable poles, unstable zeros and integrators as long as the original system matrices (A_m, B_m) are stabilizable (we can relax the controllability requirement for the original system) and the original system transfer function $C_m(zI - A_m)^{-1}B_m$ has no zeros equal to the zeros in the disturbance model $D(z^{-1})$. The requirement on the zeros of the system transfer function $C_m(zI - A_m)^{-1}B_m$ is to avoid a pole-zero cancellation due to the augmentation of the disturbance model to the A and B matrices. In a more general case, the number of outputs is assumed to be less or equal to the number of inputs. In any case, it is a good practice to check the controllability of the augmented system matrices $(A,\ B)$ in (6.13) before proceeding to the controller design.

6.3.3 Observer Design and Implementation

An observer is required if the plant state variable vector $x_m(k)$ is not measurable. The filtered state space equation (6.4) and the filtered output equation (6.5) will be used in the observer design, which are

$$x_s(k+1) = A_m x_s(k) + B_m u_s(k) + B_m \epsilon(k) \tag{6.24}$$

$$y_s(k) = C_m x_s(k) \tag{6.25}$$

where filtered state $x_s(k)$, $u_s(k)$ and output $y_s(k)$ are defined as

$$x_s(k) = D(q^{-1})x_m(k),\ \ u_s(k) = D(q^{-1})u(k),\ \ y_s(k) = D(q^{-1})y(k).$$

An observer is designed using the original system matrices A_m, C_m, to estimate the filtered state variable vector $x_s(k)$. Based on (6.24) and (6.25), the estimated state vector $\hat{x}_s(k)$ is written as:

$$\hat{x}_s(k+1) = A_m\hat{x}_s(k) + B_m u_s(k) + K_{ob}(y_s(k) - C_m\hat{x}_s(k)). \tag{6.26}$$

The estimated state calculation using (6.26) does not include the reference signal $r(k)$. This means that the observer has a neglected impact on the closed-loop response to a reference change. In general, this implementation will lead to a slower closed-loop response speed to reference signal in comparison to disturbance response speed, which might be desirable in many control applications.

However, if one wishes to have the same response speed for both reference tracking or disturbance rejection, then the reference signal $r(k)$ will enter the computation of the estimated state via the following equation:

$$\hat{x}_s(k+1) = A_m\hat{x}_s(k) + B_m u_s(k) + K_{ob}(y_s(k) - r_s(k) - C_m\hat{x}_s(k)) \tag{6.27}$$

where $r_s(k)$ is the filtered reference signal defined as

$$r_s(k) = D(q^{-1})r(k).$$

Two comments follow.

1. It is important to note that the disturbance term $B_m\epsilon(k)$ does not play any role in the observer design because $\epsilon(k)$ is assumed to be zero-mean white noise. This assumption also leads to an unbiased estimate of $x_s(k)$ and makes the task of observer design simplified. Discussions on the effect of input disturbance on the state estimate can be found in Section 5.3 of Chapter 5.
2. The reference signal $r(k)$ is not used in the calculation of the filtered estimated state $\hat{x}_s(k)$ when using (6.26). This type of implementation leads to two-degree-of-freedom control system because the observer plays a reduced role in the reference tracking. The two-degree-of-freedom control system configuration means that once the controller and the observer are designed, we have a flexibility in the implementation stage to adjust the response speed of reference tracking and disturbance rejection. For this type of control system, we can focus on tuning the state feedback controller K to improve the closed-loop dynamic response to reference change. If one wishes, the filtered reference $r_s(k)$ could be used in the observer implementation using (6.27), leading to one-degree-of-freedom control system.

6.3.4 MATLAB Tutorial for Control Implementation

The implementation of the controller and observer is slightly more complicated than the disturbance-observer-based control system introduced in Chapter 5, because one needs to carefully consider how the filtered signals are created efficiently for real-time control system computation. The following MATLAB program is created to demonstrate how the closed-loop system can be implemented and simulated for disturbance rejection and reference tracking.

Tutorial 6.2 *Write a MATLAB function that will implement the controller and observer as well as perform closed-loop simulation for reference tracking and disturbance rejection of a multi-input and multi-output system. We assume that we have the discrete-time system matrices, A_m, B_m, and C_m and have designed the state feedback controller K and the observer gain K_{ob}. In the implementation, we do not need the augmented system matrices A, B, and C, however they are required in the design of the state feedback controller K. To avoid confusion, we only implement the two-degree-of-freedom control system.*

Step by Step

1. *Create a new file called* **simu4csd.m**.
2. *The input variables to the function are the plant system matrices: A_m, B_m, C_m, the disturbance model Dmodel, the number of samples for the simulation (Nsim), controller K, and observer gain K_{ob}, reference signal sp and disturbance signal dis. The output variables from this function are output y_1 and control signal u_1. The dimension of sp is $m_1 \times Nsim$ and the dimension of dis is $n_{in} \times Nsim$ where m_1 is the number of outputs and n_{in} is the number of inputs. Enter the following program into the file:*

```
function [y1,u1]=simu4csd(Am,Bm,Cm,Dmodel,Nsim,K,Kob,
sp,dis)
```

3. *Check number of states, number of outputs and number of inputs, and the order of the disturbance model. Continue entering the following program into the file:*

```
[m1,n]=size(Cm);
[n,n_in]=size(Bm);
gamma=length(Dmodel)-1;
```

4. *Set initial conditions for the controller implementation and simulation. All initial conditions are chosen to be zero for simplicity. Continue entering the following program into the file:*

```
us=zeros(n_in,1);
xs=zeros(n,1);
```

```
y=zeros(m1,1);
xm=zeros(n,1);
```

5. *Generate the data matrices based on D(q) polynomial for implementation of the filtered output signal and the control signal, where FM will be used for calculation of filtered output and FU for calculation of control signal. Continue entering the following program into the file:*

```
FM=eye(m1);
for ii=2:gamma+1
FM=[FM Dmodel(ii)*eye(m1)];
end
FU=FM(:,m1+1:end);
```

6. *Define the following data vectors:*

$$YM = \left[y(k)^T \quad y(k-1)^T \quad \dots \quad y(k-\gamma)^T\right]^T$$
$$YME = \left[e(k)^T \quad e(k-1)^T \quad \dots \quad e(k-\gamma+1)^T\right]^T$$
$$UM = \left[u(k-1)^T \quad u(k-2)^T \quad \dots \quad u(k-\gamma)^T\right]^T$$

and choose their initial conditions to be zero. Continue entering the following program into the file:

```
YM=zeros((gamma+1)*m1,1);
YME=zeros(gamma*m1,1);
UM=zeros(gamma*m1,1);
```

7. *Perform the simulation iteratively, which corresponds to a real-time control system. Continue entering the following program into the file:*

```
 for kk=1:Nsim;
```

8. *At the sampling instant kk, with measured output y(kk) and the reference signal sp(kk), we update the data vector YM and YME. Continue entering the following program into the file:*

```
YM=[y;YM(1:end-m1,1)];
YME=[y-sp(:,kk);YME(1:end-m1,1)];
```

9. *With the estimated state variable $x_s(k)$, form the augmented state variable for feedback. Continue entering the following program into the file:*

```
Xf=[xs;YME];
```

10. *Calculate the filtered control signal $u_s(k)$ and construct the control signal using (6.26) with the definitions of FU and UM. Continue entering the following program into the file:*

```
us=-K*Xf;
u=us-FU*UM;
```

11. *Update data vector UM with the current control signal. Continue entering the following program into the file:*

```
UM=[u;UM(1:end-m1,1)];
```

12. *Compute the filtered output signal in preparation for updating the estimated state variable. Continue entering the following program into the file:*

```
ys=FM*YM;
```

13. *Update the estimated state variable with the filtered output signal and control signal in preparation for the next sampling instant. Continue entering the following program into the file:*

```
xs=Am*xs+Bm*us+Kob*(ys-Cm*xs);
```

14. *The above steps 8–13 capture the implementation of the control system. We save the control signal and output signal from this function. Continue entering the following program into the file:*

```
u1(:,kk)=u;
y1(:,kk)=y;
```

15. *Since we are in a simulation environment, the control signal calculated is sent to the plant model for closed-loop simulation. In the closed-loop simulation, we add the input disturbance to the system. This completes one-cycle of computation in the closed-loop simulation. Continue entering the following program into the file:*

```
xm=Am*xm+Bm*(u+dis(:,kk));
y=Cm*xm;
end
```

16. *Test the function using Example 6.2.*

Note that if one wishes to add output disturbance and measurement noise, the plant output y calculated in the final step should be modified to include these variables.

The following example will show the state feedback controller design, observer design and the control system implementation. One can use it to verify the MATLAB simulation program **simu4csd.m**.

Example 6.2

A discrete-time system is described by the following state space model:

$$x_m(k+1) = A_m x_m(k) + B_m u(k);\ y(k) = C_m x_m(k)$$

where the system matrices are

$$A_m = \begin{bmatrix} 1.8713 & -0.8752 \\ 1 & 0 \end{bmatrix}; \; B_m = \begin{bmatrix} 1 \\ 0 \end{bmatrix}; \; C_m = \begin{bmatrix} 0.0016 & 0.0015 \end{bmatrix}$$

where the state variables are not measured. The reference signal for the system comprises of ramp signals and constants. Design a state feedback control system to track the reference signal.

Solution

Since the reference signal comprises of ramp signals and constants, the disturbance model $D(q^{-1})$ is chosen to be

$$D(q^{-1}) = (1 - q^{-1})^2,$$

leading to the design of state feedback controller with double integrator. By using the MATLAB function **am4csd.m** *written in Tutorial 6.1, we obtain the following augmented state space model:*

$$A = \begin{bmatrix} 1.8713 & -0.8752 & 0 & 0 \\ 1 & 0 & 0 & 0 \\ 0.0045 & -0.0014 & 2 & -1 \\ 0 & 0 & 1 & 0 \end{bmatrix}; \; B = \begin{bmatrix} 1 \\ 0 \\ 0.0016 \\ 0 \end{bmatrix}; \; C = \begin{bmatrix} 0 & 0 & 1 & 0 \\ 0 & 0 & 0 & 1 \end{bmatrix}.$$

The output matrix C corresponding to the $y(k)$ and $y(k-1)$ is a good candidate for the choice of Q matrix, where we select $Q = C^T C$. Using this choice together with $R = 1$, the application of MATLAB function **dlqr.m** *results in the following state feedback controller K:*

$$K = \begin{bmatrix} 0.5423 & -0.3769 & 11.3088 & -10.2417 \end{bmatrix}.$$

For the observer design, we choose Q to be identity matrix and $R = 1$, which gives

$$K_{ob} = \begin{bmatrix} 4.3728 & 4.3649 \end{bmatrix}^T.$$

The reference signal $r(k)$ is shown in Figure 6.1(a), which is identical to that used in Example 5.6 where the MATLAB code can be found to generate $r(k)$. Closed-loop simulation of the state feedback control system is performed with disturbance signal equal to 0 by using the function **simu4csd.m**, *which is produced in Tutorial 6.2. Figure 6.1 shows the closed-loop control system response with the controller and observer, where we used the two-degree-of-freedom implementation. It is seen that the output signal closely tracks the reference signal. We can verify that the maximum track error is within ± 0.12 and the mean squared error calculated as $\frac{1}{M}\sum_{k=1}^{M} e(k)^2 = 3.3238 \times 10^{-4}$, where M is the data length. If we wish to improve the tracking performance, we can reduce the R value in the controller design to 0.01.*

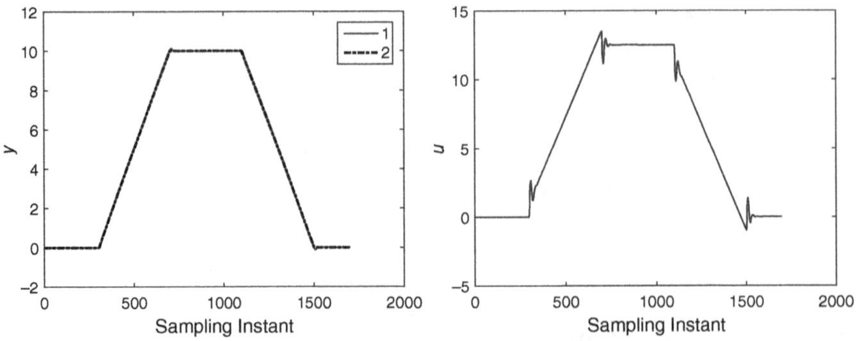

(a) Output (b) Control

Figure 6.1 Closed-loop control system response using state estimate feedback controller (Example 6.2). Key: line (1) the output signal; line (2) the reference signal.

This will increase the controller gain to

$$K = \begin{bmatrix} 0.9931 & -0.5815 & 50.9902 & -42.7982 \end{bmatrix}.$$

where the maximum tracking error is within ± 0.075 *and the mean squared error is* 6.1174×10^{-5}. *Clearly, the controller gain has increased significantly for the tracking performance improvement. If the system had measurement noise, then this higher performance demand might not be possible because the controller would amplify the noise significantly.*

6.3.5 Food for Thought

1. The state feedback control system is designed to reject input disturbance. Do you think that control system will reject an output disturbance with the same characteristic as the input disturbance? Why?
2. When we choose $Q = C^T C$ in the discrete-time LQR design where C is the output matrix from the augmented model, what is the objective function J in **dlqr.m** we try to minimize, in terms of state, output and control variables?
3. A discrete-time system is described by the difference equations:

$$x_m(k+1) = 0.6x_m(k) + u(k) + \mu(k); \; y(k) = 0.5x_m(k).$$

   If we wish the output to track a step reference signal, how do we choose the disturbance model $D(q^{-1})$? What are the augmented triplet $(A, \; B, \; C)$ matrices?
4. Continue from Problem 3. If we choose $Q = C^T C$ and $R = 1$ in the controller design, what is the objective function J with respect to the feedback error $r(k) - y(k)$ and the control signal $u(k)$?

5. Continue from Problem 3. Supposing that we need to use an observer to estimate the state, how many closed-loop poles are in the state estimate feedback system?
6. If the output $y(k)$ is severely corrupted by measurement noise and if we wish to reduce the effect of noise in the control system, would you increase the diagonal elements in R matrix when designing the controller with **dlqr.m**? Would you do the same for the observer design?

6.4 Practical Issues

6.4.1 Reducing Overshoot in Reference Tracking

In control applications, the required control system response speed for disturbance rejection is as fast as possible because the disturbances cause disruptions to plant operations. However, the requirement for reference tracking is not necessarily the same because a fast reference tracking response is often accompanied by an overshoot to the reference signal. This overshoot could cause accident in plant operation. The two-degree-of-freedom control system provides a simple remedy to reduce overshoot with a slower dynamic response to reference changes while maintaining fast response speed to disturbance signal, where the observer is implemented via the following equation:

$$\hat{x}_s(k+1) = A_m\hat{x}_s(k) + B_m u_s(k) + K_{ob}(y_s(k) - C_m\hat{x}_s(k)). \tag{6.28}$$

In contrast, the one-degree-of-freedom control system is realized by using the filtered error signal in the calculation of estimated states, which is

$$\hat{x}_s(k+1) = A_m\hat{x}_s(k) + B_m u_s(k) + K_{ob}(y_s(k) - r_s(k) - C_m\hat{x}_s(k)). \tag{6.29}$$

This means that a fast disturbance rejection speed will also lead to a fast reference response speed.

The following example is to show that the two-degree-of-freedom control system reduces the overshoot in reference response. We use the quadruple tank system in Example 5.7 for the demonstration.

Example 6.3
The linearized transfer function model for a quadruple-tank apparatus (Johansson [2000],Goodwin et al. [2000]) has the following form:

$$G(s) = \begin{bmatrix} \frac{3.7\gamma_1}{62s+1} & \frac{3.7(1-\gamma_2)}{(23s+1)(62s+1)} \\ \frac{4.7(1-\gamma_1)}{(30s+1)(90s+1)} & \frac{4.7\gamma_2}{90s+1} \end{bmatrix} \tag{6.30}$$

where $\gamma_1 = \gamma_2 = 0.8$. Design state feedback control system for the outputs to follow step reference signals and reject low frequency disturbances. Evaluate the closed-loop performances for the one-degree-of-freedom and two-degree-of-freedom control systems.

Solution

The discrete-time state space model for the water tank can be found in Example 5.7. In order for the closed-loop output to follow a step reference signal, an integrator is used in the controller. For this purpose, the disturbance model is chosen to be $D(q^{-1}) = (1 - q^{-1})$. With the discrete-time state space system matrices A_m, B_m, C_m as in Example 5.7, together with the disturbance model, the application of MATLAB function **am4csd.m** *leads to the augmented system matrices A, B, C. Both controller and observer are designed using MATLAB function* **dlqr.m**. *For the controller design, the Q matrix is chosen to be $Q = C^T C$, and R is $0.1I$ where I is identity matrix. For the observer design, the weighting matrix Q_{ob} is identity matrix and R_{ob} is $0.1I$.*

In the simulation studies, two step reference signals are chosen. The closed-loop performance for disturbance rejection is also evaluated, where the disturbance on u_1 is a white noise with $\sigma^2 = 1$ passing through a low pass filter with the transfer function $\frac{0.02}{1-0.98z^{-1}}$ and the disturbance on u_2 is a constant disturbance with amplitude of -0.5. Both reference signals and disturbance signals are identical to those used in Example 5.7. The application of the MATLAB program **simu4csd.m** *produced in Tutorial 6.2 leads to the closed-loop simulation results for the two-degree-of-freedom control system. With a small modification by adding the filtered reference signal $r_s(k)$ to the observer, we obtain the simulation results for the one-degree-of-freedom control system. In the simulation studies, we assume at the steady-state operation, 4 units of the control signals correspond to 6 units of the output signals. Closed-loop simulation starts after 16 seconds. The simulation results are discussed as follows.*

1. *Figures 6.2(a)–(b) compare the output responses for reference tracking and disturbance rejection while Figures 6.3(a)–(b) compare the control responses for reference tracking and disturbance rejection. From these responses, it is evident that the two-degree-of-freedom control system implementation has overcome the overshoot problem that occurred with the one-degree-of-freedom control system.*
2. *The advantage of the two-degree-of-freedom control system becomes more relevant when the system has an overshoot problem with the reference changes. Additionally, because the observer dynamics have a very small impact on the reference changes in this configuration, the task of tuning the dynamic performance for the reference change is simplified by focusing on tuning the state feedback controller performance. This means that if we wish to have a faster or slower closed-loop response to reference changes, we will leave the observer unchanged and change the design for the state feedback controller K.*

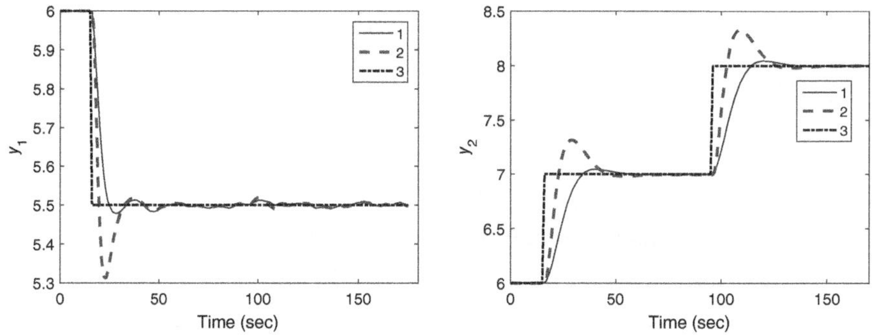

(a) Output y_1 (b) Output y_2

Figure 6.2 Closed-loop output response to reference and disturbance (Example 6.3). Key: line (1) output (two-degree-of-freedom); line (2) output (one-degree-of-freedom); line (3) reference.

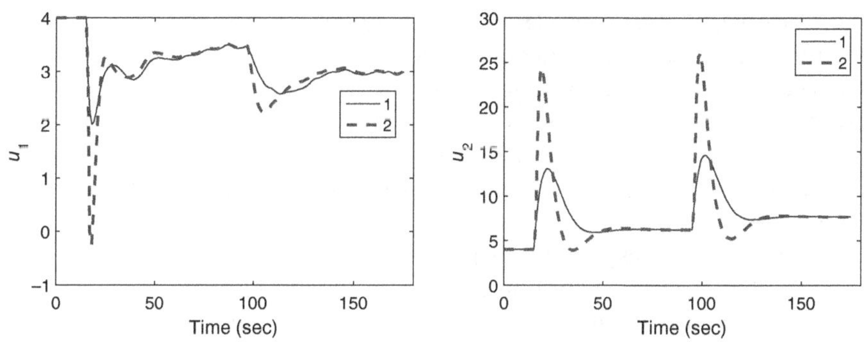

(a) Control u_1 (b) Control u_2

Figure 6.3 Closed-loop control response to reference and disturbance (Example 6.3). Key: line (1) control (two-degree-of freedom); line (2) control (one-degree-of- freedom).

3. *For this particular system, in the controller design, we can use the same $Q = C^T C$ matrix, but increase or decrease the weighting matrix R to obtain slower or faster closed-loop response speed to the reference changes. As an illustration, Figures 6.4(a)–(b) show the closed-loop responses to the same reference changes as before, however with three different performance levels where line (1) corresponds to $R = 0.1I$, line (2) corresponds to a faster response speed ($R = 0.01I$) and line (3) corresponds to a slower response speed ($R = I$).*

In summary, the two-degree-of-freedom control system configuration plays an important role in tuning the closed-loop reference responses because the observer dynamics have less influence and the matter is simpler. In practice,

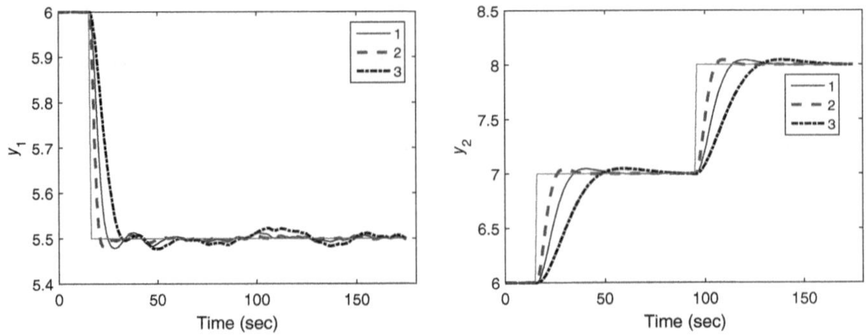

(a) Output y_1 (b) Output y_2

Figure 6.4 Closed-loop output response to reference and disturbance. Key: line (1) output ($R = 0.1I$); line (2); output ($R = 0.01I$); line (3) output ($R = I$).

the choice of $Q = C^T C$ is a good starting point to obtain a good reference response. For the cases where an overshoot in reference response is not an issue, then the one-degree-of-freedom configuration may provide a smaller reference tracking error because the observer detects the change earlier.

For disturbance rejection and measurement noise attenuation, the one-degree-of-freedom and two-degree-of-freedom control systems have identical performance, meaning that both controller and observer play important roles in these tasks. The general rules for reducing the effect of noise are to reduce the controller gain and the observer gain via the selection of desired closed-loop performance. However, to reduce the effect of disturbance, the controller gain and observer gain are to be increased. The disturbance rejection and measurement noise attenuation are two competing objectives.

6.4.2 Anti-windup Implementation

The anti-windup implementation of the multi-input and multi-output control system is outlined as follows. Suppose that the control signal $u(k)$ is constrained between u^{min} and u^{max}, where u^{min} and u^{max} contain the saturation limits for all input signals. Assume that the past output signals and the reference signals are zero vectors as

$$y(-1) = y(-2) = \ldots = y(-\gamma + 1) = 0$$

$$r(-1) = r(-2) = \ldots = r(-\gamma + 1) = 0.$$

The initial state $\hat{x}_s(0)$ is also assumed to be zero. The past control values, $u(-1),\ u(-2),\ \ldots,\ u(-\gamma)$ are given, which may not be zero, depending on the operation of the plant. The anti-windup implementation is given by the following procedures, beginning with $k = 0$.

1. Use the current measurement $y(k)$ and reference signal $r(k)$ to form

$$x(k) = \left[\hat{x}_s(k)^T \quad e(k)^T \quad e(k-1)^T \quad \dots \quad e(k-\gamma+1)^T\right]^T$$

   where $e(k) = y(k) - r(k)$.
2. Calculate the filtered control signal

$$u_s(k) = -Kx(k). \tag{6.31}$$

3. Calculate the control signal

$$u(k) = u_s(k) - d_1 u(k-1) - d_2 u(k-2) - \dots - d_\gamma u(k-\gamma).$$

4. Update $y_s(k)$ using the current output measurement $y(k)$:

$$y_s(k) = y(k) + d_1 y(k-1) + d_2 y(k-1) + \dots + d_\gamma y(k-\gamma).$$

5. Impose the saturation limits on the control signal with the following computations.
   (a) For the ith control signal, saturation limits are imposed:

$$u_i(k)^{act} = \begin{cases} u_i^{min}, & \text{if } u_i(k) < u_i^{min}; \\ u_i(k), & \text{if } u_i^{min} \le u_i(k) \le u_i^{max}; \\ u_i^{max}, & \text{if } u_i(k) > u_i^{max}, \end{cases}$$

   where $i = 1, 2, \dots, m$.
   (b) Update $u_s(k)$ using the actual control signal $u(k)^{act}$ that contains the saturated elements:

$$u_s(k)^{act} = u(k)^{act} + d_1 u(k-1) + d_2 u(k-1) + \dots + d_\gamma u(k-\gamma).$$

6. Update the estimated states for sample time $k+1$ using the $u_s(k)^{act}$ that contains the information of saturation as

$$\hat{x}_s(k+1) = A_m \hat{x}_s(k) + B_m u_s(k)^{act} + K_{ob}(y_s(k) - C_m \hat{x}_s(k)).$$

7. As the sample time progresses to $k+1$, repeat the computation from the first step.

The following tutorial illustrates the implementation of the discrete-time state feedback control algorithm in real-time simulation with the control signal constraints.

Tutorial 6.3 *An MATLAB embedded function will be written for Simulink simulation as well as for xPC Target implementation. Because the embedded functions require the dimensions of the variables to be specified, in the tutorial we* *will consider a second order system model with a second order disturbance model. For consistency, we will define all variables in the same way as those presented in Tutorial 6.2. The entire embedded MATLAB function completes one cycle of computation for the control signal. For every sampling period, it will repeat the same computation procedure.*

Step by Step

1. *Create a new Simulink file called* **dsfcI.slx**
2. *In Simulink's directory of User-Defined Functions, find the icon of the embedded MATLAB function and copy it to the dsfcI model.*
3. *Click on the icon of the embedded function, and define the input and output variables and the controller parameters to the model so that the embedded function has the following form:*

   ```
   function u=dsfcI(r,y,Am,Bm,Cm,K,Kob,Umin,Umax,FM,FU)
   ```

   where u is the calculated control signal at the sampling time k, the first two elements (r and y) among the input variables are the reference signal and the output signal at sampling time k, A_m, B_m, and C_m are the plant system matrices used in the state estimation, K is the controller gain and K_{ob} is the observer gain, and Umin and Umax are the upper and lower limits imposed on the control signal. For a single-input and single-output system, FM is equal to the disturbance model and FU is the disturbance model from its second element to its last element (see Tutorial 6.2). We calculate these parameters in MATLAB workspace in order to save computational time.
4. *We need to edit the input and output data ports in order to let the embedded function know which input ports are the real-time variables and which are the parameters. This editing task is performed using Model Explorer.*
   - *click on 'r', on Scope, select 'input', assign port '1' and size '-1', complexity 'Inherited', type 'Inherit: Same as Simulink'. Repeat the same editing procedure for the output signal 'y'.*
   - *The rest of the inputs to the embedded function are the parameters required in the computation. Click on 'Am', on Scope, select 'Parameter' and click 'Tunable' and click 'Apply' to save the changes. Repeat the same editing procedure for the rest of the parameters.*
   - *To edit the output port from the embedded function, click on 'u', on Scope, select 'Output', Port '1', Size '-1', Sampling Model 'Sample based', Type 'Inherit: Same as Simulink', and click on 'Apply' to save the changes.*
5. *In the following, the program will declare the variables stored in the embedded function during each iteration for their dimensions and initial values. For this control algorithm, the estimated state variables, and the input and output data vectors will be stored in the embedded function. Their dimensions and initial*

conditions need to be defined precisely. As an example, we consider a single input and single output system with two state variables, and a second order disturbance model. We follow Tutorial 6.2 to define the variables that need to be stored in the real-time implementation. Continue entering the following program into the file:

```
persistent xs
if isempty(xs)
xs=zeros(2,1);
end
persistent YM
if isempty(YM)
    YM=zeros(3,1);
end
persistent YME
if isempty(YME)
YME=zeros(2,1);
end
persistent UM
if isempty(UM)
UM=zeros(2,1);
end
```

6. *After defining the variables to be stored, the calculation begins for the control signals at the sampling time k. We will first prepare the data vectors for computation of control signal and the filtered output signal. Continue entering the following program into the file:*

```
YM=[y;YM(1:2,1)];
YME=[y-r;YME(1,1)];
```

7. *Calculate the filtered control signal. Continue entering the following program into the file:*

```
us=-K*[xs;YME];
```

8. *Calculate the control signal. Continue entering the following program into the file:*

```
u=us-FU*UM;
```

9. *If the control signal exceeds its maximum or minimum values, impose constraints on the amplitude of the control signal and update the filtered control signal us with the constrained control signal. Continue entering the following program into the file:*

```
if (u<Umin) u=Umin; us=u+FU*UM; end
if (u>Umax) u=Umax; us=u+FU*UM; end
```

10. *Calculate the filtered output signal in preparation for the state estimation. Continue entering the following program into the file:*

```
ys=FM*YM;
```

11. *Estimate the filtered state variable for k + 1. Continue entering the following program into the file:*

```
xs=Am*xs+Bm*us+Kob*(ys-Cm*xs);
```

12. *Shift the data vector UM in preparation for the next sampling period and complete the embedded function. Continue entering the following program into the file:*

```
UM=[u;UM(1,1)];
```

13. *Validate this embedded function by building a Simulink simulation program based on Example 6.2 and compare the simulation results. In the Simulink program, we may need to add a discrete-time delay to connect the plant output to y in this program.*

One of the advantages of this state space formulation is that the initial conditions of the estimated state could be assigned to zero because it is the filtered state variable. Although the control system design is applicable to unstable systems that have eigenvalues outside the unit circle of the complex plane, one must be extremely careful in the implementation of constraints for these systems. Due to the nature of controlling unstable systems, which requires a minimum control amplitude for stabilization, control amplitude constraints could easily lead to instability of the closed-loop systems. The anti-windup mechanisms proposed here work effectively for stable systems and systems with integrators.

6.4.3 Control System using NMSS Realization

When the disturbance model $D(q^{-1})$ is embedded into the controller design and if the state variables are measured, the design method introduced in this chapter can be directly applied without the utilization of an observer. We assume that a discrete-time system with the transfer function model:

$$G(z) = \frac{b_1 z^{n-1} + b_2 z^{n-2} + \ldots + b_n}{z^n + a_1 z^{n-1} + a_2 z^{n-2} + \ldots + a_n} z^{-n_d} \tag{6.32}$$

where n_d is the number of samples in time delay. The corresponding difference equation is:

$$\begin{aligned} y(k+1) = &-a_1 y(k) - a_2 y(k-1) - \ldots - a_n y(k-n+1) \\ &+ b_1 u(k-n_d) + b_2 u(k-n_d-1) + \ldots + b_n u(k-n-n_d+1). \end{aligned} \tag{6.33}$$

The difference equation (6.33) is converted into state space model by choosing the following state vector:

$$x_m(k) = \left[y(k) \ \ldots \ y(k-n+1) \ \ u(k-1) \ \ldots \ u(k-n-n_d+1)\right]^T,$$

which gives

$$x_m(k+1) = A_m x_m(k) + B_m u(k) \tag{6.34}$$

$$y(k) = C_m x_m(k) \tag{6.35}$$

where the system matrices A_m, B_m, and C_m are

$$A_m = \begin{bmatrix} -a_1 & -a_2 & \ldots & b_1 & \ldots & b_n \\ 1 & 0 & 0 & 0 & 0 & 0 \\ 0 & \ddots & 0 & 0 & 0 & 0 \\ 0 & 0 & 1 & 0 & 0 & 0 \\ 0 & 0 & 0 & 0 & 0 & 0 \\ 0 & 0 & \ldots & 1 & \ldots & 0 \\ \vdots & \vdots & \ddots & 0 & \ddots & 0 \\ 0 & 0 & \ldots & 0 & 1 & 0 \end{bmatrix}; \ B_m = \begin{bmatrix} 0 \\ 0 \\ \vdots \\ 0 \\ 1 \\ 0 \\ \vdots \\ 0 \end{bmatrix};$$

$$C_m = \begin{bmatrix} 1 & 0 & 0 & \ldots & 0 & 0 \end{bmatrix}.$$

Upon formulation of the non-minimal state space (NMSS) model, the next step is to augment the system model (A_m, B_m, C_m) with the disturbance model $D(z)$ to obtain (A, B, C), which was discussed in Section 6.2. The assumption is that the zeros of the transfer function $G(z)$ are not equal to the zeros of the disturbance model $D(z)$ so that the augmented system matrices (A, B) are controllable. The controller design for the non-minimal state space model follows from the design method in Section 6.3. The controller implementation with the non-minimal state space model is similar to that introduced in Tutorial 6.2, however the estimated state vector being replaced by the measured state vector $x_m(k)$.

The following two examples are aimed to highlight the features of the control system with the non-minimal state space realization. In order to do so, comparative studies are performed against the control systems with minimal state space realization.

Example 6.4

Consider the discrete-time system described by the following z-transfer function:

$$G(z) = \frac{z^3 - 1.1z^2 + 0.31z - 0.021}{z^4 - 3.0228z^3 + 3.8630z^2 - 2.6426z + 0.8084}. \tag{6.36}$$

This system is unstable, having four poles located at

$$0.5162 \pm j0.7372, \;\; 0.9952 \pm j0.0877.$$

However, it has three stable zeros located at

$$0.1, \;\; 0.3, \;\; 0.7.$$

The control system is required to track a step reference signal and to reject low frequency disturbance. Design and simulate a control system with the non-minimal state space realization.

Solution

The non-minimal state space realization leads to 7 state variables with A_m, B_m, C_m given as below:

$$A_m = \begin{bmatrix} 3.0228 & -3.8630 & 2.6426 & -0.8084 & -1.10 & 0.31 & -0.021 \\ 1 & 0 & 0 & 0 & 0 & 0 & 0 \\ 0 & 1 & 0 & 0 & 0 & 0 & 0 \\ 0 & 0 & 1 & 0 & 0 & 0 & 0 \\ 0 & 0 & 0 & 0 & 0 & 0 & 0 \\ 0 & 0 & 0 & 0 & 1 & 0 & 0 \\ 0 & 0 & 0 & 0 & 0 & 1 & 0 \end{bmatrix};$$

$$B_m = \begin{bmatrix} 1 \\ 0 \\ 0 \\ 0 \\ 1 \\ 0 \\ 0 \end{bmatrix};$$

$$C_m = \begin{bmatrix} 1 & 0 & 0 & 0 & 0 & 0 & 0 \end{bmatrix}.$$

Note that the time delay $n_d = 0$, and the first coefficient in B_m equals to 1 corresponding to the first coefficient in the numerator of the transfer function.

In order for the output to track a step signal and to reject low frequency disturbance, an integrator is required in the control system design. For this purpose, we choose the disturbance model $D(q^{-1}) = 1 - q^{-1}$ and produce the augmented state space model for the controller design using the MATLAB function **am4csd.m** *written in Tutorial 6.1.*

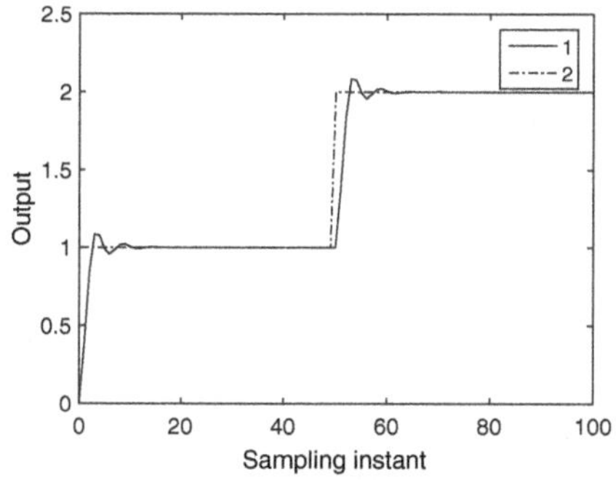

(a) Output response

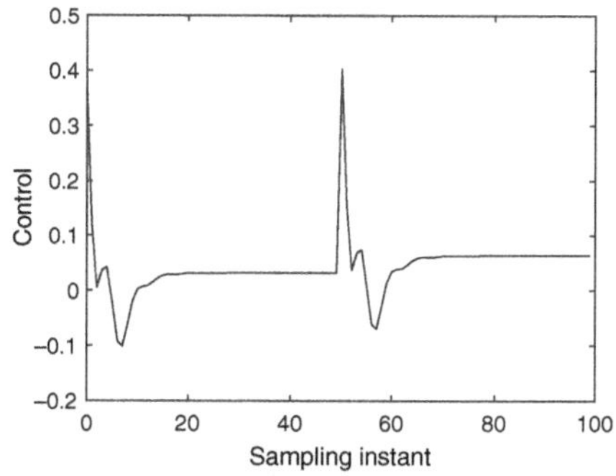

(b) Control signal

Figure 6.5 Closed-loop response to reference signal (Example 6.4). Key: line (1) output; line (2) reference.

By choosing the weighting matrices $Q = C^T C$ and $R = 1$, where the matrix C is the output matrix in the augmented model produced by **am4csd.m**, *the following state feedback controller is obtained using MATLAB function* **dlqr.m**:

$$K = \begin{bmatrix} 2.360 & -3.544 & 2.626 & -0.845 & -1.096 & 0.320 & -0.022 & 0.199 \end{bmatrix}.$$

The eight closed-loop eigenvalues are $0.2898 \pm j0.5803$, 0.7262, $0.6274 \pm j0.1558, 0, 0, 0$. *The closed-loop eigenvalues at zero correspond to the three zero eigenvalues from the non-minimal state space model, meaning that the linear quadratic controller did not alter them, which is ideal.*

Figure 6.5(a) shows the closed-loop output response to a step reference change, and Figure 6.5(b) shows the control signal response. It is seen from these figures that it takes about 10 samples for the output to reach the reference signal and 20 samples for the control signal to reach steady-state. To evaluate the closed-loop response to disturbance rejection, the reference signal is set to zero, and a series of step signals with amplitude ± 1 *with the interval of 25 samples are introduced at the input of the system. Figure 6.6(a) shows that the closed-loop output to the disturbance and Figure 6.6(b) shows the control signal response. It is seen from these two figures that the control system has responded to the step disturbances very fast.*

This section is completed with the following comparative study.

Example 6.5

We continue from Example 6.4. For this example, we will design the state feedback control system with minimal state space realization, which means that an observer is

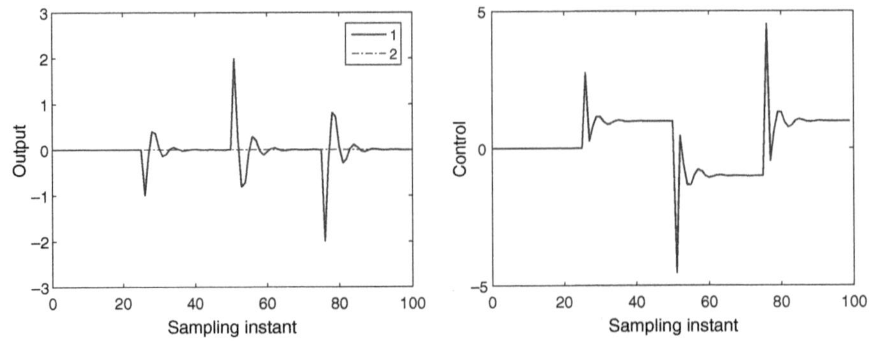

(a) Output response (b) Control signal

Figure 6.6 Closed-loop response to disturbance signal (Example 6.4). Key: line (1) output; line (2) reference.

required for the control system. We will compare the closed-loop responses with those obtained from non-minimal state space realization in Example 6.4.

Solution

From the transfer function model (6.36) with the application of MATLAB function **tf2ss.m**, *the minimal realization of the state space model is obtained with the following system matrices:*

$$A_m = \begin{bmatrix} 3.0228 & -3.8630 & 2.6426 & -0.8084 \\ 1 & 0 & 0 & 0 \\ 0 & 1 & 0 & 0 \\ 0 & 0 & 1 & 0 \end{bmatrix}; B_m = \begin{bmatrix} 1 \\ 0 \\ 0 \\ 0 \end{bmatrix};$$

$$C_m = \begin{bmatrix} 1.0000 & -1.1000 & 0.3100 & -0.0210 \end{bmatrix}.$$

For the minimal state space realization, the number of state variable is 4, which are not corresponding to the input and output measurements. Hence, an observer is required for the state feedback control.

For the controller design, the triplets (A_m, B_m, C_m) are augmented with the disturbance model $D(q^{-1}) = 1 - q^{-1}$ using the MATLAB function **am4csd.m** *written in Tutorial 6.1. By choosing the weighting matrices $Q = C^T C$, and $R = 1$, MATLAB function* **dlqr.m** *produces the controller gain K where*

$$K = \begin{bmatrix} 1.6642 & -3.1734 & 2.4328 & -0.7733 & 0.6585 \end{bmatrix}.$$

The closed-loop eigenvalues are: $0.1044 \pm j0.5183$, $0.3953 \pm j0.1507$, 0.7007. For observer design and implementation, the triplets (A_m, B_m, C_m) are used.

By choosing the weighting matrices $Q = I$ and $R = 0.1$, the following observer gain is obtained using MATLAB function **dlqr.m***:*

$$K_{ob} = \begin{bmatrix} 2.8535 & 0.7175 & -0.2694 & -0.0605 \end{bmatrix}^T$$

corresponding to the eigenvalues of the closed-loop error system: $0.6290 \pm j0.0978, -0.1086 \pm j0.0904$.

To evaluate the closed-loop performance with respect to reference changes, the same reference signal as in Example 6.4 is used. With the application of MATLAB function **simu4csd.m** *written in Tutorial 6.2, the closed-loop system responses are computed as shown in Figures 6.7(a)–(b). The comparative results are discussed as follows.*

1. *By comparing the output and control signal presented in Figure 6.7 to those in Figure 6.5 where a non-minimal state space realization was used, we can see that the closed-loop output responses are almost identical for the two different designs. This is because the implementation with minimal state space realization is a two-degree-of freedom configuration and observer plays little role in the output response to reference change.*
2. *To evaluate the closed-loop performance with respect to disturbance rejection, the same disturbance signal as in Example 6.4 is used. Figure 6.8(a) compares the output response from the minimal realization to that from non-minimal state space realization, it is seen that the control system using the non-minimal state space realization has significantly improved closed-loop performance for disturbance rejection. This is because by directly using the input and output signals as feedback signals with the non-minimal state space realization, the effect of the disturbance is detected much faster and the controller acts very*

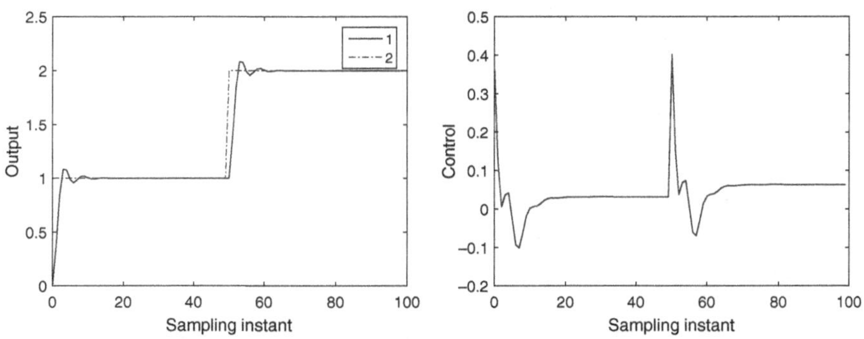

(a) Output response (b) Control signal

Figure 6.7 Closed-loop response to reference signal (Example 6.5). Key: line (1) output; line (2) reference.

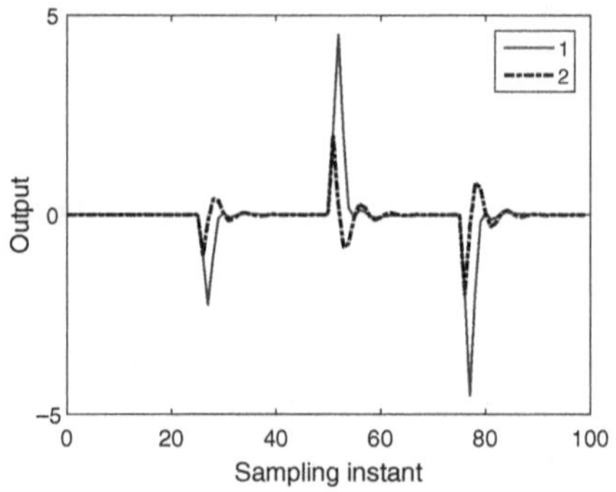

(a) Output response

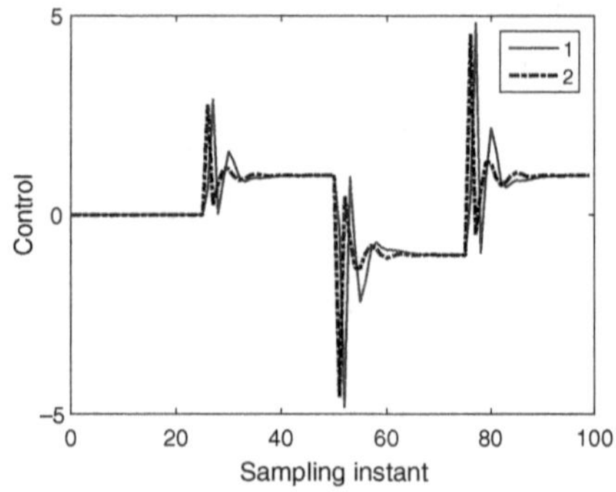

(b) Control signal

Figure 6.8 Closed-loop response to disturbance signal (Example 6.5). Key: line (1) output from minimal realization; line (2) output from non-minimal realization.

quickly to suppress it. Consequently, the amplitude of the output response to the disturbance signal is much smaller, furthermore, the amplitude of the control signal is also smaller.

One might ask what if the closed-loop error system for the observer has all eigenvalues at origin of the complex plane. We can verify that even with this type of observers, the performance of disturbance rejection from the non-minimal state space realization is still much superior than the design from the minimal state space realization. We leave this comparison as an exercise.

One cautious note follows. For complex systems, such as unstable systems with unstable zeros, the dynamic responses for disturbance rejection could be very complicated. The utilization of an observer provides us with an additional means to reach satisfactory control results. We leave this topic as exercises for practice.

6.4.4 Food for Thought

1. In the two-degree-of-freedom control system implementation, if you need to adjust the reference response speed, would you be focusing on the controller design?
2. For a state estimate feedback control system, if we use the same controller and observer, do you think that the one-degree-of-freedom implementation would produce a faster reference response speed than the one produced from the two-degree-of-freedom implementation? Why?
3. A discrete-time system designed to track a step reference signal has the state feedback controller K and the observer gain K_{ob}. Assuming the augmented

system matrices are A, B, and C, what are the key steps for the implementation of the controller?

4. Continue from Problem 3. If the control signal is constrained between u^{min} and u^{max}, how do we impose the constraints?
5. For a first order system with time delay, which has the transfer function:

$$G(z^{-1}) = \frac{bz^{-2}}{z - a},$$

what is the NMSS state space representation for this system? Do you think that the control system based the NMSS model would work for the system with time delay? Why?

6. Continue from Problem 5. If the control system is designed to track a step reference signal, for the NMSS based control system, what would you suggest for an implementation scheme that will include an anti-windup mechanism?

6.5 Repetitive Control

In this section, we will extend the results from the current chapter and the previous chapter to cover the topics of reference tracking and disturbance rejection of complex periodic exogenous signals. This type of control systems is called repetitive control systems.

6.5.1 Basic Ideas about Repetitive Control

Repetitive control systems are the classical approach to control systems with periodic exogenous signals. Originated from the seminal papers by Inoue et al. [1981], Nalano and Hara [1986], and Hara et al. [1988], in the continuous time, the repetitive control systems embed the Laplace transfer function model $\frac{1}{1-e^{-Ts}}$ into the controller structure for asymptotic tracking of a periodic signal with period T. The transfer function $\frac{1}{1-e^{-Ts}}$ is regarded as the signal generator for a periodic signal with known period T. As pointed out in Hara et al. [1988], this irrational transfer function factor has an infinite number of equally spaced poles on the imaginary axis at $0, \pm j\frac{2\pi}{T}, \pm j\frac{4\pi}{T}, \dots, \pm j\frac{k2\pi}{T}, \dots$. Hence according to the internal model principle (Francis and Wonham [1976]), the desired properties for reference tracking and disturbance rejection of the signal with the same period T can be achieved in a stable closed-loop control system.

Even though the repetitive control systems were derived in the continuous-time domain, it is the discrete-time version that has offered advantageous platforms for digital implementations (Longman [2000]). The signal generator in discrete-time for a periodic signal is $\frac{1}{1-z^{-N}}$, where N is the number of samples within a period. For the discrete-time case, the irrational transfer function e^{-Ts} is replaced by z^{-N} where $N = \frac{T}{\Delta t}$ and Δt is the sampling interval. It is interesting to note that in the frequency domain, defining the fundamental frequency $\omega_d = \frac{2\pi}{N}$, the magnitude of the signal generator $|\frac{1}{1-e^{-jN\omega}}| \rightarrow \infty$ at $\omega = 0, \omega = \pm\omega_d, \pm 2\omega_d, \ldots, \pm\frac{N-1}{2}\omega_d$ assuming that N is an odd number.

The novelty of using the factor $\frac{1}{1-e^{-Ts}}$ or $\frac{1}{1-z^{-N}}$ for repetitive control system lies in their easy realizations through a positive feedback configuration (Hara et al. [1988]). However, these factors are ideal in theory. In practice, due to the difficulties in finding the stabilizing controllers, a cut-off filter, $q(s)$ in continuous-time or $Q(z)$ in discrete-time, was used, leading to the modified repetitive control systems with the transfer functions expressed as $\frac{1}{1-q(s)e^{-Ts}}$ in the continuous-time or $\frac{1}{1-Q(z)z^{-N}}$ in the discrete-time. Figures 6.9 and 6.10 show the essential controller structures for both continuous-time and discrete-time repetitive control systems. In both control system configurations, after incorporating the signal generators, stabilization controllers $C(s)$ or $C(z)$ are required to produce stable closed-loop systems. From a historical perspective, at the time when the computational cost was expensive,

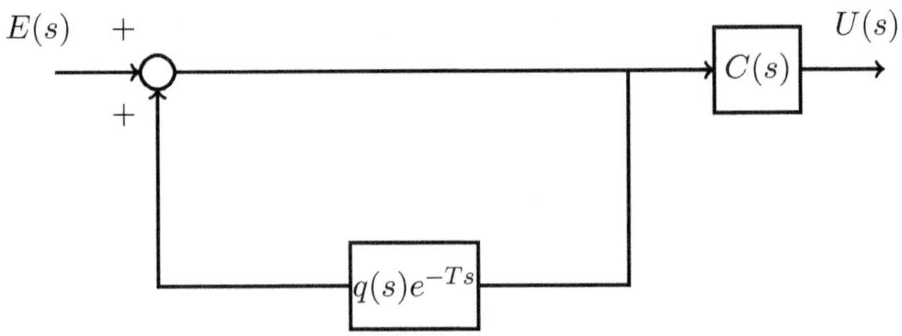

Figure 6.9 Realization of continuous-time repetitive control system.

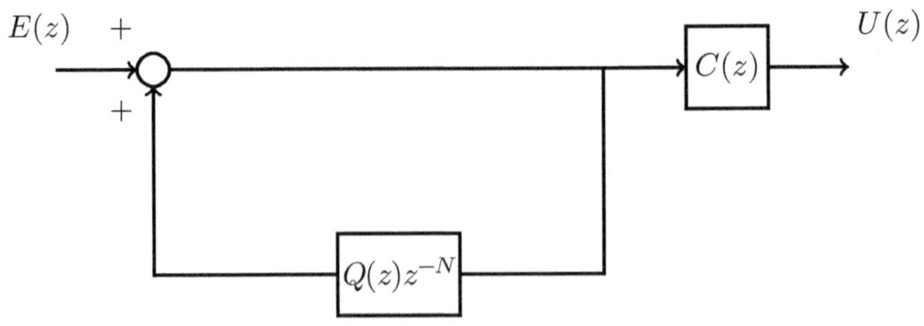

Figure 6.10 Realization of discrete-time repetitive control system.

an easy implementation of repetitive control system paved the way for its wide applications.

There are three aspects of the repetitive control systems that could be improved as outlined below.

1. The repetitive controller order is too high (at least Nth order due to the embedded signal generator) causing complications in the implementations due to noise and model uncertainty.
2. Because the repetitive controllers have unstable poles embedded in their structures, when the control signals are saturated, anti-windup mechanisms are required. This aspect is particularly important for practical applications as the control signal amplitude constraints are necessary for protecting the equipment.
3. The transfer function based design was convenient for frequency response analysis, but it is difficult to extend the design methodologies to a multi-input and multi-output system.

The first two issues imply that a repetitive controller may work perfectly for a design model, however, it may fail to produce a stable closed-loop system for the actual plant when it is implemented. The third issue limits the applications of repetitive control systems to the systems with single-input and single-output.

Can the control system design methods in Chapter 5 and this chapter be naturally extended to the design and implementation of repetitive control systems? Can this extension overcome the shortcomings of the classical repetitive control systems as listed above? Clearly the key to the answers of these questions lies in the determination of a disturbance model $D(q^{-1})$ with a limited complexity for the reason that the control system design will follow naturally for a given $D(q^{-1})$. How to find a disturbance model $D(q^{-1})$ with a limited complexity is presented in the next section.

6.5.2 Determining the Disturbance Model $D(z)$

The question is how we identify the frequency characteristics of a given periodic reference signal. We will use the Fourier analysis for this application.

Assume that a discrete periodic reference signal $r(k)$ is given with N number of samples within one period. Supposing that the fundamental frequency is expressed as $\omega_d = \frac{2\pi}{N}$, using Fourier analysis, this discrete periodic signal within one period can be uniquely represented by the inverse discrete Fourier transform (IDFT) as

$$r(k) = \frac{1}{N} \sum_{l=-\frac{N-1}{2}}^{\frac{N-1}{2}} H(e^{jl\omega_d})e^{jl\omega_d k} \tag{6.37}$$

where $H(e^{jl\omega_d})$ is the lth frequency component of the periodic reference signal $r(k)$, $l = 0, \pm 1, \pm 2, \dots, \pm \frac{N-1}{2}$, which can be calculated using Fourier analysis. N is assumed to be an odd integer in order to capture the zero frequency component. Furthermore, the z-transform of the periodic signal $r(k)$ is defined by

$$R_m(z) = \sum_{k=0}^{N-1} r(k) z^{-k}. \tag{6.38}$$

By substituting (6.37) into (6.38) and interchanging the summation, it was shown in Wang and Cluett [2000] that

$$R_m(z) = \sum_{l=-\frac{N-1}{2}}^{\frac{N-1}{2}} H(e^{jl\omega_d}) \frac{1}{N} \frac{1 - z^{-N}}{1 - e^{jl\omega_d} z^{-1}} \tag{6.39}$$

$R_m(z)$ is the frequency sampling filter (FSF) model (Bitmead and Anderson [1981], Wang and Cluett [2000]). Note that with this expression, the z-transfer function model for a periodic signal has N poles on the unit circle of the complex plane, which are e^{j0}, $e^{\pm j\omega_d}$, $e^{\pm j2\omega_d}$, …, $e^{\pm j\frac{N-1}{2}\omega_d}$. These poles are equally spaced on the unit circle.

The FSF model gives a parametric description of a periodic signal using its Fourier coefficients. There are two steps related to transform this full order model to an approximate model so that it can be used in the design of a repetitive control system. First, for the majority of the applications, when the reference signal $r(k)$ is transformed into the frequency domain, the magnitude of the frequency components, $|H(e^{jl\omega_d})|$, $l = 0, 1, 2, \dots, N-1$, decreases rapidly as l increases. Examination of the magnitudes of $|H(e^{jl\omega_d})|$ with respect to the parameter l will reveal the significant frequencies contained in a given reference signal $r(k)$. By removing the insignificant frequency coefficients from the frequency sampling filter model (6.39), which are set to be zero, the number of terms is reduced. Because the poles and the coefficients of the FSF model in (6.39) are complex parameters, they need to be changed into the equivalent expression with real parameters.

Assuming that there are $1 + \frac{n-1}{2}$ significant frequency components in a periodic signal, where n is an odd number, by taking the complex conjugate nature of the model into consideration, the z-transform of $R_m(z)$ is approximated as:

$$R_m(z) \approx R_a(z) = \frac{1}{N} \frac{1 - z^{-N}}{1 - z^{-1}} H(e^{j0}) + \sum_{l=1}^{\frac{n-1}{2}} \left[\mathrm{Re}(H(e^{jl\omega_d})) F_R^l(z) + \mathrm{Im}(H(e^{jl\omega_d})) F_I^l(z) \right] \tag{6.40}$$

where Re(.) and Im(.) denote the real and imaginary parts of their complex number argument, respectively; $F^l_R(z)$ and $F^l_I(z)$ are the second order filters, defined as

$$F^l_R(z) = \frac{1}{N}\frac{2(1-\cos(l\omega_d)z^{-1})(1-z^{-N})}{1-2\cos(l\omega_d)z^{-1}+z^{-2}} \tag{6.41}$$

$$F^l_I(z) = \frac{1}{N}\frac{2\sin(l\omega_d)z^{-1}(1-z^{-N})}{1-2\cos(l\omega_d)z^{-1}+z^{-2}}. \tag{6.42}$$

The common denominator of the frequency sampling filter model is

$$(1-z^{-1})\Pi_{l=1}^{\frac{n-1}{2}}(1-2\cos(l\omega_d)z^{-1}+z^{-2}).$$

This is the discrete-time version of the disturbance model.

The following example is to illustrate how to determine the number of frequencies required for a repetitive control system.

Example 6.6

A robot arm is undertaking a 'pick and place' task in a horizontal plane using two joints, as shown in Figure 6.11. Its end-effector travels between the pick and place locations in a straight line using joint reference trajectories which minimize the end-effector acceleration. Having reached the place location, the robot then returns back to the starting location. The reference trajectory for the anthropomorphic robot arm is shown in a two dimensional diagram as illustrated in Figure 6.12(a) as the forward journey and Figure 6.13(b) as the return journey. The trajectories on the horizontal plane are decomposed into the desired trajectory for the x-axis (see Figure 6.13(a)) and for the y-axis (see Figure 6.13(b)). Both reference signals have the period of 401 samples for completing the task of pick and place. The repetitive control system with two position outputs (x position variable and y position variable) is designed to precisely follow the desired trajectories in the x and y axes. In this

Figure 6.11 Picture of the robot arm showing pick and place locations. This robot arm was located at the University of Southampton, UK.

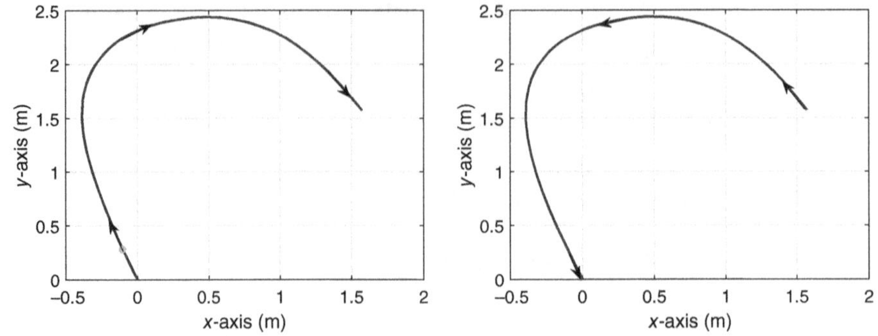

(a) Forward trajectory (b) Return trajectory

Figure 6.12 The trajectory for a robotic arm for a complete period (Example 6.6).

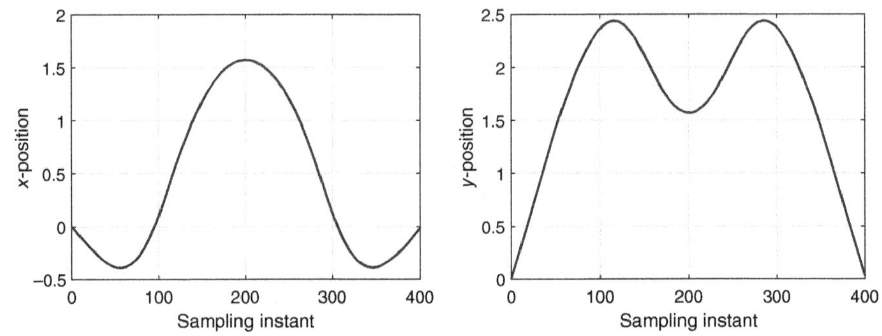

(a) x-axis trajectory (b) y-axis trajectory

Figure 6.13 The trajectories for x and y axes (Example 6.6).

example, we will find the dominant frequencies for both reference signals using Discrete Fourier analysis and analyze their effects on the re-construction of the reference signals.

Solution

We will use the MATLAB function **fft.m** *to calculate the discrete Fourier transform, and scale the Fourier coefficients against its maximum amplitude.*

For the reference signal $r_1(k)$ to the x-axis, the maximum value of $|H(e^{jl\omega_d})|$ occurs at $l = 1$. Thus, the normalized Fourier coefficients are calculated as,

$$|H(e^{jl\omega_d})/H(e^{j\omega_d})|$$

where $\omega_d = \frac{2\pi}{401}$. Figure 6.14(a) shows the magnitude of 11 Fourier coefficients. The rest of the coefficients are almost zero. From Figure 6.14(a) it is clearly seen that with a threshold 0.2, there are five Fourier coefficients selected, which then gives the following disturbance model:

$$D_1(z) = (1 - z^{-1})(1 - 2\cos(\omega_d)z^{-1} + z^{-2})(1 - 2\cos(2\omega_d)z^{-1} + z^{-2}). \quad (6.43)$$

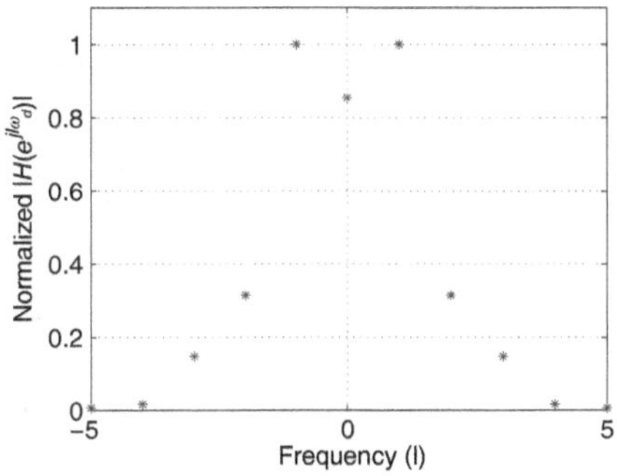

(a) Normalized Fourier coefficients of $r_1(k)$

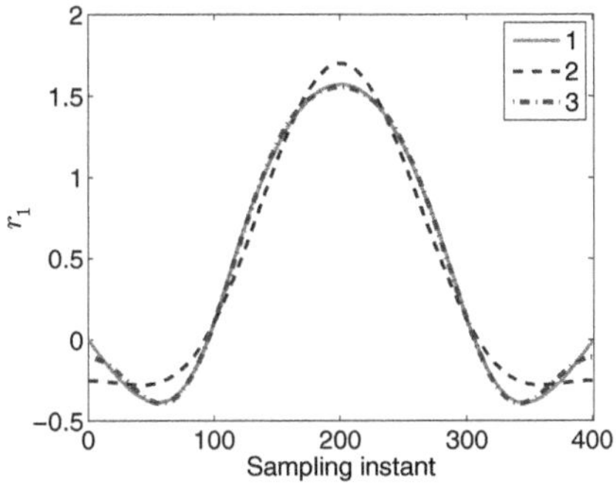

(b) Reconstructed reference signal $r_1(k)$

Figure 6.14 Reconstructed reference signal of $r_1(k)$ (Example 6.6). Key: line (1) $r_1(k)$; line (2) reconstructed signal using the five Fourier coefficients; line (3) reconstructed signal using the seven Fourier coefficients.

By decreasing the threshold to 0.1 *(see Figure 6.14(a)), two more Fourier coefficients are selected, leading to the disturbance model:*

$$\overline{D}_1(z) = (1 - z^{-1})(1 - 2\cos(\omega_d)z^{-1} + z^{-2})(1 - 2\cos(2\omega_d)z^{-1} + z^{-2})$$
$$\times (1 - 2\cos(3\omega_d)z^{-1} + z^{-2}). \tag{6.44}$$

With the MATLAB function **ifft.m**, *we reconstruct the time domain signal by neglecting the Fourier coefficients below the threshold. The reconstructed reference signals using the five and seven Fourier coefficients are compared with the original reference signal $r_1(k)$ in Figure 6.14(b). It is seen that the reconstructed reference signal using the five Fourier coefficients has already captured the general shape of the original reference signal $r_1(k)$.*

The MATLAB program for finding the Fourier coefficients and reconstructing the time-domain signal is given as below. Here, we assume that the reference signal sp is available and predefined, and calculate its Fourier coefficients followed by normalization. The Fourier coefficients are taken to be zero if they are below the threshold of 0.1.

```
N=401;
Fsp=fft(sp,N);
Sca=max(abs(Fsp));
Fsp0=Fsp/Sca;
for kk=1:N
if abs(Fsp0(kk))<0.1;
    Fsp(kk)=0;
end
end
```

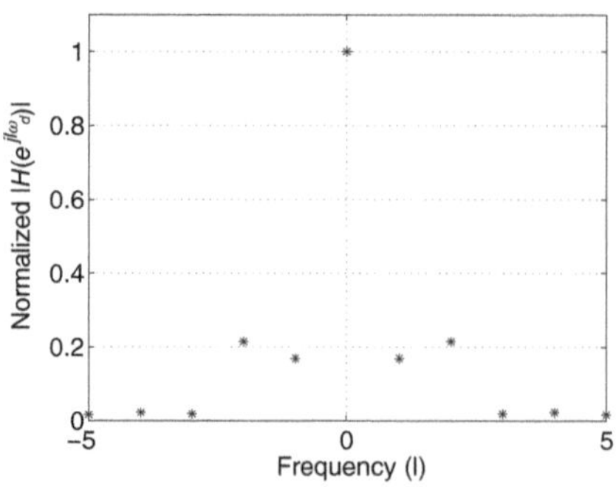

(a) Normalized Fourier coefficients of $r_2(k)$

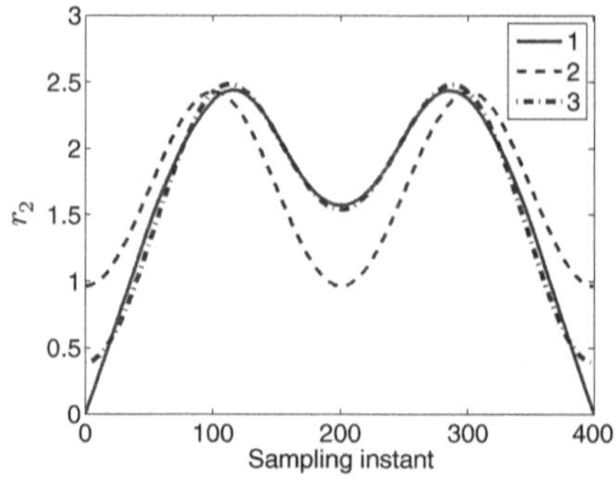

(b) Reconstructed reference signal $r_2(k)$

Figure 6.15 Reconstructed reference signal of $r_2(k)$ (Example 6.6). Key: line (1) $r_2(k)$; line (2) reconstructed signal using the three Fourier coefficients; line (3) reconstructed signal using the five Fourier coefficients.

With the inverse Fourier transform, we reconstruct a time domain signal using the dominant frequencies from the code below.

```
spA=ifft(Fsp,N);
```

The signal spA is the approximated time domain signal in samples as shown in Figure 6.14(b).

For the second reference signal $r_2(k)$ (see Figure 6.15(b)), the maximum value of $|H(e^{jl\omega_d})|$ occurs at $l = 0$. It is seen from Figure 6.15(a) that there are three Fourier coefficients above the threshold 0.2 and five coefficients above the threshold 0.1. Therefore, with the three Fourier coefficients, the disturbance model $D_2(z^{-1})$ is selected as

$$D_2(z) = (1 - z^{-1})(1 - 2\cos(2\omega_d)z^{-1} + z^{-2}) \tag{6.45}$$

and with the five Fourier coefficients,

$$\overline{D}_2(z) = (1 - z^{-1})(1 - 2\cos(\omega_d)z^{-1} + z^{-2})(1 - 2\cos(2\omega_d)z^{-1} + z^{-2}) \tag{6.46}$$

Figure 6.15(b) illustrates the original reference signal $r_2(k)$ and the reconstructed signals using the three and five Fourier coefficients. It is seen that the reconstructed signal using five Fourier coefficients has captured the general shape of the reference signal $r_2(k)$.

6.5.3 Robotic Arm Control

This case study continues from Example 6.6 for a robot arm. The robotic arm has been used in experimental tests previously in Wang et al. [2013].

The overall system model has been identified by combining experimentally-derived models of its constituent components as described by the transfer-function matrix:

$$\begin{bmatrix} y_1 \\ y_2 \end{bmatrix} = \begin{bmatrix} \frac{B_{11}(s)}{A_{11}(s)} & \frac{B_{12}(s)}{A_{12}(s)} \\ \frac{B_{21}(s)}{A_{21}(s)} & \frac{B_{22}(s)}{A_{22}(s)} \end{bmatrix} \begin{bmatrix} u_1 \\ u_2 \end{bmatrix}, \tag{6.47}$$

where the sampling interval is 0.05 sec and the transfer function elements are

$$A_{11}(s) = 5.25e - 5s^{12} + 0.01463s^{11} + 0.91s^{10} + 31.2s^9 + 714.1s^8 + 1.19e4s^7$$
$$+ 1.45e5s^6 + 1.4e6s^5 + 1.01e7s^4 + 5.7e7s^3 + 2.3e8s^2 + 5.9e8s + 7.6e8$$
$$B_{11}(s) = 0.16s^9 + 14.51s^8 + 578.2s^7 + 1.392e4s^6 + 2.26e5s^5 + 2.58e6s^4$$
$$+ 2.09e7s^3 + 1.17e8s^2 + 4.21e8s + 7.6e8$$
$$A_{12}(s) = 5.25e - 5s^10 + 0.014s^9 + 0.72s^8 + 20s^7$$
$$+ 363s^6 + 4645s^5 + 4.3e4s^4 + 2.9e005s^3 + 1.4e6s^2 + 4.18e6s + 6.323e6$$
$$B_{12}(s) = -0.022s^7 - 3.24s^6 - 88.3s^5 - 1347s^4 - 1.06e4s^3 - 4.52e4s^2$$
$$A_{21}(s) = 5.25e - 5s^{10} + 0.014s^9 + 0.67s^8 + 17.9s^7$$
$$+ 316s^6 + 3963s^5 + 3.6e4s^4 + 2.42e5s^3 + 1.1e6s^2 + 3.5e6s + 5.3e6$$
$$B_{21} = -0.16s^7 - 8.7s^6 - 194s^5 - 2498s^4 - 1.78e4s^3 - 6.64e4s^2$$
$$A_{22}(s) = 5.25e - 5s^{12} + 0.014s^{11} + 0.9s^{10} + 31s^9 + 714.1s^8 + 1.19e4s^7$$
$$+ 1.48e5s^6 + 1.4e6s^5 + 1.04e7s^4 + 5.7e7s^3 + 2.3e8s^2 + 5.9e8s + 7.6e8$$
$$B_{22}(s) = 0.027s^9 + 4.95s^8 + 264s^7 + 7394s^6 + 1.3e5s^5$$
$$+ 1.69e6s^4 + 1.5e7s^3 + 9.4e7s^2 + 3.8e8s + 7.6e8$$

In the design of repetitive control system, MATLAB functions are used to convert the transfer function model into a continuous-time state space model followed by its discretization with the sampling interval $\Delta t = 0.05$ (sec). Both repetitive controller and the observer are designed using MATLAB **dlqr.m** function. For the controller design, the diagonal elements in Q matrix are selected to be 1, if these elements correspond to the output components in state variable $x(k)$. The R matrix is chosen to be $R = 0.5I$ where I is an identity matrix. All other elements in Q are set to zero. With the choices of Q and R matrices, the largest magnitude of the closed-loop eigenvalues is about 0.93, which means that the closed-loop repetitive control system is stable. For the observer design, Q_{ob} is selected to be an identity matrix and $R_{ob} = 0.1I$. The largest magnitude of the closed-loop eigenvalues for the observer error system is 0.92, which means that the observer error system is also stable. The closed-loop eigenvalues for the entire repetitive control system consist of the eigenvalues of closed-loop control system and those of the observer error system.

In the simulation studies, there are measurement noises added to both output signals. The noise signals have zero mean with $\sigma = 0.001$. There are no control amplitude constraints because the control signals naturally satisfy all the constraints.

Case A. In the first case, we will use the disturbance model given by (6.43), which is

$$D(z^{-1}) = 1 - 4.9988z^{-1} + 9.9963z^{-2} - 9.9963z^{-3} + 4.9988z^{-4} - z^{-5}.$$

Figures 6.16(a)–(c) show the closed-loop output responses, the error signals and the control signals. The mean squared errors between the reference signals and the output signals are calculated as

$$\frac{1}{M}\sum_{k=1}^{M} e_1(k)^2 = 2.775 \times 10^{-4};\ \frac{1}{M}\sum_{k=1}^{M} e_2(k)^2 = 7.776 \times 10^{-4}$$

where $e_1(k) = r_1(k) - y_1(k)$, $e_2(k) = r_2(k) - y_2(k)$, M is the data length.

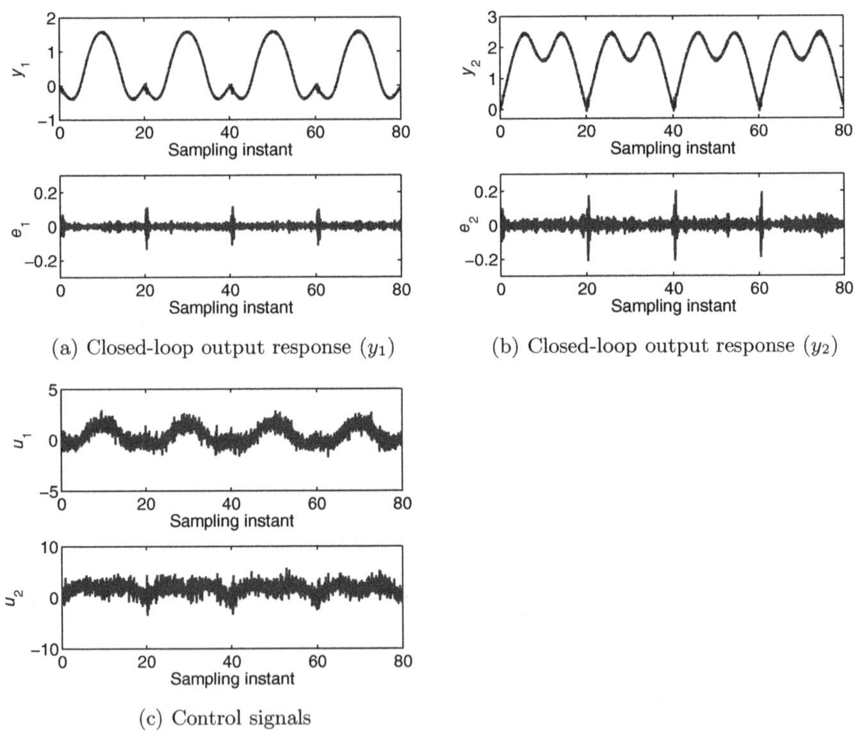

(a) Closed-loop output response (y_1)

(b) Closed-loop output response (y_2)

(c) Control signals

Figure 6.16 Reference tracking using repetitive controller (Case A).

Case B. In the second case, we will use the disturbance model given by (6.45) that includes three Fourier coefficients in the modelling of the reference signals. In this case, the disturbance model is third order given by

$$D(z^{-1}) = 1 - 2.9998z^{-1} + 2.9998z^{-2} - z^{-3}.$$

Figures 6.17(a)–(c) show the closed-loop output responses, the error signals and the control signals. The mean squared errors between the reference signals and the outputs are calculated, which are

$$\frac{1}{M}\sum_{k=1}^{M} e_1(k)^2 = 0.513 \times 10^{-4}; \quad \frac{1}{M}\sum_{k=1}^{M} e_2(k)^2 = 1.807 \times 10^{-4}$$

It is interesting to note that the mean squared errors calculated from **Case B** are actually about 5.4 ($e_1(k)$) and 4.3 ($e_2(k)$) times smaller than those calculated in **Case A**. One may wonder what causes the closed-loop control performance deterioration when the disturbance model is chosen to be a higher order. This performance deterioration is caused by the existence of measurement

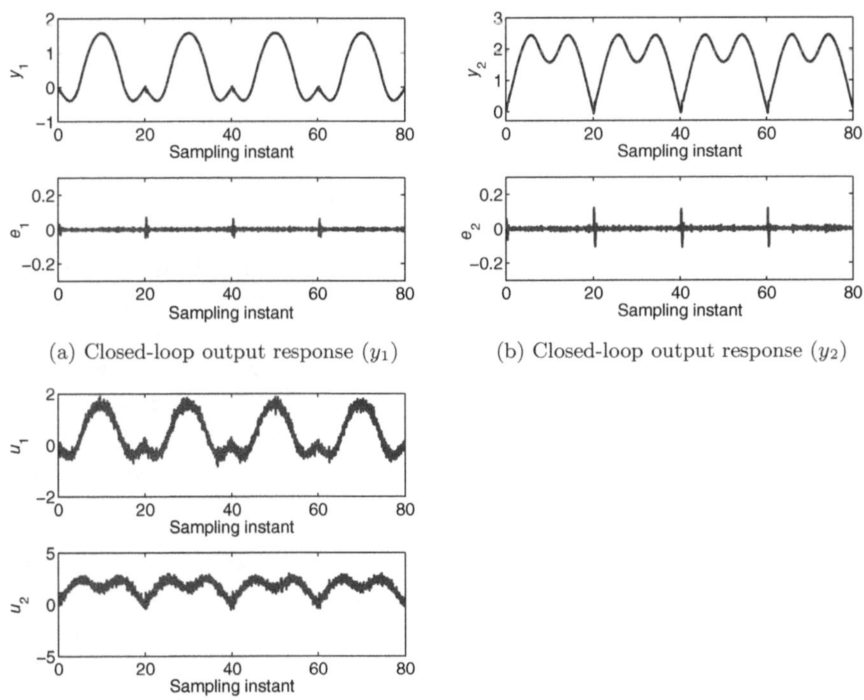

(a) Closed-loop output response (y_1)

(b) Closed-loop output response (y_2)

(c) Control signals

Figure 6.17 Reference tracking using repetitive controller (Case B).

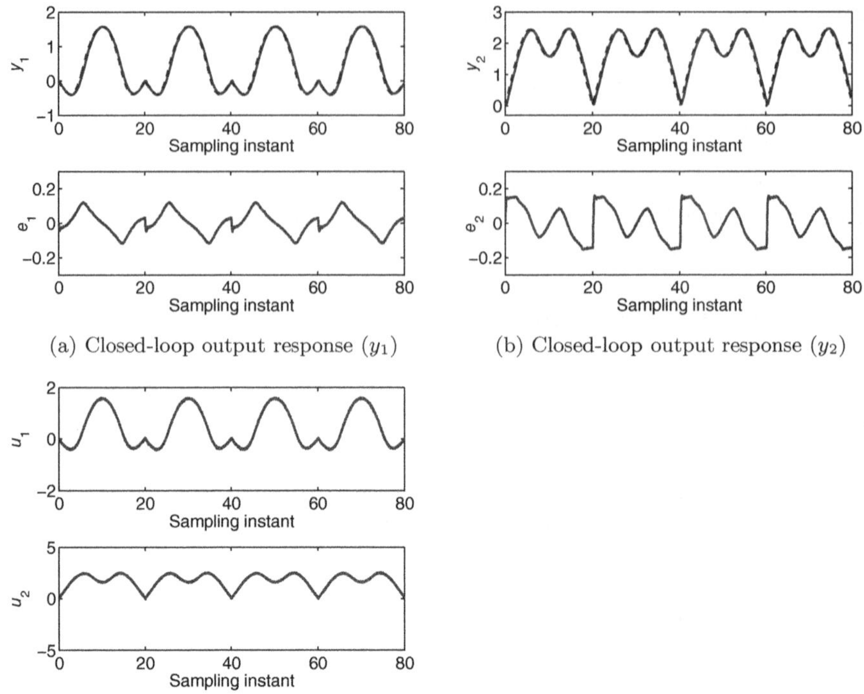

(a) Closed-loop output response (y_1)

(b) Closed-loop output response (y_2)

(c) Control signals

Figure 6.18 Reference tracking using repetitive controller (Case C).

noise in the output signals. In the same principle to the classical repetitive control systems (Inoue et al. [1981], Nalano and Hara [1986], and Hara et al. [1988]), the proposed repetitive controller has an infinite controller gain at the frequencies corresponding to the zeros of the disturbance model $D(z^{-1})$. For **Case A**, the infinite controller gains occur at $\omega = 0, \pm\omega_d, \pm 2\omega_d$. However, for **Case B**, the infinite controller gains only occur at $\omega = 0, \pm\omega_d$. Thus, the amplification of measurement noise is worse when the repetitive controller is designed using the higher order disturbance model (6.43) as in **Case A**.

Case C. It is interesting to investigate what happens if we further simplify the structure of the disturbance model $D(z^{-1})$ by including the first term only, which is

$$D(z^{-1}) = 1 - z^{-1}.$$

Figures 6.18(a)–(c) show the closed-loop output responses, the error signals and the control signals. It is clearly seen from these figures that the closed-loop control performance for reference tracking is poor and there are steady-state errors in tracking the reference signals. We have also calculated the mean squared errors for this case, which are

$$\frac{1}{M}\sum_{k=1}^{M} e_1(k)^2 = 0.0038; \; \frac{1}{M}\sum_{k=1}^{M} e_2(k)^2 = 0.0097.$$

The mean squared errors are approximately 74 and 54 times larger than those calculated in **Case B**. This investigation confirms the importance of using the correct disturbance model for the repetitive control system design.

6.5.4 Food for Thought

1. If the reference signal is a ramp signal, what is the disturbance model $D(q^{-1})$?
2. If the measurement noise is severe, would you reduce or increase the number of higher frequency components in the disturbance model $D(q^{-1})$ for the repetitive control system?
3. If the plant model is not accurate, would you reduce or increase the number of higher frequency components in the disturbance model $D(q^{-1})$?

6.6 Summary

This chapter has discussed design, simulation and implementation of state feedback control for disturbance rejection and reference tracking where the disturbance model is embedded into the state feedback controller. The essential steps in the design are: (i) to capture the characteristics of the disturbance signal with a disturbance model; (ii) to augment the plant model (A_m, B_m, C_m) with the disturbance model to obtain the augmented model (A, B, C) where the input to the augmented model is the filtered control signal $u_s(k)$; (iii) to design the state feedback controller using the augmented model (A, B); (iv) to design an observer to estimate the filtered state vector $x_s(k)$ using the plant model (A_m, B_m) if the state vector $x(k)$ is not measured.

The other important aspects of this chapter are summarized as follows.

1. The choice of a disturbance model is dependent on control system specifications of reference tracking and disturbance rejection. A common choice is to select the disturbance model $D(q^{-1}) = 1 - q^{-1}$ to ensure that the state feedback control system has integral action. The control system with integral action will follow a step reference signal and reject a step disturbance or a low frequency disturbance.
2. Anti-windup mechanism for the state feedback controller is introduced, and the implementation is straightforward.
3. One can choose to implement the state feedback control system with either one-degree-of-freedom or two-degree-of-freedom, dependent on how the

reference signal enters the controller implementation. The two-degree-of-freedom control system realization reduces overshoot in the output response to a reference signal.
4. The approach by embedding the controller with a disturbance model is particularly simple when the state variables are measured because it does not require an observer. In particular, when the discrete-time systems are represented by transfer function models, the state variables can be chosen to correspond to the plant input and output variables, which are measured. As a result, a state feedback control system can be designed and implemented without an observer.
5. The control systems for reference tracking and disturbance rejection presented in Chapter 5 and this chapter are extended to the repetitive control systems where the disturbance model $D(q^{-1})$ is selected according to the periodic reference or disturbance signal.

6.7 Further Reading

1. Repetitive predictive control system with application to a robotic arm is discussed in Wang et al. [2011a], Wang et al. [2013], Wang et al. [2011b]. A paper in tutorial style is presented in Wang [2016b].
2. When using a non-minimal state space realization, the state variables are chosen to correspond to plant input and output variables, observers are not needed for the state space design methods. Controller design methods with non-minimal state space realization were proposed in Wang and Young [1988], Taylor et al. [2000b], Wang et al. [2009] and a book Taylor et al. [2013]. The non-minimal state space realization was also used in the design of model predictive controllers (Wang [2009]).
3. An approach to harmonic disturbance rejection is presented in Goodwin et al. [1986]
4. Books for repetitive control and iterative learning control include Rogers et al. [2007], Moore [2012], Owens [2015].

Problems

Problem 6.1 Design state feedback controller K for the system that is described by the state space model:

$$x_m(k+1) = A_m x_m(k) + B_m u(k);\; y(k) = C_m x_m(k),$$

where we assume that the state variables are measured.

1. The control system is to reject a constant disturbance and track a step reference signal. The system matrices are

$$A_m = \begin{bmatrix} 0.9 & 1 \\ 0 & 0.8 \end{bmatrix};\ B_m = \begin{bmatrix} 1 \\ 1 \end{bmatrix};\ C_m = \begin{bmatrix} 1 & 2 \end{bmatrix}.$$

   With the desired closed-loop eigenvalues chosen to be 0.7, 0.71 and 0.72, find the state feedback controller K using MATLAB program **place.m**.
2. The control system is to reject a sinusoidal disturbance with the frequency $\omega_d = 2\pi f \Delta t$ where f is 1 Hz and $\Delta t = 0.1$ second while tracking a step reference signal. The system matrices are

$$A_m = \begin{bmatrix} 0.8 & 0 \\ -0.6 & 0.6 \end{bmatrix};\ B_m = \begin{bmatrix} 1 & 0 \\ 0 & 0.5 \end{bmatrix};\ C_m = \begin{bmatrix} 1 & 1 \\ 0 & 1 \end{bmatrix}.$$

   Use MATLAB program **dlqr.m** to find the state feedback controller K where both Q and R are identity matrices with compatible dimensions.

Problem 6.2 Design a state feedback controller K for a discrete-time system with actuator time-delay, which is described by the state space model:

$$x_m(k+1) = A_m x_m(k) + B_m u(k - n_d);\ y(k) = C_m x_m(k),$$

where the state vector $x_m(k)$ is measured.

1. The time-delay n_d is 3 samples and the system matrices are

$$A_m = \begin{bmatrix} 1 & 1 \\ 0 & 1 \end{bmatrix};\ B_m = \begin{bmatrix} 0.5 \\ 1 \end{bmatrix};C_m = \begin{bmatrix} 1 & 1 \end{bmatrix}.$$

   The control system is to reject a random walk type of disturbance while tracking a step reference signal. Use MATLAB program **dlqr.m** to find the state feedback controller K where the weighting matrices are chosen to be $Q = C^T C$, $R = 0.1I$ and C being the output matrix from the augmented state space model. Where are the closed-loop eigenvalues?
2. The time-delay n_d is 2 samples and the system matrices are:

$$A_m = \begin{bmatrix} 0.9 & -1 \\ 1 & 0.8 \end{bmatrix};\ B_m = \begin{bmatrix} 1 & 0 \\ 0 & 1 \end{bmatrix};\ C_m = \begin{bmatrix} 1 & 0 \\ 1 & 1 \end{bmatrix}.$$

   The control system is to track a sinusoidal reference signal with frequency $\omega_d = 2\pi f \Delta t$ ($\Delta t = 1$(sec), $f = 2$ Hz) for output $y_1(k)$, and a ramp reference signal for output $y_2(k)$. Use MATLAB program **dlqr.m** to find the state feedback controller K where both Q and R are identity matrices with compatible dimensions. Where are the closed-loop eigenvalues?

Problem 6.3 This exercise is to design a state feedback control system for a hot air balloon using the technique embedding the disturbance model into the controller. One can compare this study with the design and simulation results obtained from solving Problem 5.3 in Chapter 5. The detailed description of the physical system can be found in Problem 5.3. For convenience, the state space model to describe the dynamics of a hot air balloon is given as

$$\begin{bmatrix} \dot{T}(t) \\ \dot{v}(t) \\ \dot{h}(t) \end{bmatrix} = \begin{bmatrix} -\frac{1}{\tau_1} & 0 & 0 \\ \beta & -\frac{1}{\tau_2} & 0 \\ 0 & 1 & 0 \end{bmatrix} \begin{bmatrix} T(t) \\ v(t) \\ h(t) \end{bmatrix} + \begin{bmatrix} \alpha \\ 0 \\ 0 \end{bmatrix} u(t) + \begin{bmatrix} 0 \\ \frac{1}{\tau_2} \\ 0 \end{bmatrix} w(t) \tag{6.48}$$

where the state variables $T(t)$, $v(t)$ and $h(t)$ are the temperature change of air in the balloon, the vertical velocity of the balloon, and the change in altitude from steady-state altitude respectively. The input variable $u(t)$ is the manipulated variable and $w(t)$ is the vertical wind velocity. The output is the altitude $h(t)$, which is measured. The normalized constants in the model are $\tau_1 = 1$, $\tau_2 = 10$, $\alpha = 1$ and $\beta = 0.01$. The system is sampled with sampling interval $\Delta t = 0.1$ (sec). Design a state estimate feedback control system that will reject the wind disturbance predominately in low frequency and will track a step reference signal for the altitude.

1. Calculate the state feedback controller gain K with closed-loop poles positioned around $e^{-0.1\Delta t}$.
2. Calculate the observer gain K_{ob} with closed-loop poles positioned around $e^{-\Delta t}$.
3. Generate a wind disturbance using the following code where *Nsim* is the number of samples for the simulation.

```
Nsim=4000;
dis=randn(1,Nsim);
dis=filter(0.01,[1 -0.99],dis);
dis=dis+2*[zeros(1,Nsim/2) ones(1,Nsim/2)];
```

4. Simulate the closed-loop response with the reference signal generated by the following code:

```
r=[ones(1,Nsim/2) zeros(1, Nsim/2)];
```

   Add the wind disturbance signal to the simulation. Present the closed-loop response of the altitude and the control signal. What are your observations?
5. Compare the simulation results with those obtained from disturbance observer based approach in Problem 5.3. What are your observations?

Problem 6.4 Continue from Problem 6.3. We will investigate the implementation of anti-windup mechanism with respect to constraints on the control signal.

1. Determine the minimum and maximum control amplitude from Problem 6.3, which are denoted as u^{min} and u^{max}. Assume that the control signal is constrained between $0.8u^{min}$ and $0.8u^{max}$. Explore the integrator windup scenario by using the constrained control signal for the plant simulation, however using the unconstrained control signal in the observer implementation. Present the closed-loop response of the altitude and the control signal. What are your observations?
2. Implement the control signal constraints with the anti-windup mechanism. Present the closed-loop response of the altitude and the control signal.
3. Compare the constrained control results with those obtained from disturbance observer based approach in Problem 5.3. What are your observations?

Problem 6.5 Continue from Problem 6.4. We will investigate the effect of time delay from the sensor and the actuator on the closed-loop control performance, where the measurement for the altitude has $2\Delta t$ seconds of delay and the actuator has $5\Delta t$ seconds of delay. The state space model with the time delays become:

$$\begin{bmatrix} \dot{T}(t) \\ \dot{v}(t) \\ \dot{h}(t) \end{bmatrix} = \begin{bmatrix} -\frac{1}{\tau_1} & 0 & 0 \\ \beta & -\frac{1}{\tau_2} & 0 \\ 0 & 1 & 0 \end{bmatrix} \begin{bmatrix} T(t) \\ v(t) \\ h(t) \end{bmatrix} + \begin{bmatrix} \alpha \\ 0 \\ 0 \end{bmatrix} u(t - 5\Delta t) + \begin{bmatrix} 0 \\ \frac{1}{\tau_2} \\ 0 \end{bmatrix} w(t) \tag{6.49}$$

$$y(t) = h(t - 2\Delta t) \tag{6.50}$$

1. Build a real-time control system in Simulink by following Tutorial 6.3, where the identical controller and observer with the anti-windup mechanism from Problem 6.4 are used. Simulate the closed-loop control system with the plant that has both actuator delay and sensor delay. Plot the closed-loop response of the control signal and output signal. Is the closed-loop system stable? In comparison to the simulation results obtained from Problem 6.4, what are your observations?
2. Consider the measurement delay and the actuator delay in the state estimate feedback control system design by choosing additional state variables to capture the delays. Present the discrete-time state space model.
3. Use MATLAB program **dlqr.m** to find the state feedback controller K where $Q = C^T C$ (C is the output matrix from the augmented model) and $R = 1$. Design the observer K_{ob} where both Q and R are identity matrices with compatible dimensions.
4. Perform closed-loop simulation studies for the time delay system with the identical conditions. Plot the closed-loop response of the control signal and output signal. What are your observations in terms of control performance improvement?

Problem 6.6 A transfer function model was given in Ralhan and Badgwell [2000] to describe the operations of a gas fired heater at a high fuel operation. The system has a substantial time delay because the temperature sensors are located away from the heating source. In Example 5.9 of Chapter 5, a disturbance- observer-based control system was designed for the gas fired heater. For convenience, the discrete-time model with time delay is presented below:

$$\begin{bmatrix} x_0(k+1) \\ y_0(k) \\ y_0(k-1) \\ y_0(k-2) \\ y_0(k-3) \\ y_0(k-4) \end{bmatrix} = \overbrace{\begin{bmatrix} A_0 & 0 & 0 & 0 & 0 & 0 \\ C_0 & 0 & 0 & 0 & 0 & 0 \\ 0 & 1 & 0 & 0 & 0 & 0 \\ 0 & 0 & 1 & 0 & 0 & 0 \\ 0 & 0 & 0 & 1 & 0 & 0 \\ 0 & 0 & 0 & 0 & 1 & 0 \end{bmatrix}}^{A_m} \begin{bmatrix} x_0(k) \\ y_0(k-1) \\ y_0(k-2) \\ y_0(k-3) \\ y_0(k-4) \\ y_0(k-5) \end{bmatrix} + \overbrace{\begin{bmatrix} B_0 \\ 0 \\ 0 \\ 0 \\ 0 \\ 0 \end{bmatrix}}^{B_m} u(k) \tag{6.51}$$

with the measured output $y(k)$ being

$$y(k) = \overbrace{\begin{bmatrix} 0 & 0 & 0 & 0 & 0 & 1 \end{bmatrix}}^{C_m} \begin{bmatrix} x_0(k) \\ y_0(k-1) \\ y_0(k-2) \\ y_0(k-3) \\ y_0(k-4) \\ y_0(k-5) \end{bmatrix} \tag{6.52}$$

where the system matrices A_0, B_0 and C_0 are

$$A_0 = \begin{bmatrix} 0.6550 & -0.0327 \\ 0.8187 & 0.9825 \end{bmatrix}; \; B_0 = \begin{bmatrix} 0.8187 \\ 0.4381 \end{bmatrix}; \; C_0 = \begin{bmatrix} 0 & 0.0400 \end{bmatrix}.$$

1. Design a state estimate feedback control system that will maintain a constant house temperature by embedding an integrator into the controller. Use MATLAB program **dlqr.m** to find the state feedback controller K where $Q = C^T C$ (C is the output matrix from the augmented model) and $R = 1$. Find the observer K_{ob} where both Q and R are identity matrices with compatible dimensions.
2. The sampling interval Δt is one minute and the incremental reference signal is 1°C to start the simulation and increased to 2°C after 50 minutes. All initial conditions are chosen to be zero. Perform closed-loop simulation for the state estimate feedback control of the gas fired heater system. Present the control signal and the output signal.

3. The control signal is constrained above zero and 3 units, which is

$$0 \leq u(k) \leq 3.$$

Simulate the closed-loop control system with the control signal constraints. Present the control signal and the output signal.
4. Comparing the simulation results with those obtained from Example 5.9 of Chapter 5, what are your observations?

Problem 6.7 Continue from Problem 6.6. In this exercise, we wish to impose a constraint on the maximum value of the desired closed-loop time constant for both observer and controller, which is to be achieved through the prescribed degree of stability in discrete-time linear quadratic regulator.

1. For the controller design, the maximum value of the desired closed-loop time constant is chosen to be 5 minutes. With sampling interval $\Delta t = 1$ (min) and the same Q and R matrices as in Problem 6.6, find the state feedback controller K.
2. For the observer design, the maximum value of the desired time constant for the error system is chosen to be 1 minute. With sampling interval $\Delta t = 1$ (min) and the same Q and R matrices as in Problem 6.6, find the state feedback controller K_{ob}.
3. Evaluate the closed-loop control performance with the same simulation conditions as outlined in Problem 6.6 with and without the control signal constraints. Comparing the simulation results with those obtained from Problem 6.6, what are your observations?

Problem 6.8 Consider the control system design for the drive train of a wind turbine. We have studied this system in the continuous-time control system design in Section 2.4 of Chapter 2 (see Figure 2.11) and in Problem 5.5 of Chapter 5 where a disturbance-observer-based control system is designed.
For convenience, defining the angular velocities of the turbine and generator as ω_t and ω_r, and the angular positions as θ_t and θ_r, the drive train of the wind turbine is described by the following differential equations:

$$2H_t \frac{d\omega_t}{dt} = T_t - T_w - K_s(\theta_r - \theta_t) - D_m(\omega_r - \omega_t) \tag{6.53}$$

$$2H_g \frac{d\omega_r}{dt} = -T_e + T_L + K_s(\theta_r - \theta_t) + D_m(\omega_r - \omega_t) \tag{6.54}$$

$$\frac{d\theta_t}{dt} = \omega_t \tag{6.55}$$

$$\frac{d\theta_r}{dt} = \omega_r \tag{6.56}$$

where H_t and H_g are the inertial time constants for the turbine and generator, K_s is the shaft stiffness constant and D_m is the mutual damping coefficient. T_w is the input disturbance generated by wind changes and T_L represents the load disturbance reflecting the demand of power output. For the current study, as in Perdana [2008], the turbine inertial constant $H_t = 2.6$ second, and the generator inertial constant $H_g = 0.22$ second, the stiffness constant $K_s = 141.0$ pu, and mutual damping coefficient $D_m = 3.0$ pu.

The manipulated variables are the turbine torque T_t and the motor torque T_e. In order to optimize the wind energy generation, the angular velocity of the generator requires to be under control. Thus, the control objective is to achieve reference tracking of ω_r while rejecting wind disturbance T_w and load disturbance T_L. To avoid causing mechanical stress to the wind turbine generator, the angular velocity of the turbine ω_t desires to be synchronized with that of the generator. Namely, in the steady-state operation, the control objective is $\omega_t = \omega_r$.

1. Convert the continuous-time state space model into a discrete-time state space model using sampling interval $\Delta t = 0.01$.
2. Assuming that θ_t and θ_r are measured, design a state estimate feedback control system by embedding integrator into the controller. For both controller design and observer design, MATLAB function **dlqr.m** is used with both Q and R being identity matrices with compatible dimensions. What is K? What is K_{ob}?
3. Simulate the closed-loop responses for reference tracking and disturbance rejection with your choice of N_{sim}. For simplicity, all initial conditions are chosen to be zero. The reference signals to θ_t and θ_r are unit ramp signals in discrete-time. The wind disturbance is a random walk disturbance existed for all the simulation time while the load disturbance T_L is a unit step disturbance entering the system at half of the simulation time. Present the control signals and the output signals θ_r and θ_t.

Problem 6.9 A mechanical system is described by the following z-transfer function (see van Donkelaar et al. [1999]):

$$G(z) = \frac{-5.7980z^3 + 19.5128z^2 - 21.6452z + 7.9547}{z^4 - 3.0228z^3 + 3.8630z^2 - 2.6426z + 0.8084}. \tag{6.57}$$

This system has four poles located at

$$0.5162 \pm j0.7372, \quad 0.9952 \pm j0.0877$$

and the three zeros at

$$1.3873, \quad 0.9891 \pm j0.1034.$$

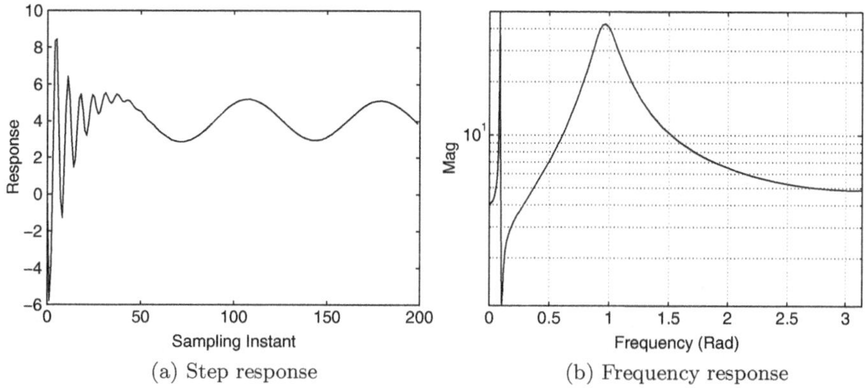

Figure 6.19 Open-loop response of mechanical system.

This system was used in the design of model predictive control in Wang [2009] where the unit step response and the frequency response of this system were shown as in Figure 6.19.

Clearly, this system is highly oscillatory with unstable zeros. Design a state estimate control system that will track a step reference signal and reject a sinusoidal disturbance with discrete-time frequency $\omega_d = 0.1$ (rad).

1. For both controller design and observer design, MATLAB function **dlqr.m** is used with both Q and R being identity matrices with compatible dimensions. What is K? What is K_{ob}?
2. Simulate the closed-loop responses for reference tracking and disturbance rejection with your choice of N_{sim}. For simplicity, all initial conditions are chosen to be zero. The reference signal to the output is a unit step signal while adding a sinusoidal disturbance with frequency ω_d to the input signal. Present the control signal and the output signal.
3. Because the system is highly oscillatory with unstable zeros, we may carefully tune the Q and R matrices in the controller design and observer design or use the prescribed degree of stability to reduce the oscillations in the closed-loop control system. Explore both approaches and compare the closed-loop control results via the simulation studies.

Problem 6.10 Continue from Problem 6.9. We will investigate the implementation of anti-windup mechanism with respect to constraints on the control signal.

1. Determine the minimum and maximum control amplitude from Problem 6.9, which are denoted as u^{min} and u^{max}. Assume that the control signal is constrained between $0.8u^{min}$ and $0.8u^{max}$. Explore the integrator windup scenario by using the constrained control signal for the plant simulation, however using the unconstrained control signal in the observer implementation. Present the closed-loop response of the output and the control signal. What are your observations?
2. Implement the control signal constraints with the anti-windup mechanism. Present the closed-loop response of the output and the control signal. Comparing the simulation results obtained from ones obtained from incorrectly implemented results in the previous step, what are your observations?

Problem 6.11 Extrusion is a continuous process in which a rotating screw is used to force the food material through the barrel of the machine and out through a narrow die opening. In this process the material is simultaneously transported, mixed, shaped, stretched and sheared under elevated temperature and pressure. Figure 6.20 shows the block diagram of a food extruder.

A continuous-time model predictive control was designed for the food extruder with experimental validation in Wang [2009]. Suppose that u_1, u_2, y_1 and y_2 represent screw speed, liquid pump speed, specific mechanical energy (SME) and motor torque, respectively. A continuous-time model for the food extruder was obtained through system identification, which is

$$\begin{bmatrix} y_1 \\ y_2 \end{bmatrix} = \begin{bmatrix} G_{11} & G_{12} \\ G_{21} & G_{22} \end{bmatrix} \begin{bmatrix} u_1 \\ u_2 \end{bmatrix} \tag{6.58}$$

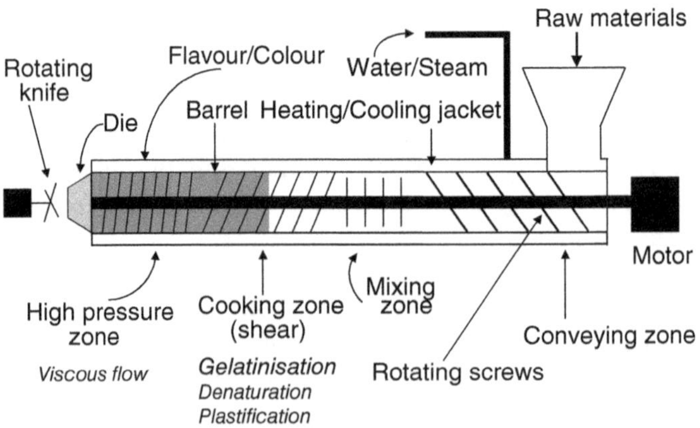

Figure 6.20 Block diagram of a food extruder.

where

$$G_{11} = \frac{0.21048s + 0.00245}{s^3 + 0.302902s^2 + 0.066775s + 0.002186}$$

$$G_{12} = \frac{-0.001313s^2 + 0.000548s - 0.000052}{s^4 + 0.210391s^3 + 0.105228s^2 + 0.00777s + 0.000854}$$

$$G_{21} = \frac{0.000976s - 0.000226}{s^3 + 0.422036s^2 + 0.091833s + 0.003434}$$

$$G_{22} = \frac{-0.00017}{s^2 + 0.060324s + 0.006836}.$$

The control objective is to maintain steady-state operation of both SME and motor torque in the presence of low frequency disturbances where a state estimate feedback control system with integral action is designed.

1. Choosing sampling interval $\Delta t = 3$ sec, obtain the discrete-time food extruder model.
2. Use MATLAB program **dlqr.m** to find the state feedback controller K where $Q = C^T C$ (C is the output matrix from the augmented model) and $R = I$. Find the observer K_{ob} where both Q and R are identity matrices with compatible dimensions.
3. In the simulation studies, we will only consider the incremental changes of the reference signals, meaning that all steady-state conditions are assumed to be zero. Additionally, we assume all zero initial conditions for simplicity. Simulate the closed-loop control system with unit step signal for the SME and zero reference for the motor torque. In the simulation studies, add random walk disturbances to both control signals. Present the control signals and the output signals. What are your observations?

Problem 6.12 The dynamics of a ground vehicle are described by the following differential equations:

$$\dot{x}(t) = v_x(t) \tag{6.59}$$

$$\dot{v}_x(t) = a_x(t) \tag{6.60}$$

$$2\dot{a}_x(t) + a_x(t) = u_x(t) \tag{6.61}$$

$$\dot{y}(t) = v_y(t) \tag{6.62}$$

$$\dot{v}_y(t) = a_y(t) \tag{6.63}$$

$$2\dot{a}_y(t) + a_y(t) = u_y(t) \tag{6.64}$$

The main control objective is for the vehicle to track a given reference trajectory shown in Figure 6.21 subject to operational constraints on the velocity and

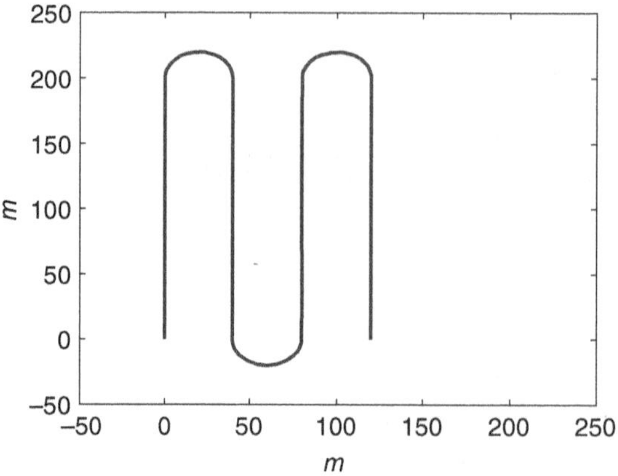

Figure 6.21 Desired trajectory in *x*-*y* plane.

acceleration on both *x* and *y* directions. More specifically the constraints are:

$$|v_x| \leq 30m/sec; \quad |a_x| \leq 5m/sec; \quad |v_y| \leq 30m/sec; \quad |a_y| \leq 5m/sec.$$

In the first step, the trajectory in *x*-*y* plane is decomposed into the two reference signals for the *x*-axis and *y*-axis separately. Clearly for the first 200 meters, the vehicle moves in a straight line in the *y*-direction. Therefore, the trajectory for the *x*-axis is a constant of zero and the trajectory for the *y*-axis is a straight line with slope ω_1 for the *y*-axis velocity. The second characteristic of the *x*-*y* plane trajectory is the semicircle, which has a radius of 20 meters. This semicircle can be described by sinusoidal functions in both *x* and *y* axes. The angular velocity for the circle ω_2 will be determined in the trajectory decomposition. Both ω_1 and ω_2 will be selected in consideration of the operational constraints.

1. Select the desired velocities ω_1 and ω_2 such that the velocity limits and acceleration limits are satisfied.
2. With $\Delta t = 0.1$ (sec), construct the desired trajectories for the *x*-axis and *y*-axis of the vehicle.
3. Without considering the vehicle load disturbance, determine the simplest disturbance model $D(z)$ that will enable the control system to track the desired trajectories in both *x*-axis and *y*-axis.
4. Assuming that the vehicle has a load that can not be neglected (a constant unknown load), determine the disturbance model $D(z)$ that will enable the repetitive control system to follow the desired trajectories while rejecting the load disturbance.
5. Design the control system by embedding the disturbance model into the controller. Adjust the desired closed-loop performance for the controller and

observer so that the constraints on the velocity and acceleration are satisfied. Perform the design and simulation for the disturbance model $D(z)$.

6. Design the control system through disturbance estimation with the full state variables estimated. Perform the design and simulation for the disturbance model $D(z)$.
7. What are your observations for the design methods?

Part III

Kalman Filtering

7

The Kalman Filter

7.1 Introduction

The Kalman filter is another one of the real-time computational algorithms for state estimation. The key similarity between state observer from the previous chapters and the Kalman filter is that they both estimate the unknown state variables via the measurement of output variables in combination with knowledge of the process. An observer computes the observer gain vector off-line for a time-invariant system. In contrast, the Kalman filter computes the Kalman filter gain in real-time and recursively, hence it can effectively estimate the state vector for time-varying systems. The Kalman filter yields optimal gain and state estimate if the process noise and measurement noise are Gaussian and the system is linear.

The chapter begins with an introduction to mathematical models used in the Kalman filter (see Section 7.2.1) followed by a derivation of the filter (see Sections 7.2.2 and 7.2.3). The derivation presented in this section follows an intuitive approach, although there are a number of ways to derive the Kalman filter. After the presentation of the Kalman filter algorithm, in Section 7.2.4, several examples are presented together with MATLAB tutorials. In Section 7.2.5 the compensation of sensor bias and load disturbance is discussed. The commonly encountered scenarios in the Kalman filter applications include multi-rate sampled data and missing measurements. These important issues are addressed in Section 7.3 with case studies and MATLAB tutorials. Section 7.4 presents the extended Kalman filter for state estimation of nonlinear systems where several examples are given together with MATLAB tutorials. Section 7.5 extends the idea of prescribed degree of stability in DLQR to the Kalman filter, leading to the Kalman filter with fading memory. The final topic of this chapter is to discuss the relationship between the Kalman filter and an observer (see Section 7.6).

State Feedback Control and Kalman Filtering with MATLAB/Simulink Tutorials, First Edition.
Liuping Wang and Robin Ping Guan.

Companion website: www.wiley.com/go/feedback19

7.2 The Kalman Filter Algorithm

7.2.1 State Space Models in the Kalman Filter

Recall that in Chapters 3- 6, we have used the following state space model for control system design:

$$x(k+1) = Ax(k) + Bu(k); \; y(k) = Cx(k),$$

where the system matrices A, B, and C are constant matrices used to describe a time-invariant system. Because the Kalman filter in this chapter is optimally derived for linear time-varying systems, the mathematical model is assumed to be described by the following difference equations:

$$x(k) = A(k-1)x(k-1) + B(k-1)u(k-1) + w(k-1) \tag{7.1}$$

$$y(k) = C(k)x(k) + v(k), \tag{7.2}$$

where $x(k)$ is the state vector that has the dimension $n \times 1$, $u(k)$ is the external input signal with dimension $p \times 1$ and $y(k)$ is the vector of measured variables with dimension $m \times 1$, $w(k)$ and $v(k)$ are the process noise and the measurement noise respectively that are assumed to be sequences of independent random variables with zero mean and covariances Q and R. Two aspects are emphasized as follows.

1. In the mathematical model (see (7.1) and (7.2)), the vector $x(k)$ is the *true* state that is unknown, the external input signal $u(k)$ is assumed to be known and the output signal $y(k)$ is measured. The time-varying system matrices $A(k)$, $B(k)$, and $C(k)$ are given. The goal for the Kalman filter is to estimate the true state $x(k)$.
2. The variables $w(k)$ and $v(k)$ satisfy the following statistical properties:

$$E[w(k)] = 0_{n\times 1} \tag{7.3}$$

$$E[v(k)] = 0_{m\times 1} \tag{7.4}$$

$$E[w(k)w(j)^T] = Q(k)\delta(k-j) \tag{7.5}$$

$$E[v(k)v(j)^T] = R(k)\delta(k-j) \tag{7.6}$$

$$E[v(k)w(j)^T] = 0_{m\times n} \tag{7.7}$$

where $\delta(k-j)$ is the Kronecker delta function, that is $\delta(k-j) = 1$ if $k = j$ and $\delta(k-j) = 0$ if $k \neq j$. Equations (7.3) and (7.4) correspond to the zero mean assumption of the noises, and (7.5)–(7.6) correspond to the assumptions of the noise being independent random variables. Equation (7.7) means that there is no correlation between the variables $v(k)$ and $w(k)$. The statistical properties of the noises (see (7.3)–(7.7)) are part of the conditions for optimality of the Kalman filter.

Note that the state space model given by (7.1) and (7.2) is also commonly expressed in the Kalman filter literature as:

$$x_k = A_{k-1}x_{k-1} + B_{k-1}u_{k-1} + w_{k-1};\ y_k = C_k x_k + v_k,$$

where the subscript k is used to denote the sampling index. For the notational consistency in this book, we will use the state space model given in (7.1) and (7.2) for the Kalman filter.

7.2.2 An Intuitive Computational Procedure

The Kalman filter algorithm optimally estimates the state vector $x(k)$ using the knowledge of process dynamics and the measured output $y(k)$ based on the statistical properties of the process noise $w(k)$ and measurement noise $v(k)$. This estimation is performed iteratively.

We start the derivation with an intuitive computational procedure for the Kalman filter.

1. At the sampling time $k = 0$, the covariance matrix for the process noise $w(0)$ is $Q(0) = E[w(0)w(0)^T]$ and for the measurement noise $v(0)$ is $R(0) = E[v(0)v(0)^T]$. We envisage that all the output measurements were utilized up to $k = 0$ to construct an initial estimate $\hat{x}(0)^+$, where the superscript '+' denotes the measurement $y(0)$ has been utilized. For this initial estimate $\hat{x}(0)^+$, there is an error: $e(0)^+ = x(0) - \hat{x}(0)^+$ and the uncertainty of the initial estimate is quantified using the covariance matrix:

$$P(0)^+ = E[\left(x(0) - \hat{x}(0)^+\right)\left(x(0) - \hat{x}(0)^+\right)^T],$$

because $x(0)$ is assumed to be a vector containing random variables.

2. At the sampling time $k = 1$, before the measurement $y(1)$ is taken, given the most recent best estimate $\hat{x}(0)^+$, we will find the best estimate $\hat{x}(1)^-$ for $x(1)$ where the superscript '−' denotes that the estimate does not utilize the measurement $y(1)$. The dynamic model (7.1) is used to calculate the best estimate $\hat{x}(1)^-$ as

$$\begin{aligned} \hat{x}(1)^- &= E[A(0)\hat{x}(0)^+ + B(0)u(0) + w(0)] \\ &= A(0)\hat{x}(0)^+ + B(0)u(0) \end{aligned} \tag{7.8}$$

where the expected value $E[w(0)]$ is zero (see (7.3)). However, there is an error due to the process noise $w(0)$, which is

$$e(1)^- = x(1) - \hat{x}(1)^- = A(0)(x(0) - \hat{x}(0)^+) + w(0). \tag{7.9}$$

This error is quantified using its covariance matrix:

$$\begin{aligned} P(1)^- &= E[(x(1) - \hat{x}(1)^-)(x(1) - \hat{x}(1)^-)^T] \\ &= A(0)E[\left(x(0) - \hat{x}(0)^+\right)\left(x(0) - \hat{x}(0)^+\right)^T]A(0)^T + E[w(0)w(0)^T]. \end{aligned}$$

Note that from the previous step,

$$E[\left(x(0) - \hat{x}(0)^+\right)\left(x(0) - \hat{x}(0)^+\right)^T] = P(0)^+$$

is the covariance of $e(0)^+$. Additionally, $E[w(0)w(0)^T] = Q(0)$ (see (7.5)). Therefore, the covariance matrix of $e(1)^-$ is equal to

$$P(1)^- = A(0)P(0)^+A(0)^T + Q(0). \tag{7.10}$$

3. When the measurement $y(1)$ becomes available, this information will be utilized to improve the estimate $\hat{x}(1)^-$. This improvement is achieved using a correction with the error between $y(1)$ and $C(1)\hat{x}(1)^-$, where $C(1)\hat{x}(1)^-$ has the dimension $m \times 1$ and is the predicted measurement taken from the point of view of the estimated state vector $\hat{x}(1)^-$. The improved estimate is calculated as

$$\hat{x}(1)^+ = \hat{x}(1)^- + K(1)(y(1) - C(1)\hat{x}(1)^-), \tag{7.11}$$

where $K(1)$ is the Kalman filter gain at the sampling instant $k = 1$, which will be optimized shortly in Section 7.2.3. If $\hat{x}(1)^-$ is perfect such that the error $y(1) - C(1)\hat{x}(1)^-$ is zero, then it is clear that $\hat{x}(1)^+$ equals $\hat{x}(1)^-$. Otherwise, a correction is made based on the error multiplied with the Kalman filter gain $K(1)$.

For the estimate $\hat{x}(1)^+$, there is a corresponding covariance matrix $P(1)^+$ of the error $x(1) - \hat{x}(1)^+$. To calculate this covariance matrix, we first note that

$$x(1) - \hat{x}(1)^+ = x(1) - \hat{x}(1)^- - K(1)C(1)(x(1) - \hat{x}(1)^-) - K(1)v(1),$$

where we have used the relation $y(1) = C(1)x(1) + v(1)$. The covariance matrix for $x(1) - \hat{x}(1)^+$ is calculated as

$$\begin{aligned} P(1)^+ &= E[(x(1) - \hat{x}(1)^+)(x(1) - \hat{x}(1)^+)^T] \\ &= (I - K(1)C(1))E[(x(1) - \hat{x}(1)^-)(x(1) - \hat{x}(1)^-)^T](I - K(1)C(1))^T \\ &\quad + K(1)E[v(1)v(1)^T]K(1)^T \\ &= (I - K(1)C(1))P(1)^-(I - K(1)C(1))^T + K(1)R(1)K(1)^T \end{aligned} \tag{7.12}$$

where we have used the following equations:

$$P(1)^- = E[(x(1) - \hat{x}(1)^-)(x(1) - \hat{x}(1)^-)^T]; \; R(1) = E[v(1)v(1)^T].$$

4. At the sampling time $k = 2$, the computational procedure repeats itself by following the same steps from Step 2.

The computational procedure essentially separates the estimation process into two steps: a time update step (see Step 2) and a measurement update step (see Step 3). This procedure allows us to combine two sources of information from process dynamics and measurements as well as to separate the effect of two noise terms (process noise $w(k)$ and measurement noise $v(k)$) on the estimate.

Associated with the time update step and the measurement update step, there are two sets of estimates and covariances.

For the time update step, at the sampling instant k, $\hat{x}(k)^-$ and $P(k)^-$ are called *a priori* state estimate and *a priori* covariance, which are denoted with the superscript '–'. For an arbitrary k, the *a priori* estimate $\hat{x}(k)^-$ and the covariance of the error $P(k)^-$ are computed by generalizing (7.8) and (7.10) as

$$\hat{x}(k)^- = A(k-1)\hat{x}(k-1)^+ + B(k-1)u(k-1) \tag{7.13}$$

$$P(k)^- = A(k-1)P(k-1)^+A(k-1)^T + Q(k-1). \tag{7.14}$$

It is seen that the *a priori* estimate utilizes the knowledge of system dynamics with the measurements $y(0)$, $y(1)$, $y(2)$, …, $y(k-1)$, but excluding the current measurement $y(k)$. Hence, the superscript '–' denotes the exclusion.

For the measurement update step, the variables are called *a posteriori* state estimate and *a posteriori* covariance, which are denoted with the superscript '+'. At the sampling instant k, the *a posteriori* estimate is $\hat{x}(k)^+$ and the covariance of the error is $P(k)^+$, which are computed by generalizing (7.11) and (7.12) for an arbitrary k as

$$\hat{x}(k)^+ = \hat{x}(k)^- + K(k)(y(k) - C(k)\hat{x}(k)^-) \tag{7.15}$$

$$P(k)^+ = (I - K(k)C(k))P(k)^-(I - K(k)C(k))^T + K(k)R(k)K(k)^T. \tag{7.16}$$

In contrast to the *a priori* estimate, the *a posteriori* estimate utilizes the output measurements $y(0)$, $y(1)$, …, $y(k)$, which includes the current measurement $y(k)$. Hence, the superscript '+' indicates the inclusion.

The remaining question is how to compute the Kalman filter gain $K(k)$ at the sampling instant k. This will be explained in Section 7.2.3.

7.2.3 Optimization of Kalman Filter Gain

At the sampling instant k, the *a posteriori* covariance matrix $P(k)^+$ plays an important role in optimally determining the Kalman filter gain $K(k)$. At the sampling instant k, the objective function is chosen as the sum of the variances of the errors between the state and the *a posteriori* estimate $\hat{x}(k)^+$:

$$J(k) = \sum_{j=1}^{n} E\left[\left(x_j(k) - \hat{x}_j(k)^+\right)^2\right] = E\left[\left(x(k) - \hat{x}(k)^+\right)^T\left(x(k) - \hat{x}(k)^+\right)\right]. \tag{7.17}$$

Note that for a vector η with dimension $n \times 1$, we have

$$\eta^T\eta = trace(\eta\eta^T),$$

where the trace of a matrix M equals to the sum of all its diagonal elements. Thus, the objective function $J(k)$ becomes the trace of the covariance matrix $P(k)^+$, which is

$$\begin{aligned}J(k) &= trace\{E[(x(k) - \hat{x}(k)^+)(x(k) - \hat{x}(k)^+)^T]\} = trace\{P(k)^+\}\\ &= trace\{(I - K(k)C(k))P(k)^-(I - K(k)C(k))^T + K(k)R(k)K(k)^T\}.\end{aligned} \tag{7.18}$$

The optimal Kalman filter gain $K(k)$ is to be found by minimizing the objective function $J(k)$, which is derived as follows.

Note that (7.18) can be expressed as

$$\begin{aligned}J(k) = trace\{&P(k)^- - 2K(k)C(k)P(k)^-\\ &+ K(k)[C(k)P(k)^-C(k)^T + R(k)]K(k)^T\}.\end{aligned} \tag{7.19}$$

We calculate the first derivative of the objective function as

$$\frac{\partial J(k)}{\partial K(k)} = -2P(k)^-C(k)^T + 2K(k)[C(k)P(k)^-C(k)^T + R(k)], \tag{7.20}$$

where we have used the following matrix equalities:

$$\frac{\partial(trace\{XYX^T\})}{\partial X} = 2XY; \quad \frac{\partial(trace\{XY\})}{\partial X} = Y^T.$$

Now, assuming that the matrix $C(k)P(k)^-C(k)^T + R(k)$ is positive definite at the sampling instant k, the Kalman filter gain $K(k)$ that minimizes the objective function $J(k)$ is obtained by solving the algebraic equation:

$$\frac{\partial J(k)}{\partial K(k)} = 0.$$

This optimization leads to the following solution:

$$K(k) = P(k)^-C(k)^T[C(k)P(k)^-C(k)^T + R(k)]^{-1}. \tag{7.21}$$

This is the Kalman filter gain that minimizes the objective function $J(k)$ when the matrix $C(k)P(k)^-C(k)^T + R(k)$ is positive definite.

The Kalman filter algorithm is summarized below. To begin the iterative computation, we will choose initial conditions for the *a posteriori* state estimate $\hat{x}(0)^+ = E(x(0))$ and the covariance of the error $P(0)^+ = E[(x(0) - \hat{x}(0)^+)(x(0) - \hat{x}(0)^+)^T]$ as well as the covariance matrices Q for the process noise and R for the measurement noise. The following computational procedure is repeated for each sampling instant k.

1. Compute the *a priori* estimate of the state and the covariance matrix:

$$\hat{x}(k)^- = A(k-1)\hat{x}(k-1)^+ + B(k-1)u(k-1) \tag{7.22}$$

$$P(k)^- = A(k-1)P(k-1)^+A(k-1)^T + Q(k-1). \tag{7.23}$$

2. Compute the Kalman filter gain:

$$K(k) = P(k)^- C(k)^T [C(k)P(k)^- C(k)^T + R(k)]^{-1}. \tag{7.24}$$

3. Compute the *a posteriori* estimate of the state and the covariance matrix:

$$\hat{x}(k)^+ = \hat{x}(k)^- + K(k)(y(k) - C(k)\hat{x}(k)^-) \tag{7.25}$$

$$\begin{aligned} P(k)^+ = &(I - K(k)C(k))P(k)^-(I - K(k)C(k))^T \\ &+ K(k)R(k)K(k)^T. \end{aligned} \tag{7.26}$$

There is a simplified expression of the *a posteriori* estimate of the covariance matrix $P(k)^+$. The simplification follows by noting that from the original expression:

$$\begin{aligned} P(k)^+ &= (I - K(k)C(k))P(k)^-(I - K(k)C(k))^T \\ &\quad + K(k)R(k)K(k)^T \\ &= (I - K(k)C(k))P(k)^- - P(k)^- C(k)^T K(k)^T \\ &\quad + K(k)C(k)P(k)^- C(k)^T K(k)^T + K(k)R(k)K(k)^T, \end{aligned} \tag{7.27}$$

where the sum of the last two terms equal to:

$$\begin{aligned} &K(k)C(k)P(k)^- C(k)^T K(k)^T + K(k)R(k)K(k)^T \\ &= K(k)[C(k)P(k)^- C(k)^T + R(k)]K(k)^T \\ &= P(k)^- C(k)^T [C(k)P(k)^- C(k)^T + R(k)]^{-1} [C(k)P(k)^- C(k)^T + R(k)]K(k)^T \\ &= P(k)^- C(k)^T K(k)^T. \end{aligned}$$

Hence, we arrive at the simplified version of the *a posteriori* estimate of the covariance matrix $P(k)^+$:

$$P(k)^+ = (I - K(k)C(k))P(k)^-. \tag{7.28}$$

The *a posteriori* covariance matrix $P(k)^+$ given by (7.28) is less computationally demanding in comparison with that given by (7.26). However it may loss its symmetric property due to numerical accuracy if one chooses to use this simpler form. The simplified expression of $P(k)^+$ will be used in Chapter 8 for the Kalman filter algorithm with orthogonal decomposition, where the symmetric property will be enforced through the associated computational procedure.

7.2.4 Kalman Filter Examples with MATLAB Tutorials

In this section, we will first produce the MATLAB function **KF.m** so that the examples in this section can be verified using simulation studies.

Tutorial 7.1 *The objective of this tutorial is to write a MATLAB function* **KF.m** *that will produce the estimated states for a multi-input and multi-output system. The input variables to the function are:*

- *the system matrices A, B, and C;*
- *the input and output data u ($p \times L$), y ($m \times L$), where p and m are the number of inputs and outputs respectively, and L is the number of the simulations in samples;*
- *the initial conditions for the Kalman filter $\hat{x}(0)^+$ and $P(0)^+$;*
- *the noise covariances Q and R.*

The output variables from the function are $\hat{x}(k)^+$ and $\hat{y}(k)$. Although the objective is to estimate $\hat{x}(k)^+$, it could be useful to have the information of estimated output $\hat{y}(k)$ ($\hat{y}(k) = C\hat{x}(k)^+$).

Step by Step

1. *Create a new file called* **KF.m**.
2. *Enter the following program into the file:*

```
function [xhat,yhat]=KF(A,B,C,u,y,x_plus,P_plus,Q,R)
```

3. *Check the data length, the number of measurements and states. Continue entering the following program into the file:*

```
L=length(y);
[m,n]=size(C);
```

4. *Assign the initial condition $\hat{x}(0)^+$ to $\hat{x}(:,1)$ and the initial condition $C\hat{x}(0)^+$ to $\hat{y}(:,1)$. Continue entering the following program into the file:*

```
xhat(:,1)=x_plus;
yhat(:,1)=C*x_plus;
```

5. *The computation will be performed recursively. Compute the* a priori *estimate of the state $\hat{x}(k)^-$ and the covariance matrix $P(k)^-$. Continue entering the following program into the file:*

```
for k=2:L
x_minus=A*x_plus+B*u(:,k-1);
P_minus=A*P_plus*A'+Q;
```

6. *Compute the Kalman filter gain. Continue entering the following program into the file:*

```
K=P_minus*C'*inv(C*P_minus*C'+R);
```

7. *Compute the* a posteriori *estimate of the state $\hat{x}(k)^+$ and the covariance matrix $P(k)^+$. Continue entering the following program into the file:*

```
x_plus=x_minus+K*(y(:,k)-C*x_minus);
P_plus=(eye(n)-K*C)*P_minus*(eye(n)-K*C)'+K*R*K';
```

8. *Save the estimated state variable and output variable, and complete one cycle of computation. Continue entering the following program into the file:*

```
xhat(:,k)=x_plus;
yhat(:,k)=C*x_plus;
end
```

9. *Test this function using Example 7.1.*

Example 7.1

A signal is known to be piece-wise constant with unknown amplitude. This signal is corrupted by zero-mean measurement noise with variance 0.01. Design a Kalman filter to extract the signal from the noise corrupted measurement. This example will demonstrate that when there is a discrepancy between the true process noise and the noise assumption used in the model, the Kalman filter can compensate this to a certain degree.

Solution

To design a Kalman filter, the first step is to build a dynamic model for the system. Being a piece-wise constant signal, we can assume that the signal could be modeled by an unknown constant c. That is at the sampling instant k, the state variable $x(k) = c$, *and* $x(k) - x(k-1) = 0$. *Because the parameter c might change to a different constant after a certain time interval, this possible variation is modeled by the process noise* $w(k)$. *Therefore, we arrive at the dynamic model for a piece-wise constant as*

$$x(k) = x(k-1) + w(k-1) \tag{7.29}$$

and the measurement model as

$$y(k) = x(k) + v(k) \tag{7.30}$$

where $v(k)$ *is white noise with zero mean and variance 0.01.*

To test the Kalman filter algorithm, the parameter c is 1 for the first 500 samples and changes to -1 *for the second 500 samples. A measurement noise* ($v(k)$) *generated using MATLAB function* **randn.m** *is added to the parameter c. Figure 7.1(a) shows the corrupted signal. Note that the process noise* $w(k)$ *is zero except at the sampling instant* $k = 500$ *where the signal is changed from 1 to* -1. *Therefore, the true process noise* $w(k)$ *is not white noise with zero mean and there is a mismatch between the true process noise and the assumed white noise in the model.*

For the Kalman filter application, $A = 1$, $B = 0$, $C = 1$, $R = 0.01$. *The initial conditions are* $\hat{x}(0)^+ = 0$, $P(0)^+ = 100$. *Because the process noise* $w(k)$ *is added to reflect the uncertainty caused by the parameter variation, its variance is unknown. We use the Q parameter as a tuning parameter. Here as an example, we select* $Q = 0.0001$. *Figure 7.1(a) shows the noise corrupted signal and Figure 7.1(b)*

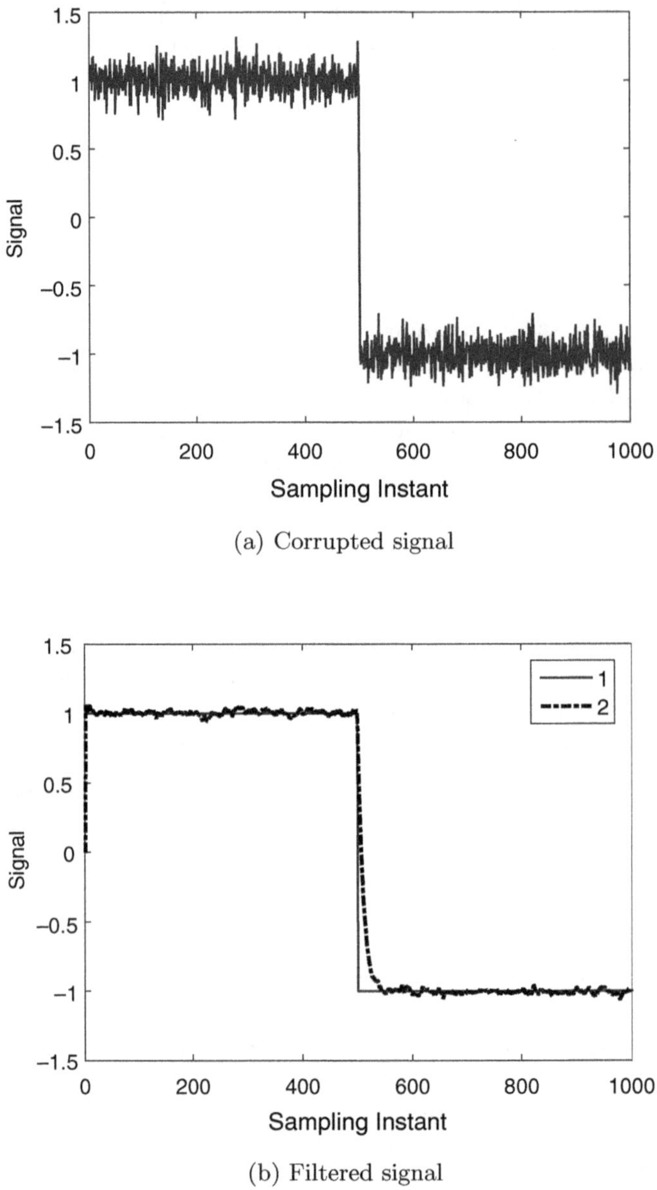

Figure 7.1 Piece-wise constant signal (Example 7.1). Key: line (1) parameter c; line (2) filtered signal.

compares the filtered signal with the parameter c. By following Tutorial 7.2, we can verify that if $Q = 0$, the filter works optimally if the parameter c is unchanged, however, it would take a long time for the Kalman filter to converge to the new parameter when c changes.

This example shows that when using the Kalman filter there is a degree of robustness with respect to the assumption of noise in the process. For instance, the assumption of $w(k)$ is mismatched with the true process noise, but the filter has tolerated this mismatch.

The modeling framework derived from this example has been used in compensation of sensor bias and load disturbance in Section 7.2.5.

Tutorial 7.2 *This tutorial is to test the MATLAB function* **KF.m** *based on Example 7.1.*

Step by Step

1. *Create a new file called* **test4KF.m**.
2. *Generate the test signal based on Example 7.1. The MATLAB function randn.m is used to generate the measurement noise with variance 0.01. By specifying the seed number, we know exactly which noise sequence used in the simulation. Enter the following program into the file:*

```
y1=[ones(1,500) -ones(1,500)];
randn('seed',0);
noi=0.1*randn(1,1000);
y=y1+noi;
u=zeros(1,1000);
```

3. *Define the system matrices for the Kalman filter algorithm. Continue entering the following program into the file:*

```
A=1;
B=0;
C=1;
```

4. *Define the Q and R matrices for the Kalman filter. Continue entering the following program into the file:*

```
Q=0.0001;
R=0.01;
```

5. *Choose initial conditions for the Kalman filter. Continue entering the following program into the file:*

```
P_plus=100;
x_plus=0;
```

6. *Call MATLAB function* **KF.m** *written in Tutorial 7.1. Continue entering the following program into the file:*

```
[xhat,yhat]=KF(A,B,C,u,y,x_plus,P_plus,Q,R);
```

7. *Compare the results with those obtained in Example 7.1.*

The following two examples are used to illustrate the role of Q matrix played on the state estimate.

Example 7.2

A linear time-varying system is described by the following state space model:

$$\begin{bmatrix} x_1(k) \\ x_2(k) \end{bmatrix} = \begin{bmatrix} \sin\left(\frac{2\pi(k-1)}{300}\right) & 1 \\ 0 & \cos\left(\frac{2\pi(k-1)}{300}\right) \end{bmatrix} \begin{bmatrix} x_1(k-1) \\ x_2(k-1) \end{bmatrix} + \begin{bmatrix} 1 \\ 1 \end{bmatrix} u(k-1) \quad (7.31)$$

$$y(k) = \begin{bmatrix} 0 & 1 \end{bmatrix} \begin{bmatrix} x_1(k) \\ x_2(k) \end{bmatrix} + v(k) \quad (7.32)$$

where the input signal $u(k) = \sin(\frac{2\pi k}{100})$ *and the measurement noise* $v(k)$ *is white noise with variance* $\sigma^2 = 1$. *Figure 7.2 shows the measured output data, where the initial condition* $x_1(0) = x_2(0) = 1$ *is used to create this set of data. Design and implement a Kalman filter to estimate the states* $x_1(k)$ *and* $x_2(k)$ *for this time-varying system.*

Solution

Because there is no process noise in the system, we choose Q *to be a zero matrix.* $R = 1$ *is selected for the given variance of the measurement noise. The initial conditions are arbitrarily chosen as* $\hat{x}(0)^+ = [0\ 0]^T,\ P(0)^+ = 100I$. *The time update equations are*

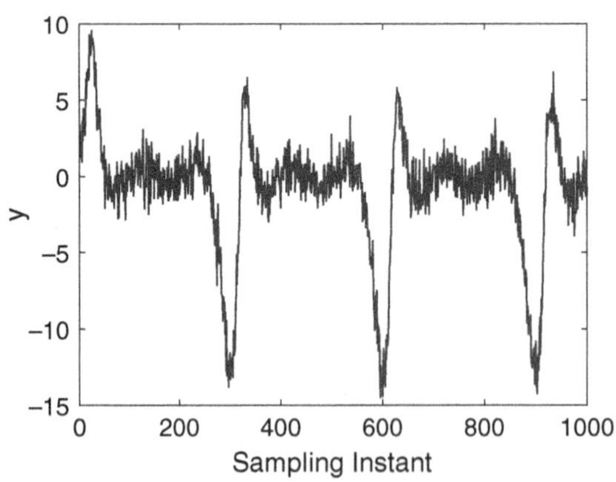

Figure 7.2 Measured output data (Example 7.2).

restated for convenience:

$$\hat{x}(k)^- = A(k-1)\hat{x}(k-1)^+ + B(k-1)u(k-1) \tag{7.33}$$

$$P(k)^- = A(k-1)P(k-1)^+A(k-1)^T + Q \tag{7.34}$$

where the following time-varying matrix $A(k-1)$ and matrix B are used:

$$A(k-1) = \begin{bmatrix} \sin\left(\frac{2\pi(k-1)}{300}\right) & 1 \\ 0 & \cos\left(\frac{2\pi(k-1)}{300}\right) \end{bmatrix}; \; B = \begin{bmatrix} 1 \\ 1 \end{bmatrix}.$$

The Kalman filter gain is updated using (7.24), the a posteriori *estimate of the state and the covariance follow from (7.25) and (7.26).*

Because it is in a simulation environment, the numerical values of the state variables $x_1(k)$ and $x_2(k)$ are precisely known. We evaluate the performance of the Kalman filter against the true state variables. In the top diagram of Figure 7.3(a), the estimated state $\hat{x}_1(k)^+$ is compared with $x_1(k)$ and $\hat{x}_1(k)^+$ is not distinguishable from $x_1(k)$. Thus, in the bottom diagram, the error $e_1 = x_1(k) - \hat{x}_1(k)^+$ is shown and it can be seen that the error converges to a small value after 40 samples. Similar conclusions are drawn for the estimated state $\hat{x}_2(k)^+$ and $e_2(k)$ as illustrated in Figure 7.3(b). Additionally, we calculate the mean squared error for the two states as $\frac{1}{M}\sum_{k=1}^{M}(e_1(k)^2 + e_2(k)^2) = 0.0077$ where M is the data length.

One can follow Tutorial 7.1 to re-create the simulation results presented in this example. A small modification to the previous program occurs at the time update step, where the time-varying system matrix is used.

The next question we ask is what it means when the Q matrix is selected to be zero. It effectively means that the dynamic model (7.31) in Example 7.2 is absolutely accurate and we trust it with 100% confidence. It also means that

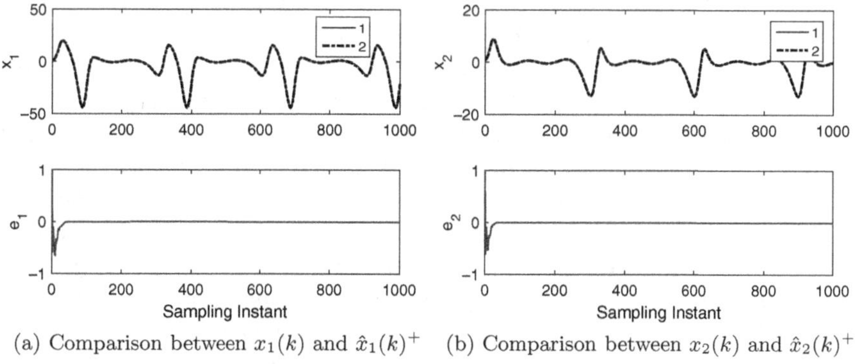

(a) Comparison between $x_1(k)$ and $\hat{x}_1(k)^+$ (b) Comparison between $x_2(k)$ and $\hat{x}_2(k)^+$

Figure 7.3 Performance evaluation of Kalman filter (Example 7.2).

the measurement does not matter to us. Therefore, when $Q = 0$ is selected, an extreme scenario is assumed. In reality, the dynamic model will not be perfect, and the choice of $Q = 0$ is not realistic. The following example is to show that when the dynamic model is not correct, the choice of Q could be used to improve the estimation.

Example 7.3

In this example, we consider the dynamic system with the following state space model:

$$\begin{bmatrix} x_1(k) \\ x_2(k) \end{bmatrix} = \begin{bmatrix} \sin\left(\frac{2\pi(k-1)}{300}\right) & 1 \\ 0 & \cos\left(\frac{2\pi(k-1)}{300}\right) \end{bmatrix} \begin{bmatrix} x_1(k-1) \\ x_2(k-1) \end{bmatrix} + \begin{bmatrix} 0.8 \\ 0.8 \end{bmatrix} u(k-1) \tag{7.35}$$

$$y(k) = \begin{bmatrix} 0 & 1 \end{bmatrix} \begin{bmatrix} x_1(k) \\ x_2(k) \end{bmatrix} + v(k) \tag{7.36}$$

where the input signal $u(k) = \sin\left(\frac{2\pi k}{100}\right)$ and the measurement noise $v(k)$ is white noise with variance $\sigma^2 = 1$. The input matrix B is different from that given in (7.31) of Example 7.2, and the rest of the model is identical. Assuming that we were not aware of this change in B matrix, and we applied the same Kalman filter to the data generated using (7.35) and (7.36), demonstrate that there are large errors between the estimated and true states. Furthermore, show that selection of Q matrix could lead to the reductions of errors.

Solution

When we apply the same Kalman filter where the B matrix is equal to $[1\ 1]^T$ as from the original model (7.31). There is a modelling error between the actual system and the model used in the Kalman filter. With $Q = 0$ and $R = 1$, the solid line in Figure 7.4 shows $e_1(k) = x_1(k) - \hat{x}_1(k)^+$ and $e_2(k) = x_2(k) - \hat{x}_2(k)^+$. We calculated the mean squared error as $\frac{1}{M}\sum_{k=1}^{M}(e_1(k)^2 + e_2(k)^2) = 8.2702$. In contrast to the previous mean squared error when the model was correct, the mean squared error has increased from 0.0077 to this large number of 8.2702. This clearly shows that model accuracy plays an important role in the Kalman filtering, and a mismatch of the system model leads to the deterioration of Kalman filter performance.

In order to take into account the possibility that there are errors in the dynamic model, either due to model imperfection or due to process noise, we choose $Q = 0.1I$ to reflect the imperfection. We apply the Kalman filter with the new selection of Q matrix to the data generated using (7.35) and (7.36). Figure 7.4 compares the errors of the two Kalman filters with different Q matrices. The plots show that when $Q = 0.1I$, there is an improvement on the errors of state estimation, however, the effect of measurement noise is also reflected in the state errors. The mean squared error for $Q = 0.1I$ is 6.968.

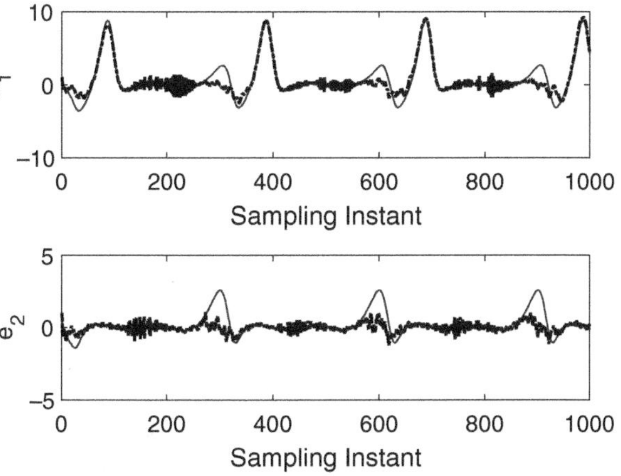

Figure 7.4 Evaluation of Kalman filter performance when B matrix is wrong (Example 7.3). Key: solid-line $Q = 0$; dashed-line $Q = 0.1I$.

The optimal performance of Kalman filter makes it an attractive choice for state estimation as demonstrated by the examples. However, we emphasize that the optimal results are obtained based on the assumption that the time-varying system model is precisely known.

7.2.5 Compensation of Sensor Bias and Load Disturbance

In engineering applications, it is common to encounter the problems of low frequency disturbances affecting the process dynamics (see Chapters 5 and 6) and sensor bias affecting the measurements. For instance, the electrical load in the drive train of a wind turbine (see Section 2.4.1). The mathematical model we used in the Kalman filter (see Section 7.2.1) has not included the descriptions for these commonly encountered signals in some engineering applications. This section will show how to modify the original mathematical model in order to include the descriptions of the process disturbance and sensor bias.

Sensor bias compensation

For simplicity, we consider the case of a system with a single output. In some engineering applications, the measurement contains a component and this component is unknown unless we put extra effort to determine it experimentally. We often call this component sensor bias. Another problem with the sensor bias is that it may change slowly with time due to temperature change or aging of the sensor electronic elements.

For a single input and a single output system, the state space model that contains the description of sensor bias is chosen to be the following form:

$$x(k) = A(k-1)x(k-1) + B(k-1)u(k-1) + \omega(k-1) \tag{7.37}$$

$$y(k) = C(k)x(k) + \upsilon_0 + \upsilon(k) \tag{7.38}$$

where v_0 is an unknown constant ($v_0 \neq 0$) to describe the sensor bias. As before, $x(k)$ is the state vector that has the dimension $n \times 1$, $u(k)$ is the external input signal and $y(k)$ is the measured variable, $w(k)$ and $v(k)$ are the process noise and the measurement noise that are assumed to be sequences of independent random variables with zero mean and covariances Q and R. Clearly, the unknown constant v_0 will cause problems to the Kalman filter algorithm because the measurement error is $v_0 + v(k)$ that is no longer zero mean. This is confirmed in Example 7.4.

For the unknown constant v_0, we describe it with the following difference equation:

$$v_0(k) = v_0(k-1) + \epsilon(k-1)$$

where $\epsilon(k)$ is a zero mean random noise with a small variance σ_ϵ^2. This is the same model used to describe the unknown constant in Example 7.1. Now we modify the state space model (7.37) and (7.38) to include $v_0(k)$ as part of the state vector:

$$\begin{bmatrix} x(k) \\ v_0(k) \end{bmatrix} = \begin{bmatrix} A(k-1) & 0 \\ 0_{1\times n} & 1 \end{bmatrix} \begin{bmatrix} x(k-1) \\ v_0(k-1) \end{bmatrix} + \begin{bmatrix} B(k-1) \\ 0 \end{bmatrix} u(k-1) + \begin{bmatrix} w(k-1) \\ \epsilon(k-1) \end{bmatrix} \tag{7.39}$$

$$y(k) = \begin{bmatrix} C(k) & 1 \end{bmatrix} \begin{bmatrix} x(k) \\ v_0(k) \end{bmatrix} + v(k). \tag{7.40}$$

It can be seen that the process noises and measurement noise from (7.39) and (7.40) are sequences of independent random variables with zero mean. The covariance for $v(k)$ remains the same R, the covariance for the process noise has changed with its dimension increased from $n \times n$ to $(n+1) \times (n+1)$ in order to include the effect of $\epsilon(k)$.

Since we are familiar with the time-varying system given in Example 7.2, we will use that system to illustrate the compensation of sensor bias based on the state space model given by (7.39) and (7.40). This is shown in Example 7.4.

Example 7.4

Consider the time-varying system described by the following state space model:

$$\begin{bmatrix} x_1(k) \\ x_2(k) \end{bmatrix} = \begin{bmatrix} \sin\left(\frac{2\pi(k-1)}{300}\right) & 1 \\ 0 & \cos\left(\frac{2\pi(k-1)}{300}\right) \end{bmatrix} \begin{bmatrix} x_1(k-1) \\ x_2(k-1) \end{bmatrix} + \begin{bmatrix} 1 \\ 1 \end{bmatrix} u(k-1) \tag{7.41}$$

$$y(k) = \begin{bmatrix} 0 & 1 \end{bmatrix} \begin{bmatrix} x_1(k) \\ x_2(k) \end{bmatrix} + v_0 + v(k) \tag{7.42}$$

where the input signal $u(k) = \sin\left(\frac{2\pi k}{100}\right)$ *and the measurement noise* $v(k)$ *is white noise with variance* $\sigma^2 = 1$. *With the initial condition* $x_1(0) = x_2(0) = 1$ *and* $v_0(k) = 1$

for $k = 0, 1, 2, \ldots, 499$ and $v_0(k) = -1$ for $k = 500, 501, \ldots, 999$, a set of simulation data is generated to illustrate the sensor bias problem and the evaluation of the proposed compensation scheme. Use this set of test data to show that the estimation errors have increased significantly if the sensor bias is not compensated and the errors will be reduced with the proposed compensation.

Solution

To illustrate the scenario without the compensation, we use the state space model by neglecting the unknown constant v_0 in the Kalman filter. By choosing $Q = 0.1I_{2\times2}$ and $R = 1$, we apply the Kalman filter algorithm to this set of test data with initial conditions $P(0)^+ = 100I$ and $\hat{x}(0)^+$ to be zero. Figures 7.5(a)–(b) show the errors $x_1(k) - \hat{x}_1(k)$ and $x_2(k) - \hat{x}_2(k)$ in line (1). It is clearly seen that the sensor bias without compensation has made the performance of the Kalman filter deteriorate and it has caused large errors for the estimated state variables. The sum of mean squared errors is calculated as

$$\frac{1}{M}\left\{\sum_{k=1}^{M}(x_1(k) - \hat{x}_1(k)^+)^2 + (x_2(k) - \hat{x}_2(k)^+)^2\right\} = 3.6287$$

where M is the data length.

To illustrate the scenario with the sensor bias compensation, we use the augmented state space model presented in (7.39) and (7.40). By choosing $Q = 0.1I_{3\times3}$ and $R = 1$, we apply the Kalman filter algorithm based on the augmented model to the same set of test data, where the initial conditions are chosen to be $P(0)^+ = 100I_{3\times3}$ and $\hat{x}(0)^+ = 0_{3\times1}$. In comparison, Figures 7.5(a)–(b) show the errors $x_1(k) - \hat{x}_1(k)$ and $x_2(k) - \hat{x}_2(k)$ in line (2). It is clearly seen that the state estimate errors have

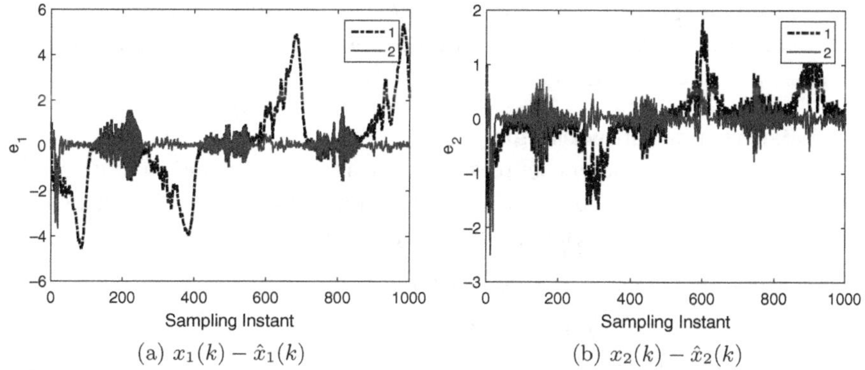

Figure 7.5 Performance evaluation of Kalman filter with compensation for sensor bias (Example 7.4). Key: line (1) error without compensation; line (2) error with compensation.

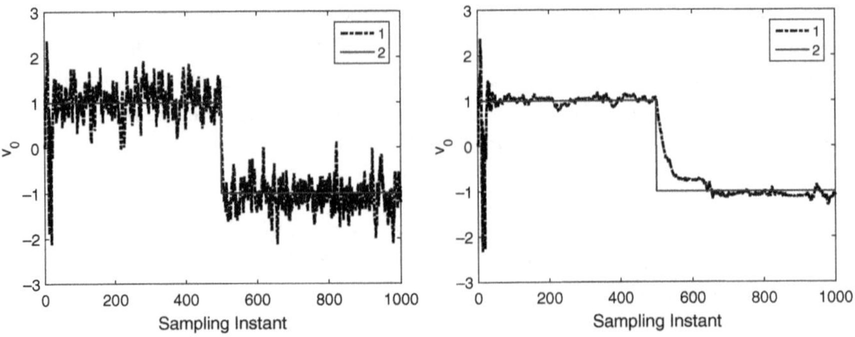

(a) $Q = 0.1I$ (b) $Q(3,3) = 0.001$

Figure 7.6 Estimate of the sensor bias (Example 7.4). Key: line (1) the estimated bias parameter; line (2) true bias parameter.

been significantly reduced with the compensation for sensor bias. Furthermore, we calculate the sum of the mean squared errors as

$$\frac{1}{M}\left\{\sum_{k=1}^{M}(x_1(k)-\hat{x}_1(k)^+)^2+(x_2(k)-\hat{x}_2(k)^+)^2\right\}=0.272$$

which is only a fraction of the error without the sensor bias compensation.

Because the bias parameter $v_0(k)$ is also estimated as part of the state vector, we can find this physical parameter from the Kalman filter without doing experiment. Figure 7.6(a) shows the estimated bias parameter with $Q = 0.1I$. Clearly, the estimate converges quickly to the bias, however, the measurement noise is amplified severely. To reduce the effect of measurement noise on the estimated bias, the third diagonal element in Q is reduced to 0.001. Figure 7.6(b) shows that the effect of measurement noise is reduced, however, it takes a longer time for the estimate to converge to the bias. The sum of mean squared errors is 0.3751 for this choice of Q, which is slightly increased from the previous case.

Load disturbance compensation

A load disturbance is another source of unknown factors that are not described in the commonly used state space model for the Kalman filter. In describing the load disturbance, the following state space model is used:

$$x(k) = A(k-1)x(k-1) + B(k-1)u(k-1) + B_L(k-1)T_L(k-1) + w(k-1) \tag{7.43}$$

$$y(k) = C(k)x(k) + v(k) \tag{7.44}$$

where $T_L(k)$ is an unknown constant representing a load disturbance in an electrical or mechanical system. The matrix $B_L(k)$ is assumed to be given. If it were unknown, we would assume that $B_L(k) = B(k)$.

To describe the unknown load $T_L(k)$, the following difference equation is used:

$$T_L(k) = T_L(k-1) + \epsilon(k-1)$$

where $\epsilon(k)$ is a zero mean random noise with a small variance σ_ϵ^2. For compensation of the load disturbance, the following state space model is used:

$$\begin{bmatrix} x(k) \\ T_L(k) \end{bmatrix} = \begin{bmatrix} A(k-1) & B_L(k-1) \\ 0_{1\times n} & 1 \end{bmatrix} \begin{bmatrix} x(k-1) \\ T_L(k-1) \end{bmatrix} + \begin{bmatrix} B(k-1) \\ 0 \end{bmatrix} u(k-1) + \begin{bmatrix} w(k-1) \\ \epsilon(k-1) \end{bmatrix} \tag{7.45}$$

$$y(k) = \begin{bmatrix} C(k) & 0 \end{bmatrix} \begin{bmatrix} x(k) \\ T_L(k) \end{bmatrix} + v(k). \tag{7.46}$$

We will use the state space model (7.45) and (7.46) to compensate the effect of load disturbance in Kalman filter.

Example 7.5

Consider that a time-varying system is described by the following state space model:

$$\begin{bmatrix} x_1(k) \\ x_2(k) \end{bmatrix} = \begin{bmatrix} \sin\left(\frac{2\pi(k-1)}{300}\right) & 1 \\ 0 & \cos\left(\frac{2\pi(k-1)}{300}\right) \end{bmatrix} \begin{bmatrix} x_1(k-1) \\ x_2(k-1) \end{bmatrix} + \begin{bmatrix} 1 \\ 1 \end{bmatrix} (u(k-1) + T_L(k-1)) \tag{7.47}$$

$$y(k) = \begin{bmatrix} 0 & 1 \end{bmatrix} \begin{bmatrix} x_1(k) \\ x_2(k) \end{bmatrix} + v(k) \tag{7.48}$$

where the input signal $u(k) = \sin\left(\frac{2\pi k}{100}\right)$ and the measurement noise $v(k)$ is white noise with variance $\sigma^2 = 1$. With the initial conditions $x_1(0) = x_2(0) = 1$ and the load $T_L(k) = 1$ for the first 500 samples and 2 for the next 500 samples, a set of simulation data is generated for the illustration of load disturbance problem and its compensation when using Kalman filter. Use this set of test data to show the efficacy of the compensation scheme when the system has an unknown disturbance.

Solution

We will first illustrate that without considering the existence of $T_L(k)$, the Kalman filter fails to correctly estimate the state variables. This is demonstrated by neglecting $T_L(k)$ in the dynamic model when we design the Kalman filter.

By selecting $Q = 0.1I_{2\times 2}$ and $R = 1$, the Kalman filter algorithm is applied using the test data that contain the load $T_L(k)$ where the initial conditions are chosen to be $P(0)^+ = 100I$ and $\hat{x}(0)^+ = 0$. In dash-dot line, Figures 7.7(a)–(b) show that the errors between the true states and the estimated states. It is seen that the Kalman filter fails

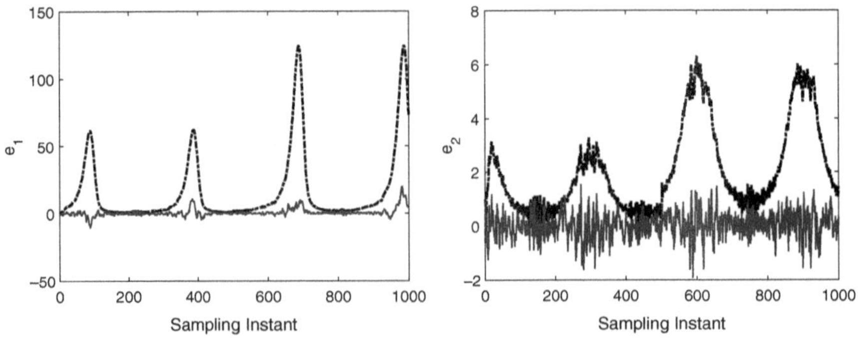

(a) $x_1(k) - \hat{x}_1(k)$ (b) $x_2(k) - \hat{x}_2(k)$

Figure 7.7 Performance evaluation of Kalman filter with compensation for load disturbance (Example 7.5). Key: solid-line error with compensation; dash-dot line error without compensation.

to correctly estimate the state variables. Furthermore, the sum of mean squared errors is calculated as

$$\frac{1}{M}\left\{\sum_{k=1}^{M}(x_1(k)-\hat{x}_1(k)^+)^2+(x_2(k)-\hat{x}_2(k)^+)^2\right\}=1035.9.$$

In compensating the load disturbance, the state space model given by (7.45) and (7.46) is used for the Kalman filter design. With this augmented model, the covariance matrices $Q = 0.1I_{3\times3}$ and $R = 1$ are selected. With the initial conditions $P(0)^+ = 100I_{3\times3}$ and $\hat{x}(0)^+$ being zero, the Kalman filter with load compensation is used to estimate the state variables. In comparison to the results obtained without the compensation, Figures 7.7(a)–(b) show the errors between the true states and the estimated states. It is seen that the errors have been significantly reduced. Furthermore, the sum of mean square errors is calculated as

$$\frac{1}{M}\left\{\sum_{k=1}^{M}(x_1(k)-\hat{x}_1(k)^+)^2+(x_2(k)-\hat{x}_2(k)^+)^2\right\}=9.57$$

It is evident that it is necessary to use load compensation in the Kalman filter estimation if the system has an existing load.

7.2.6 Food for Thought

1. A state space model is described by

$$x(k) = 0.9x(k-1) + 0.1u(k-1) + c_0 + w(k-1); y(k) = 2x(k) + v(k),$$

where c_0 is a constant and is assumed known. What is the *a priori* estimate $\hat{x}(k)^-$ assuming a given *a posteriori* estimate $\hat{x}(k-1)^+$? Does the constant c_0 play a role here?

2. Continue from 1. What is the *a posteriori* estimate $\hat{x}(k)^+$? How does the constant c_0 affect $\hat{x}(k)^+$?
3. For the same system as in 1, however with unknown constant c_0, if we neglect the unknown constant c_0, what are the *a priori* estimate $\hat{x}(k)^-$ and the *a posteriori* estimate $\hat{x}(k)^+$?
4. Continue from 3. Do you think that neglecting the unknown constant c_0 is a correct way to obtain the *a priori* estimate $\hat{x}(k)^-$ and the *a posteriori* estimate $\hat{x}(k)^+$? If not, what would you propose to do?
5. What is the objective function we used to obtain the Kalman filter gain K?
6. In the Kalman filter algorithm, we have assumed that Q and R matrices are known. If they are unknown, what would you do?

7.3 The Kalman Filter in Multi-rate Sampling Environment

7.3.1 KF Algorithm for Missing Data Scenarios

From Section 7.2.1, in a uniformly sampled environment, we assume that the mathematical model is:

$$x(k) = A(k-1)x(k-1) + B(k-1)u(k-1) + w(k-1) \tag{7.49}$$

$$y(k) = C(k)x(k) + v(k). \tag{7.50}$$

Supposing that the system has m output measurements, we denote the components in (7.50) as

$$y(k) = \begin{bmatrix} y_1(k) \\ \vdots \\ y_i(k) \\ \vdots \\ y_m(k) \end{bmatrix}; \; C(k) = \begin{bmatrix} C_1(k) \\ \vdots \\ C_i(k) \\ \vdots \\ C_m(k) \end{bmatrix}; \; v(k) = \begin{bmatrix} v_1(k) \\ \vdots \\ v_i(k) \\ \vdots \\ v_m(k) \end{bmatrix}$$

where $C_i(k)$ with dimension $1 \times n$ is the ith row in the output matrix $C(k)$. In order to accommodate the systems with either missing measurements or in a multi-rate sampling environment, we will modify the above state space model and the Kalman filter accordingly.

Time update with missing input data

We assume that the process model (7.49) has the sampling interval Δt and the time update step in the Kalman filter is performed in the uniformly sampled environment. Namely, we will compute the *a priori* estimate of the state and the covariance matrix:

$$\hat{x}(k)^- = A(k-1)\hat{x}(k-1)^+ + B(k-1)u(k-1) \tag{7.51}$$

$$P(k)^- = A(k-1)P(k-1)^+A(k-1)^T + Q(k-1). \tag{7.52}$$

However, if the input signal $u(k-1)$ is unavailable, we will replace $u(k-1)$ by the most recent past sample available, for example, $u(k-2)$, and compute the *a priori* estimate of the state:

$$\hat{x}(k)^- = A(k-1)\hat{x}(k-1)^+ + B(k-1)u(k-2) \tag{7.53}$$

while $P(k)^-$ from (7.52) is unchanged.

Measurement update with missing measurement data

It often occurs in sensor fusion applications that the measurements have different sampling rates. The question arises as how to adapt the Kalman filter derived from a uniformly sampling environment to a multi-rate sampling environment. The key is to capture the characteristics of the multi-rate sampled measurements with a time-varying measurement matrix $C(k)$. We will consider three scenarios.

Scenario I: all measurements are available. At the sampling instant k, when all measurements are available, we will use the measurement equation (7.50) for calculating the Kalman filter gain and the *a posteriori* estimate of the state and the covariance matrix (see (7.24)–(7.26)). Thus, there is no change on the measurement update step.

Scenario II: partial measurements are available. At the sampling instant k, when partial measurements are available, the measurement equation (7.50) is modified to accommodate this situation. As an illustration, we assume that at the sampling instant k, $y_1(k)$ is available from the measurement, but $y_i(k)$, $i = 2, \ldots, m$ are not available. Then the measurement equation is reduced to:

$$y(k) = C_1(k)x(k) + v_1(k) \tag{7.54}$$

where $C_1(k)$ is the first row of the original $C(k)$ matrix. Accordingly, $R(k)$ becomes the scalar variance of $v_1(k)$. By using the measurement equation (7.54) and the scalar $R(k)$, we will calculate the Kalman filter gain $K(k)$, and the *a posteriori* estimate of the state and the covariance matrix (see (7.24)–(7.26)).

Scenario III: no measurement is available. At the sampling instant k, if no measurements are available, the $C(k)$ matrix becomes $0_{m\times n}$. The *a posteriori* estimate of the state and the covariance matrix become:

$$\hat{x}(k)^+ = \hat{x}(k)^-$$
$$P(k)^+ = P(k)^-.$$

7.3.2 Case Studies with MATLAB Tutorial

The following example illustrates the applications of Kalman filter in a multi-rate sampling environment.

Example 7.6

The movement of an object is measured by an acceleration and a GPS measurement. The measurement from the acceleration has the sampling rate of 100 Hz and the GPS measurement has a sampling rate of 1 Hz. Additionally, there are a small number of missing data from the acceleration measurements. Design a Kalman filter to estimate its position, velocity and acceleration. Evaluate the performance of the filter through simulation studies.

Solution

We will first establish the dynamic equations of a moving object. For simplicity, we will consider the $x - y$ plane movement and neglect the rotational dynamics.

Assuming that the acceleration at the x-axis is constant a_x, the velocity along the x-axis at time t is

$$v_x(t) = a_x t + v_x(0) \tag{7.55}$$

where $v_x(0)$ is its initial velocity. The integration of (7.55) gives the position of the object as

$$p_x(t) = \frac{1}{2}a_x t^2 + v_x(0)t + p_x(0) \tag{7.56}$$

where $p_x(0)$ is its initial position.

Now, let us consider the incremental expressions of (7.55) and (7.56). At the sampling instant k, the time t is chosen to be the sampling interval Δt, the initial velocity $v_x(0)$ is $v_x(k-1)$, the initial position $p_x(0)$ is $p_x(k-1)$ and the constant acceleration a_x is $a_x(k-1)$. With these selections of variables, (7.55) and (7.56) become:

$$v_x(k) = a_x(k-1)\Delta t + v_x(k-1) \tag{7.57}$$

$$p_x(k) = \frac{1}{2}a_x(k-1)\Delta t^2 + v_x(k-1)\Delta t + p_x(k-1). \tag{7.58}$$

In a similar procedure, we establish the incremental motion equations for the y-axis:

$$v_y(k) = a_y(k-1)\Delta t + v_y(k-1) \tag{7.59}$$

$$p_y(k) = \frac{1}{2}a_y(k-1)\Delta t^2 + v_y(k-1)\Delta t + p_y(k-1). \tag{7.60}$$

Combining together (7.57)–(7.60) yields the following state space model in discrete-time:

$$\overbrace{\begin{bmatrix} p_x(k) \\ v_x(k) \\ p_y(k) \\ v_y(k) \end{bmatrix}}^{x(k)} = \overbrace{\begin{bmatrix} 1 & \Delta t & 0 & 0 \\ 0 & 1 & 0 & 0 \\ 0 & 0 & 1 & \Delta t \\ 0 & 0 & 0 & 1 \end{bmatrix}}^{A} \overbrace{\begin{bmatrix} p_x(k-1) \\ v_x(k-1) \\ p_y(k-1) \\ v_y(k-1) \end{bmatrix}}^{x(k-1)} + \overbrace{\begin{bmatrix} \frac{1}{2}\Delta t^2 & 0 \\ \Delta t & 0 \\ 0 & \frac{1}{2}\Delta t^2 \\ 0 & \Delta t \end{bmatrix}}^{B} \overbrace{\begin{bmatrix} a_x(k-1) \\ a_y(k-1) \end{bmatrix}}^{u(k-1)}. \tag{7.61}$$

The position measurement equation corresponds to the GPS measurement as

$$y(k) = \begin{bmatrix} y_1(k) \\ y_2(k) \end{bmatrix} = \overbrace{\begin{bmatrix} 1 & 0 & 0 & 0 \\ 0 & 0 & 1 & 0 \end{bmatrix}}^{C(k)} \begin{bmatrix} p_x(k) \\ v_x(k) \\ p_y(k) \\ v_y(k) \end{bmatrix} + \begin{bmatrix} v_1(k) \\ v_2(k) \end{bmatrix} \tag{7.62}$$

where $v_1(k)$ and $v_2(k)$ are the measurement noise assumed to be independent with zero mean for simplicity.

The dynamic model given by (7.61) and (7.62) is based on the assumption that the system is uniformly sampled with the sampling interval Δt. For this application, the input variables to the state equation (7.61) are the measured accelerations $a_x(k)$ and $a_y(k)$ that have a faster sampling rate 100 Hz, that is $\Delta t = 0.01$ (sec). For every sampling instant k, when the acceleration measurements are available, we will compute the a priori *estimate of the state and the covariance matrix:*

$$\hat{x}(k)^- = A(k-1)\hat{x}(k-1)^+ + B(k-1)u(k-1) \tag{7.63}$$

$$P(k)^- = A(k-1)P(k-1)^+A(k-1)^T + Q(k-1). \tag{7.64}$$

However, when the acceleration measurements are not available from the previous sample, we will use the most recent past input signal (for instance $u(k-2)$) to compute the a priori *estimate of the state:*

$$\hat{x}(k)^- = A(k-1)\hat{x}(k-1)^+ + B(k-1)u(k-2) \tag{7.65}$$

while $P(k)^-$ is unchanged.

There are two cases for the a posteriori *estimate of the state and the covariance matrix.*

Case A. *When there is no GPS measurement, the measurement matrix $C(k)$ is zero. The* a posteriori *estimate of the state and the covariance matrix become*

$$\hat{x}(k)^+ = \hat{x}(k)^- \tag{7.66}$$

$$P(k)^+ = P(k)^-. \tag{7.67}$$

Case B. *When there is a GPS measurement at the sampling instant k, the measurement matrix C is*

$$C(k) = \begin{bmatrix} 1 & 0 & 0 & 0 \\ 0 & 0 & 1 & 0 \end{bmatrix}.$$

We compute the Kalman filter gain:

$$K(k) = P(k)^{-}C(k)^{T}[C(k)P(k)^{-}C(k)^{T} + R(k)]^{-1} \tag{7.68}$$

and the a posteriori *estimate of the state and the covariance matrix:*

$$\hat{x}(k)^{+} = \hat{x}(k)^{-} + K(k)(y(k) - C(k)\hat{x}(k)^{-}) \tag{7.69}$$

$$P(k)^{+} = (I - K(k)C(k))P(k)^{-}(I - K(k)C(k))^{T} + K(k)R(k)K(k)^{T}. \tag{7.70}$$

To evaluate the performance of the Kalman filter, we generate a set of simulation data based on the uniformly sampled model given by (7.61) and (7.62). With the accelerations $a_x(k) = 0.1\sin\left(\frac{2\pi k}{6000}\right)$ *and* $a_y(k) = 0.1\cos\left(\frac{2\pi k}{6000}\right)$, *and* $x(0)$ *being zero vector, the true state vector* $x(k)$ *is created. To evaluate the performance of the Kalman filter, we add white noise* $w_x(k)$ *and* $w_y(k)$ *to the accelerations* $a_x(k)$ *and* $a_y(k)$, *where both* $w_x(k)$ *and* $w_y(k)$ *are zero mean with variance* $\sigma^2 = 1$ *that are generated with the seed being 2 based on the MATLAB function* **randn.m**. *Both measurement noise* $v_1(k)$ *and* $v_2(k)$ *are white noise with* $\sigma^2 = 1$, *which are generated with the seed being* 1. *For the acceleration measurements, there is one measurement missing out of six samples.*

Note that the process noise $w(k)$ *acting on the state vector* $x(k)$ *is:*

$$w(k) = B\begin{bmatrix} w_x(k) \\ w_y(k) \end{bmatrix}$$

where the matrix B is based on the process model (7.61). We choose the process noise covariance:

$$Q = E[w(k)w(k)^{T}] = BE[\begin{bmatrix} w_x(k) \\ w_y(k) \end{bmatrix} \begin{bmatrix} w_x(k) & w_y(k) \end{bmatrix}]B^{T} = BB^{T}$$

because we assume that the variances of $w_x(k)$ *and* $w_y(k)$ *are known to be* 1. *More specifically,* $Q(1,1) = Q(3,3) = 2.5 \times 10^{-9}$, $Q(2,2) = Q(4,4) = 10^{-4}$; $Q(2,1) = Q(1,2) = Q(4,3) = Q(3,4) = 5 \times 10^{-7}$, *and the remaining elements in Q are zero. The measurement covariance matrix R is an identity matrix. The initial conditions for the Kalman filter are selected as* $P(0)^{+} = I$ *and* $\hat{x}(0) = 0$. *The simulation results are summarized as follows.*

Without GPS measurement. *The first part of the simulation studies is to establish the importance of the GPS measurement. Figures 7.8(a)–(d) show the comparison*

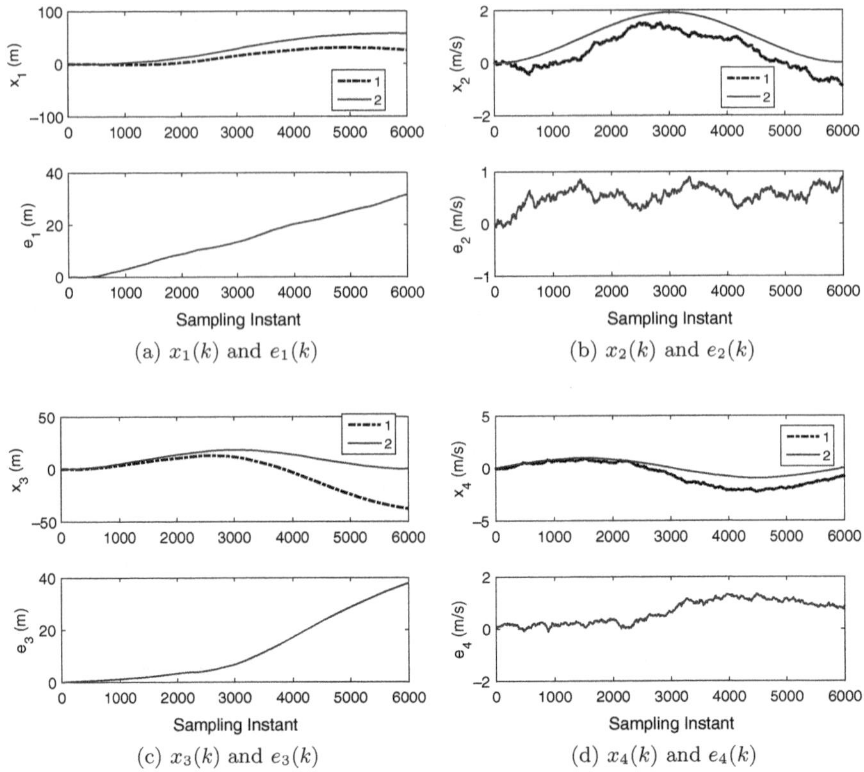

Figure 7.8 Comparison between true and estimated state without GPS measurement (Example 7.6). Key: line (1) estimated state; line (2) true state.

between true and the estimated state without the GPS measurement. It is seen that all state errors increase as k increases. Figure 7.9 confirms that without the GPS measurement, the estimated trajectory could not converge to the true trajectory. The sums of the mean squared error for the position and velocity are

$$\frac{1}{M}\sum_{k=1}^{M}\left(e_1(k)^2 + e_3(k)^2\right) = 601;\ \frac{1}{M}\sum_{k=1}^{M}\left(e_2(k)^2 + e_4(k)^2\right) = 0.9023.$$

With GPS measurement. *The second part of the simulation studies shows that the GPS measurement plays an important role in the estimation. In the simulation studies, the Kalman filter is implemented in the multi-rate sampling environment, where the measurement from the acceleration has the sampling rate of 100 Hz and the GPS measurement has a sampling rate of 1 Hz. This means that the GPS measurement update occurs at* $k = 0,\ 100,\ 200,\ 300, \ldots$*. Figures 7.10(a)–(d) present the*

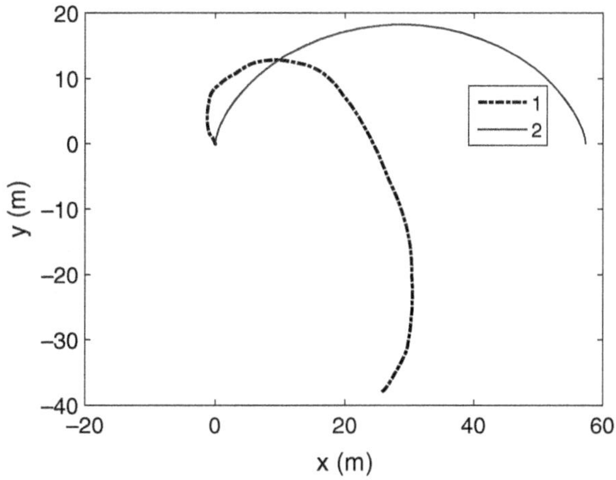

Figure 7.9 Trajectory of the moving object (Example 7.6). Key: line (1) the estimated trajectory without GPS; line (2) true trajectory.

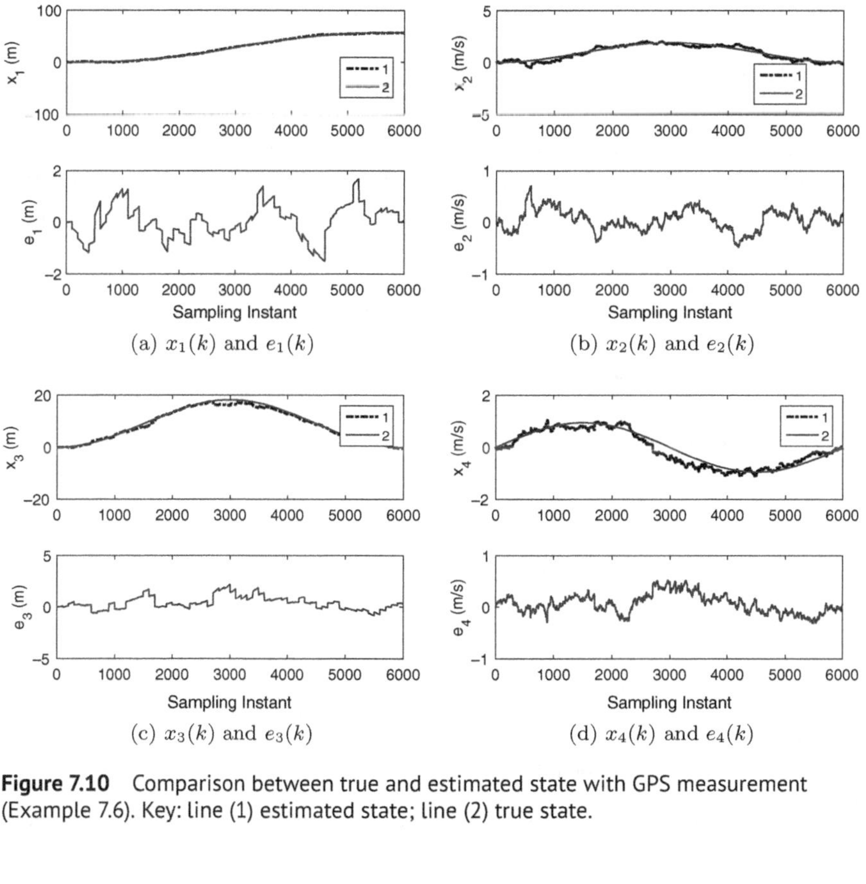

Figure 7.10 Comparison between true and estimated state with GPS measurement (Example 7.6). Key: line (1) estimated state; line (2) true state.

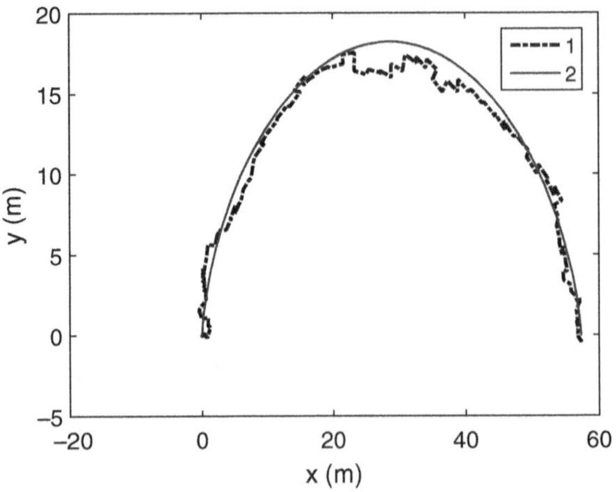

Figure 7.11 Trajectory of the moving object (Example 7.6). Key: line (1) the estimated trajectory with GPS; line (2) true trajectory.

comparisons between true and individual states and their errors. The sums of the mean squared error for the position and velocity are

$$\frac{1}{M}\sum_{k=1}^{M}\left(e_1(k)^2 + e_3(k)^2\right) = 0.9503; \quad \frac{1}{M}\sum_{k=1}^{M}\left(e_2(k)^2 + e_4(k)^2\right) = 0.0829.$$

Figure 7.11 compares the estimated trajectory of the moving object with the true trajectory. Figure 7.12 shows two diagonal elements of $P(k)^+$ where $300 \le k \le 900$. It is seen that when there is a measurement from the GPS at $k = 300, 400, \ldots, 800$, the variances of the a posteriori *estimate reduce. However, when there is no measurement from the GPS, the variances increase.*

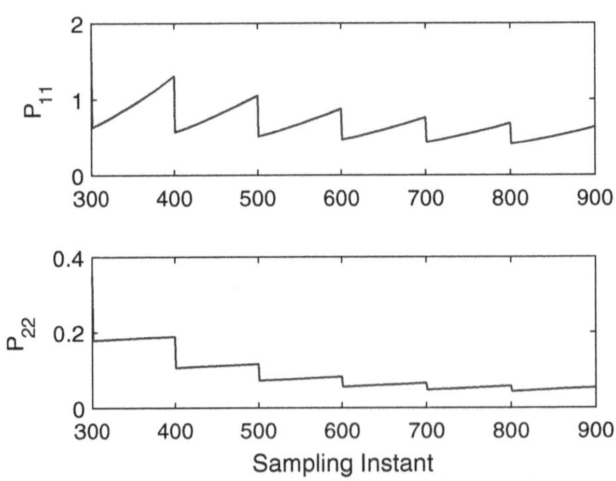

Figure 7.12 Diagonal element of $P(k)^+$ (Example 7.6). Top figure: $P_{11}(k)^+$; bottom figure: $P_{22}(k)^+$.

This section is completed with the following example and the corresponding MATLAB tutorial for practicing the Kalman filter with missing data and in a multi-rate sampling environment.

Example 7.7

In this example, we will demonstrate how to use two GPS measurements for improving the accuracy of state estimation using the Kalman filter with missing data.

Solution

While the state space equation for the moving object is the same as in (7.61), the position measurement equation corresponding to the two GPS measurements is changed as

$$y(k) = \begin{bmatrix} y_1(k) \\ y_2(k) \\ y_3(k) \\ y_4(k) \end{bmatrix} = \begin{bmatrix} 1 & 0 & 0 & 0 \\ 0 & 0 & 1 & 0 \\ 1 & 0 & 0 & 0 \\ 0 & 0 & 1 & 0 \end{bmatrix} \begin{bmatrix} p_x(k) \\ v_x(k) \\ p_y(k) \\ v_y(k) \end{bmatrix} + \begin{bmatrix} v_1(k) \\ v_2(k) \\ v_3(k) \\ v_4(k) \end{bmatrix}. \tag{7.71}$$

If the noises from the two GPS measurements are independent, we expect improvements from using the second GPS measurement. In any case, the two GPS measurements offer a redundancy in the situation when one of the GPS signals is lost. When one of the GPS measurements is not available, this corresponds to the case that a partial measurement is available. For this case, we will choose the correct $C(k)$ and $R(k)$ to reflect the lack of measurements from one of GPS measurements. For example, if the first GPS measurement is available, then the output equation becomes:

$$y(k) = \begin{bmatrix} y_1(k) \\ y_2(k) \end{bmatrix} = \begin{bmatrix} 1 & 0 & 0 & 0 \\ 0 & 0 & 1 & 0 \end{bmatrix} \begin{bmatrix} p_x(k) \\ v_x(k) \\ p_y(k) \\ v_y(k) \end{bmatrix} + \begin{bmatrix} v_1(k) \\ v_2(k) \end{bmatrix}. \tag{7.72}$$

and the noise covariance matrix $R(k)$ has the dimension of 2×2.

Assuming that the first GPS has a higher quality, we generate the independent noises $v_1(k)$ and $v_2(k)$ with $\sigma_1^2 = \sigma_2^2 = 0.25$ and choose the covariance of measurement noise: $R(k) = 0.25I$. We evaluate the performance of the Kalman filter in a multi-rate sampling environment with missing data. All the other simulation conditions are the same as those given in Example 7.6.

The higher quality second GPS has significantly improved the accuracy of the estimated states. The sums of the mean squared error for the position and velocity are

$$\frac{1}{M}\sum_{k=1}^{M} \left(e_1(k)^2 + e_3(k)^2\right) = 0.2862; \quad \frac{1}{M}\sum_{k=1}^{M} \left(e_2(k)^2 + e_4(k)^2\right) = 0.0576.$$

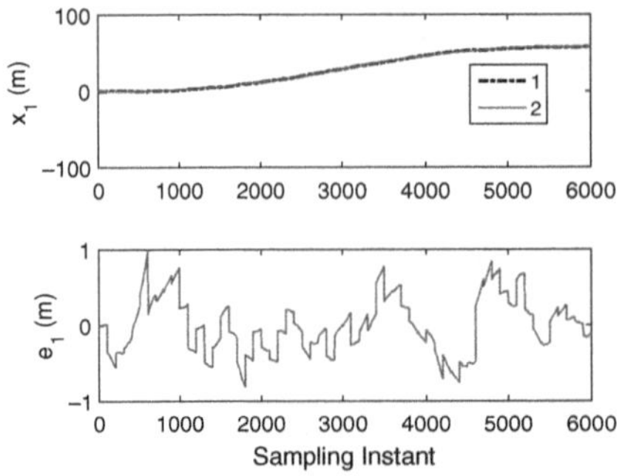

(a) $x_1(k)$ and $e_1(k)$

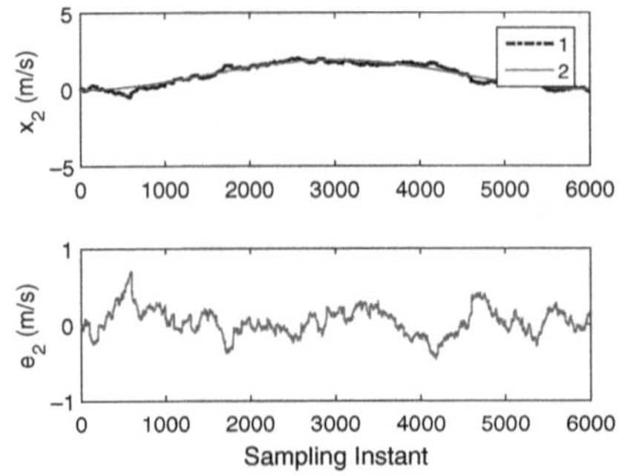

(b) $x_2(k)$ and $e_2(k)$

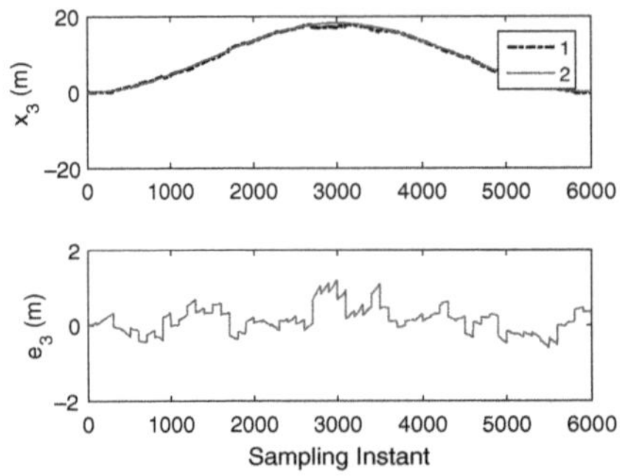

(c) $x_3(k)$ and $e_3(k)$

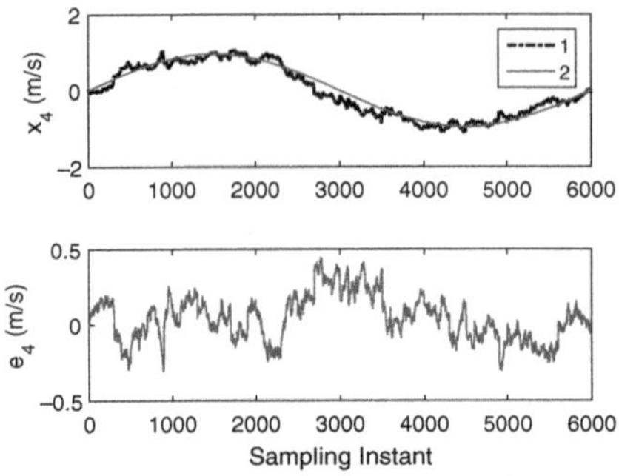

(d) $x_4(k)$ and $e_4(k)$

Figure 7.13 Comparison between true and estimated state with two GPS measurements (Example 7.7). Key: line (1) estimated state; line (2) true state.

In Figures 7.13(a)–(d) the comparisons between true and individual states and their errors are presented. Additionally, Figure 7.14 compares the estimated trajectory of the moving object with the true trajectory.

Tutorial 7.3 *This tutorial is to implement the Kalman filter in a multi-rate sampling environment with missing measurements. It is based on Examples 7.6 and 7.7.*

Step by Step

1. *Create a new file called* **KFMultiRate.m**.
2. *Enter the following system matrices based on (7.61) and (7.71):*

```
Deltat=0.01;
A=[1 Deltat 0 0;0 1 0 0;0 0 1 Deltat;0 0 0 1];
B=[0.5*Deltat^2 0;Deltat 0;0 0.5*Deltat^2;0 Deltat];
C=[1 0 0 0;0 0 1 0;1 0 0 0;0 0 1 0];
[m,n]=size(C);
```

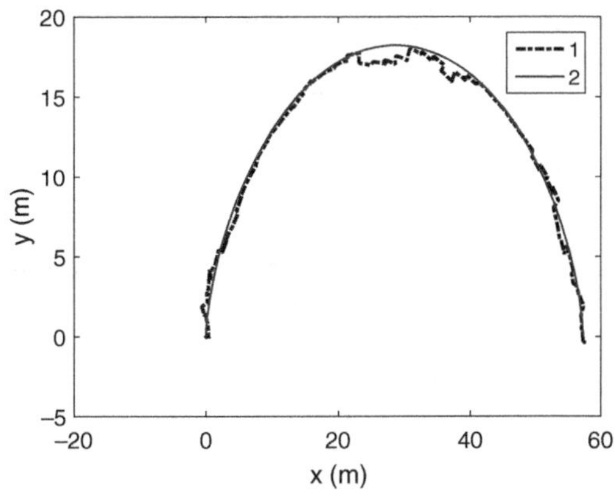

Figure 7.14 Trajectory of the moving object (Example 7.7). Key: line (1) the estimated trajectory with two GPS; line (2) true trajectory.

3. *Generate the input signals, which are the accelerations for the x and y axes. Continue entering the following program into the file:*

```
k=1:6000;
u1=0.1*sin(2*pi*k/6000);
u2=0.1*cos(2*pi*k/6000);
u=[u1;u2];
L=length(u1);
```

4. *Generate the true state variables and the output signals for simulation study. The true state variables will be used in the evaluation of the Kalman filter. Continue entering the following program into the file:*

```
x=[0;0;0;0];
xmodel(:,1)=x;
for kk=2:L;
    x=A*x+B*u(:,kk-1);
    xmodel(:,kk)=x;
    y(:,kk)=C*x;
end
```

5. *Generate the measurement noises and add to the outputs. Enter the following program into the file:*

```
randn('seed',1);
noi1=0.5*randn(1,L);
noi2=0.5*randn(1,L);
noi3=1*randn(1,L);
noi4=1*randn(1,L);
y=y+[noi1;noi2;noi3;noi4];
```

6. *Generate the process noises to be used in the simulation study:*

```
randn('seed',2);
w1=1*randn(1,L);
w2=1*randn(1,L);
w=[w1;w2];
```

7. *Define the Q and R matrices for the Kalman filter. Continue entering the following program into the file:*

```
Q=B*B';
R=eye(m);
R(1,1)=0.25;
R(2,2)=0.25;
```

8. *Choose initial conditions for the Kalman filter. Continue entering the following program into the file:*

```
P_plus=eye(n);
x_plus=zeros(n,1);
xhat(:,1)=x_plus;
```

9. *Define the triggers for the multi-rate sampled data and the missing data, where nratio is the ratio between the fast and slow sampling rates, 'nys=600' corresponds to the case where there is one sample missing from the second GPS measurement among every 600 samples, and 'nxs=6' corresponds to the case where there is one sample missing from the acceleration measurements every six samples.*

```
nratio=100;
nys=600;
nxs=6;
```

10. *Activate the trigger signals:*

```
kstar=0;
kstar1=0;
kstar2=0;
```

11. *The computation will be performed recursively. We also update the trigger signals.*

```
for k=2:L
kstar=kstar+1;
kstar1=kstar1+1;
kstar2=kstar2+1;
```

12. *Compute the* a priori *estimate of the state $\hat{x}(k)^-$ and the covariance matrix $P(k)^-$ with the measured acceleration signals that have noises. If the acceleration signal is missing, then we replace it with the most past sampled signals. Continue entering the following program into the file:*

```
if kstar1==nxs;
kstar1=0;
x_minus=A*x_plus+B*(u(:,k-2)+w(:,k-2));
P_minus=A*P_plus*A'+Q;
else
x_minus=A*x_plus+B*(u(:,k-1)+w(:,k-1));
P_minus=A*P_plus*A'+Q;
end
```

13. *Compute the Kalman filter gain, the* a posteriori *estimate of the state $\hat{x}(k)^+$ and the covariance matrix $P(k)^+$ according to the slower sampling rate. Continue entering the following program into the file:*

```
if kstar==nratio;
      kstar=0;
    C=[1 0 0 0;0 0 1 0;1 0 0 0;0 0 1 0];
    R=eye(4);
    R(1,1)=0.25;
    R(2,2)=0.25;
    K=P_minus*C'*inv(C*P_minus*C'+R);
    x_plus=x_minus+K*(y(:,k)-C*x_minus);
    P_plus=(eye(n)-K*C)*P_minus*(eye(n)-K*C)'+K*R*K';
    else
      x_plus=x_minus;
      P_plus=P_minus;
    end
```

14. *The final case corresponds to the situation when the second GPS measurement is not available. Continue entering the following program into the file:*

```
if kstar2==nys;
kstar2=0;
C=[1 0 0 0;0 0 1 0];
R=0.25*eye(2);
 K=P_minus*C'*inv(C*P_minus*C'+R);
 x_plus=x_minus+K*(y(1:2,k)-C*x_minus);
 P_plus=(eye(n)-K*C)*P_minus*(eye(n)-K*C)'+K*R*K';
 end
```

15. *Save the estimated state variable and diagonal elements of $P(k)^+$ for performance evaluation, and complete one cycle of computation:*

```
  xhat(:,k)=x_plus;
  Pk(:,k)=diag(P_plus);
end
```

16. *Test the program based on Example 7.7.*

7.3.3 Food for Thought

1. Can you think of an application with a multi-rate sampled data system?
2. A state space model is described by

$$x(k) = 0.6x(k-1) + 0.1u(k-1) + w(k-1); y(k) = 6x(k) + v(k),$$

assuming that the process is sampled with $\Delta t = 0.01$ (sec) and the measurement is sampled with $10\Delta t$, what are the key equations of the Kalman filter for the multi-rate sampled data system?

7.4 The Extended Kalman Filter (EKF)

The Kalman filter assumes a linear time-varying model in its design. However, many applications go beyond linear systems and are nonlinear in nature. For these nonlinear systems, the extended Kalman filter is an effective tool to estimate their states.

7.4.1 Linearization in Extended Kalman Filter

The core idea in the extended Kalman filter lies in the concept of linearization of a nonlinear system. In order to understand how the extended Kalman filter works, we will examine two cases of linearization of a nonlinear function. Assume that there is a nonlinear function $f(x)$ with a single variable x and the function $f(x)$ is smooth and its derivatives exist for all the orders. The nonlinear function $f(x)$ can be expressed in terms of a Taylor series expansion at $x = x^0$, where x^0 is a constant, as

$$f(x) = f(x^0) + \frac{df(x)}{dx}|_{x=x^0}(x - x^0) + \frac{1}{2!}\frac{d^2f(x)}{dx^2}|_{x=x^0}(x - x^0)^2 + \dots \quad (7.73)$$

All the derivatives are evaluated at $x = x^0$. In the linearization of the nonlinear function $f(x)$, we will only use the first two terms in the Taylor series expansion leading to the approximation of the nonlinear function $f(x)$ at the specific point x^0:

$$f(x) \approx f(x^0) + \frac{df(x)}{dx}|_{x=x^0}(x - x^0). \tag{7.74}$$

It is clearly seen that the first order Taylor series approximates the original nonlinear function $f(x)$ using the function evaluated at x^0 and its first derivative at $x = x^0$ with multiplication of $(x - x^0)$. When $x = x^0, f(x) = f(x^0)$.

In the multivariable case, x is a vector with dimension $n \times 1$ that contains n variables, where

$$x = \begin{bmatrix} x_1 & x_2 & \dots & x_n \end{bmatrix}^T,$$

and the nonlinear function $f(x)$ is also a vector with dimension $p \times 1$ that contains p nonlinear functions, where

$$f(x) = \begin{bmatrix} f_1(x) & f_2(x) & \dots & f_p(x) \end{bmatrix}^T.$$

The ith nonlinear function $f_i(x)$ is approximated around a constant vector x^0, using the first $n + 1$ terms in the multivariable first order Taylor series expansion as

$$\begin{aligned} f_i(x) \approx f_i(x^0) &+ \left.\frac{\partial f_i(x)}{\partial x_1}\right|_{x=x^0} (x_1 - x_1^0) \\ &+ \left.\frac{\partial f_i(x)}{\partial x_2}\right|_{x=x^0} (x_2 - x_2^0) + \dots + \left.\frac{\partial f_i(x)}{\partial x_n}\right|_{x=x^0} (x_n - x_n^0). \end{aligned} \tag{7.75}$$

Note that we have assumed the existence of the partial derivatives of the nonlinear function $f_i(x)$ against all its variables. We emphasize that at $x = x^0, f_i(x) = f_i(x^0)$.

In the extended Kalman filter design, we assume that a discrete-time nonlinear system is described by the nonlinear difference equation:

$$x(k) = f(x(k-1), u(k-1), w(k-1)), \tag{7.76}$$

where the state vector $x(k)$, the nonlinear function $f(x(k-1), u(k-1), w(k-1))$ and the process noise $w(k)$ have the dimension $n \times 1$. More specifically, the nonlinear function $f(x(k-1), u(k-1), w(k-1))$ is expressed as:

$$f(x(k-1), u(k-1), w(k-1)) = \begin{bmatrix} f_1(x(k-1), u(k-1), w(k-1)) \\ f_2(x(k-1), u(k-1), w(k-1)) \\ \vdots \\ f_n(x(k-1), u(k-1), w(k-1)) \end{bmatrix}.$$

Additionally, we assume that the measurement equation is given by

$$y(k) = h(x(k), v(k)) \tag{7.77}$$

where the output vector $y(k)$ and the measurement noise $v(k)$ have the dimension $m \times 1$. The nonlinear function $h(x(k), v(k))$ is written as:

$$h(x(k), v(k)) = \begin{bmatrix} h_1(x(k), v(k)) \\ h_2(x(k), v(k)) \\ \vdots \\ h_m(x(k), v(k)) \end{bmatrix}.$$

Both process noise $w(k)$ and measurement noise $v(k)$ are zero mean and independent, and they have the covariance matrices $Q(k)$ and $R(k)$ respectively.

The key to the extended Kalman filter lies in two linearization procedures.

1. The nonlinear function $f(x(k-1), u(k-1), w(k-1))$ in (7.76) is linearized with respect to the state vector $x(k-1)$ and the process noise $w(k-1)$ at the operating conditions: $x(k-1) = \hat{x}(k-1)^+$ and $w(k-1) = 0$.
2. The nonlinear function $h(x(k), v(k))$ in (7.77) is linearized with respect to the state vector $x(k)$ and the measurement noise $v(k)$ at the operating conditions: $x(k) = \hat{x}(k)^-$ and $v(k) = 0$. Here, the same as in the Kalman filter algorithm, $\hat{x}(k-1)^+$ is the *a posteriori* estimate of the state at the sampling instant $k-1$ and $\hat{x}(k)^-$ is the *a priori* estimate of the state at the sampling instant k.

A. Linearization of $f(x(k-1), u(k-1), w(k-1))$

At the sampling time k, we are given the *a posteriori* estimate $\hat{x}(k-1)^+$, where

$$\hat{x}(k-1)^+ = \left[\hat{x}_1(k-1)^+ \quad \hat{x}_2(k-1)^+ \quad \dots \quad \hat{x}_n(k-1)^+\right]^T.$$

The first order Taylor series expansion of the ith nonlinear function $f_i(x(k-1), u(k-1), w(k-1))$ with respect to the state vector $x(k-1)$ and process noise vector $w(k-1)$ at $x(k-1) = \hat{x}(k-1)^-$ and $w(k-1) = 0$ gives the following approximation:

$$\begin{aligned} & f_i(x(k-1), u(k-1), w(k-1)) \approx f_i(\hat{x}(k-1)^+, u(k-1), 0) \\ & + \frac{\partial f_i(x(k-1), u(k-1), w(k-1))}{\partial x_1(k-1)}|_{(\hat{x}(k-1)^+,0)}(x_1(k-1) - \hat{x}_1(k-1)^+) \\ & + \frac{\partial f_i(x(k-1), u(k-1), w(k-1))}{\partial x_2(k-1)}|_{(\hat{x}(k-1)^+,0)}(x_2(k-1) - \hat{x}_2(k-1)^+) + \dots \\ & + \frac{\partial f_i(x(k-1), u(k-1), w(k-1))}{\partial x_n(k-1)}|_{(\hat{x}(k-1)^+,0)}(x_n(k-1) - \hat{x}_n(k-1)^+) \\ & + \frac{\partial f_i(x(k-1), u(k-1), w(k-1))}{\partial w_1(k-1)}|_{(\hat{x}(k-1)^+,0)} w_1(k-1) \end{aligned}$$

$$+ \frac{\partial f_i(x(k-1), u(k-1), w(k-1))}{\partial w_2(k-1)}|_{(\hat{x}(k-1)^+,0)} w_2(k-1) + \dots$$
$$+ \frac{\partial f_i(x(k-1), u(k-1), w(k-1))}{\partial w_n(k-1)}|_{(\hat{x}(k-1)^+,0)} w_n(k-1) \tag{7.78}$$

where $\frac{\partial f_i(x(k-1),u(k-1),w(k-1))}{\partial x_i(k-1)}|_{(\hat{x}(k-1)^+,0)}$ denotes that the partial derivative is evaluated at $x(k-1) = \hat{x}(k-1)^+$ and $w(k-1) = 0$. After applying the first order Taylor series expansion to all $f_i(x(k-1), u(k-1), w(k-1))$ where $i = 1, 2, \dots, n$, we obtain the linearized state space model at the given state vector $\hat{x}(k-1)^+$:

$$x(k) \approx f(\hat{x}(k-1)^+, u(k-1), 0) + A(k-1)(x(k-1) - \hat{x}(k-1)^+)$$
$$+ M_w(k-1)w(k-1) \tag{7.79}$$

where the matrices $A(k-1)$ and $M_w(k-1)$ have the dimension $n \times n$ and are called Jacobian matrices. More specifically, the variables in (7.79) are computed using the following expressions:

$$A(k-1) = \overbrace{\begin{bmatrix} \frac{\partial f_1(x(k-1),u(k-1),w(k-1))}{\partial x_1(k-1)} & \cdots & \frac{\partial f_1(x(k-1),u(k-1),w(k-1))}{\partial x_n(k-1)} \\ \frac{\partial f_2(x(k-1),u(k-1),w(k-1))}{\partial x_1(k-1)} & \cdots & \frac{\partial f_2(x(k-1),u(k-1),w(k-1))}{\partial x_n(k-1)} \\ \vdots & \vdots & \vdots \\ \frac{\partial f_n(x(k-1),u(k-1),w(k-1))}{\partial x_1(k-1)} & \cdots & \frac{\partial f_n(x(k-1),u(k-1),w(k-1))}{\partial x_n(k-1)} \end{bmatrix}}^{\text{Jacobian}} |_{(\hat{x}(k-1)^+,0)} \tag{7.80}$$

$$M_w(k-1) = \overbrace{\begin{bmatrix} \frac{\partial f_1(x(k-1),u(k-1),w(k-1))}{\partial w_1(k-1)} & \cdots & \frac{\partial f_1(x(k-1),u(k-1),w(k-1))}{\partial w_n(k-1)} \\ \frac{\partial f_2(x(k-1),u(k-1),w(k-1))}{\partial w_1(k-1)} & \cdots & \frac{\partial f_2(x(k-1),u(k-1),w(k-1))}{\partial w_n(k-1)} \\ \vdots & \vdots & \vdots \\ \frac{\partial f_n(x(k-1),u(k-1),w(k-1))}{\partial w_1(k-1)} & \cdots & \frac{\partial f_n(x(k-1),u(k-1),w(k-1))}{\partial w_n(k-1)} \end{bmatrix}}^{\text{Jacobian}} |_{(\hat{x}(k-1)^+,0)} \tag{7.81}$$

and $f(\hat{x}(k-1)^+, u(k-1), 0)$ is given as

$$f(\hat{x}(k-1)^+, u(k-1), 0) = \begin{bmatrix} f_1(\hat{x}(k-1)^+, u(k-1), 0) \\ f_2(\hat{x}(k-1)^+, u(k-1), 0) \\ \vdots \\ f_n(\hat{x}(k-1)^+, u(k-1), 0) \end{bmatrix}.$$

B. Linearization of $h(x(k), v(k))$

At the sampling time k, we are given the *a priori* estimate $\hat{x}(k)^-$, where

$$\hat{x}(k)^- = \left[\hat{x}_1(k)^- \;\; \hat{x}_2(k)^- \;\; \dots \;\; \hat{x}_n(k)^-\right]^T.$$

Following the same linearization procedure with the first order Taylor series expansion, the nonlinear measurement equation (7.77) is linearized against $x(k)$ and $v(k)$ at the operating conditions $x(k) = \hat{x}(k)^-$ and $v(k) = 0$. The linearized output equation at the sampling instant k is written as

$$y(k) \approx h(\hat{x}(k)^-, 0) + C(k)(x(k) - \hat{x}(k)^-) + M_v(k)v(k) \tag{7.82}$$

where the output measurement matrix $C(k)$ and matrix $M_v(k)$ are computed as the corresponding Jacobian:

$$C(k) = \overbrace{\begin{bmatrix} \frac{\partial h_1(x(k),v(k))}{\partial x_1(k)} & \frac{\partial h_1(x(k),v(k))}{\partial x_2(k)} & \cdots & \frac{\partial h_1(x(k),v(k))}{\partial x_n(k)} \\ \frac{\partial h_2(x(k),v(k))}{\partial x_1(k)} & \frac{\partial h_2(x(k),v(k))}{\partial x_2(k)} & \cdots & \frac{\partial h_2(x(k),v(k))}{\partial x_n(k)} \\ \vdots & \vdots & \vdots & \vdots \\ \frac{\partial h_m(x(k),v(k))}{\partial x_1(k)} & \frac{\partial h_m(x(k),v(k))}{\partial x_2(k)} & \cdots & \frac{\partial h_m(x(k),v(k))}{\partial x_n(k)} \end{bmatrix}}^{\text{Jacobian}} \Bigg|_{(\hat{x}(k)^-,0)}$$

$$M_v(k) = \overbrace{\begin{bmatrix} \frac{\partial h_1(x(k),v(k))}{\partial v_1(k)} & \frac{\partial h_1(x(k),v(k))}{\partial v_2(k)} & \cdots & \frac{\partial h_1(x(k),v(k))}{\partial v_m(k)} \\ \frac{\partial h_2(x(k),v(k))}{\partial v_1(k)} & \frac{\partial h_2(x(k),v(k))}{\partial v_2(k)} & \cdots & \frac{\partial h_2(x(k),v(k))}{\partial v_m(k)} \\ \vdots & \vdots & \vdots & \vdots \\ \frac{\partial h_m(x(k),v(k))}{\partial v_1(k)} & \frac{\partial h_m(x(k),v(k))}{\partial v_2(k)} & \cdots & \frac{\partial h_m(x(k),v(k))}{\partial v_m(k)} \end{bmatrix}}^{\text{Jacobian}} \Bigg|_{(\hat{x}(k)^-,0)},$$

and $h(\hat{x}(k)^-, 0)$ is computed as:

$$h(\hat{x}(k)^-, 0) = \begin{bmatrix} h_1(\hat{x}(k)^-, 0) \\ h_2(\hat{x}(k)^-, 0) \\ \vdots \\ h_m(\hat{x}(k)^-, 0) \end{bmatrix}.$$

7.4.2 The Extended Kalman Filter Algorithm

The extended Kalman filter algorithm is based on the linearized state space model (see (7.79) and (7.82)) and in spirit, follows the standard Kalman filter that was introduced in the previous sections (see Sections 7.2.1–7.2.4).

There are three key elements in the Kalman filter for a time-varying system, which are the computations of the *a priori* estimate $\hat{x}(k)^-$ and the covariance $P(k)^-$, the *a posteriori* estimate $\hat{x}(k)^+$ and the covariance $P(k)^+$, as well as the Kalman filter gain $K(k)$. In the following, we will show how these computations can be naturally extended to the nonlinear case based on the approximate models given by (7.79) and (7.82).

A. The *a priori* estimate $\hat{x}(k)^-$ and the covariance $P(k)^-$

For the first order Taylor series approximation of the original nonlinear model, the best estimate of the $\hat{x}(k)^-$ is the first term in (7.79), which is

$$\hat{x}(k)^- = f(\hat{x}(k-1)^+, u(k-1), 0). \tag{7.83}$$

The error between $x(k)$ given by (7.79) and $\hat{x}(k)^-$ is calculated as,

$$x(k) - \hat{x}(k)^- \approx A(k-1)(x(k-1) - \hat{x}(k-1)^+) + M_w(k-1)w(k-1). \tag{7.84}$$

From (7.84), we calculate its covariance matrix $P(k)^-$ as,

$$\begin{aligned} P(k)^- &= E[(x(k) - \hat{x}(k)^-)(x(k) - \hat{x}(k)^-)^T] \\ &= A(k-1)P(k-1)^+A(k-1)^T + M_w(k-1)Q(k-1)M_w(k-1)^T \end{aligned} \tag{7.85}$$

where $Q(k-1) = E[w(k-1)w(k-1)^T]$. Note that the computation of $P(k)^-$ is derived by following the same steps as in the standard Kalman filter (see, for example, (7.9)–(7.10)).

B. The *a posteriori* estimate $\hat{x}(k)^+$ and the covariance $P(k)^+$

When the measurement $y(k)$ becomes available, this information will be used to improve the estimate $\hat{x}(k)^-$ and yield the *a posteriori* estimate $\hat{x}(k)^+$. In the extended Kalman filter, differing from the standard Kalman filter, this improvement is achieved using a correction with the error between $y(k)$ and $h(\hat{x}(k)^-, 0)$ from the first term of the Taylor series expansion (see (7.82)), where $h(\hat{x}(k)^-, 0)$ has the dimension $m \times 1$ and is the predicted measurement taken from the point of view of the *a priori* estimate $\hat{x}(k)^-$ based on the linearized measurement model. The *a posteriori* estimate $\hat{x}(k)^+$ is calculated as

$$\hat{x}(k)^+ = \hat{x}(k)^- + K(k)(y(k) - h(\hat{x}(k)^-, 0)). \tag{7.86}$$

When using the linearized measurement model (7.82), the error between $x(k)$ and $\hat{x}(k)^+$ becomes:

$$\begin{aligned} x(k) - \hat{x}(k)^+ &\approx x(k) - \hat{x}(k)^- \\ &- K(k)\left(h(\hat{x}(k)^-, 0) + C(k)(x(k) - \hat{x}(k)^-) + M_v(k)v(k) - h(\hat{x}(k)^-, 0)\right) \\ &= (I - K(k)C(k))(x(k) - \hat{x}(k)^-) - K(k)M_v(k)v(k). \end{aligned} \tag{7.87}$$

Hence, we have the *a posteriori* estimate of the covariance matrix as

$$\begin{aligned}P(k)^+ &= E[(x(k)-\hat{x}(k)^+)(x(k)-\hat{x}(k)^+)^T] \\ &= (I-K(k)C(k))P(k)^-(I-K(k)C(k))^T \\ &\quad + K(k)M_v(k)R(k)M_v(k)^TK(k)^T. \end{aligned} \tag{7.88}$$

C. The optimal Kalman filter gain $K(k)$

Note that by considering $M_v(k)R(k)M_v(k)^T$ as the new noise term, $P(k)^+$ has the identical form as that in the standard Kalman filter (see (7.16)). Thus, by following the same derivation procedure outlined in Section 7.2.3, the extended Kalman filter gain $K(k)$ is optimized and has the following form:

$$K(k) = P(k)^-C(k)^T[C(k)P(k)^-C(k)^T + M_v(k)R(k)M_v(k)^T]^{-1}. \tag{7.89}$$

For clarity, the extended Kalman filter algorithm is summarized as follows.

To begin the iterative computation, we will choose initial conditions for the *a posteriori* state estimate $\hat{x}(0)^+ = E[x(0)]$ and the covariance of the error $P(0)^+ = E[(x(0)-\hat{x}(0)^+)(x(0)-\hat{x}(0)^+)^T]$ as well as the covariance matrices Q for the process noise $w(k)$ and R for the measurement noise $v(k)$. The following computational procedure is repeated for each sampling instant k based on the linearized state space model (see (7.79) and (7.82)).

1. Compute the *a priori* estimate of the state using $f(\hat{x}(k-1)^+, u(k-1), 0)$ from (7.79):

$$\hat{x}(k)^- = f(\hat{x}(k-1)^+, u(k-1), 0). \tag{7.90}$$

2. Compute the *a priori* estimate of the covariance matrix:

$$P(k)^- = A(k-1)P(k-1)^+A(k-1)^T + M_w(k-1)Q(k-1)M_w(k-1)^T. \tag{7.91}$$

3. Compute the Kalman filter gain:

$$K(k) = P(k)^-C(k)^T[C(k)P(k)^-C(k)^T + M_v(k)R(k)M_v(k)^T]^{-1}. \tag{7.92}$$

4. Compute the *a posteriori* estimate of the state:

$$\hat{x}(k)^+ = \hat{x}(k)^- + K(k)(y(k) - h(\hat{x}(k)^-, 0)). \tag{7.93}$$

5. Compute the *a posteriori* covariance matrix:

$$\begin{aligned}P(k)^+ &= (I-K(k)C(k))P(k)^-(I-K(k)C(k))^T \\ &\quad + K(k)M_v(k)R(k)M_v(k)^TK(k)^T. \end{aligned} \tag{7.94}$$

The *a posteriori* covariance matrix also has the following simplified form:

$$P(k)^+ = (I-K(k)C(k))P(k)^- \tag{7.95}$$

7.4.3 Case Studies with MATLAB Tutorial

The following two examples are to illustrate the estimation of unknown state using the extended Kalman filter.

Example 7.8

A continuous-time system is described by the following differential equations:

$$\begin{bmatrix} \dot{x}(t) \\ \dot{\theta}(t) \end{bmatrix} = \begin{bmatrix} x(t) + \cos\theta(t) + w_1(t) \\ -x(t) + \sin\theta(t) + w_2(t) \end{bmatrix} \tag{7.96}$$

where the process noises $w_1(t)$ and $w_2(t)$ are known to be uncorrelated with zero mean and variance 0.1.

The variable $x(t)$ is measured, which gives the output equation:

$$y(t) = \begin{bmatrix} 1 & 0 \end{bmatrix} \begin{bmatrix} x(t) \\ \theta(t) \end{bmatrix} + v(t) \tag{7.97}$$

Design an extended Kalman filter to find the estimated states $\hat{x}(t)$ and $\hat{\theta}(t)$ with sampling interval $\Delta t = 0.01$. Although $x(t)$ is measured, the estimated state $\hat{x}(t)$ is a filtered signal and $\hat{\theta}(t)$ is an estimate of the unknown state $\theta(t)$.

Solution

With sampling interval Δt, the continuous-time system is discretized using the following approximations:

$$\dot{x}(t) \approx \frac{x(k) - x(k-1)}{\Delta t} \tag{7.98}$$

$$\dot{\theta}(t) \approx \frac{\theta(k) - \theta(k-1)}{\Delta t}. \tag{7.99}$$

Here, the corresponding discrete-time system has two states: $[x(k)\ \theta(k)]^T$, and process noise $w(k) = [w_1(k)\ w_2(k)]^T$. The measurement $y(k)$, is a scalar with measurement noise $v(k)$. Following discretization, the discrete-time system is:

$$\begin{bmatrix} x(k) \\ \theta(k) \end{bmatrix} = \Delta t \begin{bmatrix} x(k-1) + \cos\theta(k-1) \\ -x(k-1) + \sin\theta(k-1) \end{bmatrix} + \begin{bmatrix} x(k-1) \\ \theta(k-1) \end{bmatrix} + \Delta t \begin{bmatrix} w_1(k-1) \\ w_2(k-1) \end{bmatrix}$$
$$= \begin{bmatrix} f_1(x(k-1), \theta(k-1), w_1(k-1)) \\ f_2(x(k-1), \theta(k-1), w_2(k-1)) \end{bmatrix}$$

The measurement is collected from the most current state, which gives the following measurement equation:

$$y(k) = \begin{bmatrix} 1 & 0 \end{bmatrix} \begin{bmatrix} x(k) \\ \theta(k) \end{bmatrix} + v(k) = h(x(k), \theta(k), v(k))$$

where $w(k-1)$ and $v(k)$ are both assumed to be independent and Gaussian distributed with zero-mean.

The system matrix $A(k-1)$ is calculated with the Jacobian matrix:

$$A(k-1) = \begin{bmatrix} \frac{\partial f_1}{\partial x} & \frac{\partial f_1}{\partial \theta} \\ \frac{\partial f_2}{\partial x} & \frac{\partial f_2}{\partial \theta} \end{bmatrix} = \begin{bmatrix} \Delta t + 1 & -\Delta t \sin\theta(k-1) \\ -\Delta t & \Delta t \cos\theta(k-1) + 1 \end{bmatrix} \tag{7.100}$$

and the output matrix C is a constant matrix:

$$C = \begin{bmatrix} \frac{\partial h_1}{\partial x} & \frac{\partial h_1}{\partial \theta} \end{bmatrix} = \begin{bmatrix} 1 & 0 \end{bmatrix} \tag{7.101}$$

The process noise $w(k-1)\Delta t$ is selected to have a covariance matrix:

$$Q = \begin{bmatrix} 0.001 & 0 \\ 0 & 0.001 \end{bmatrix}$$

and the measurement noise $v(k)$ is assumed to have a scalar covariance i.e. variance, $R = 0.1$, and the diagonal elements in the Q matrix are selected to corresponding to the noise in continuous-time with the multiplication of Δt. We can tune these parameters for the extended Kalman filter if they are unknown to us. Figures 7.15(a)–(b) compare the estimated states to the true states. The underlying nonlinear system is unstable, where we have seen that the states are growing with respect to time and the estimated states $\hat{x}$ and $\hat{\theta}$ track the true states very well. The instability can be confirmed with a longer simulation time. Figures 7.16(a)–(b) further confirm that the errors between the true states and the estimated states are stable.

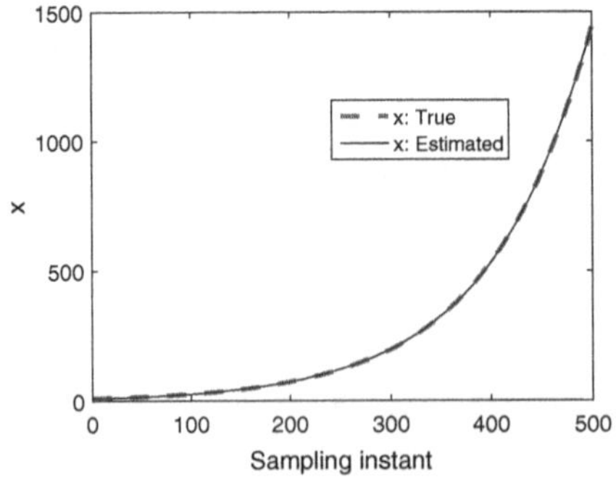

(a) Estimated and true x

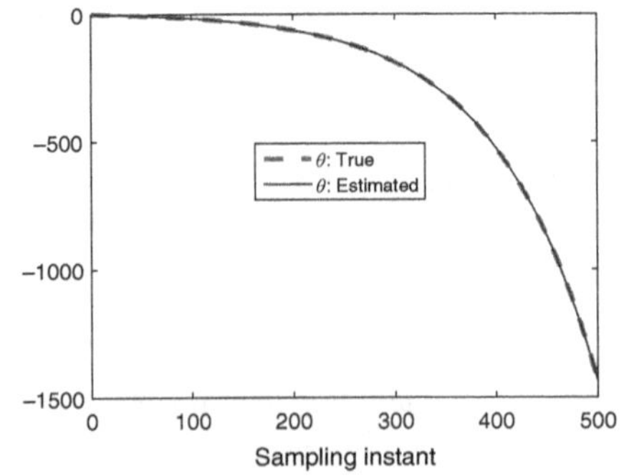

(b) Estimated and true θ

Figure 7.15 Estimation results for EKF (Example 7.8). $Q = 0.001I$ and $R = 0.1$.

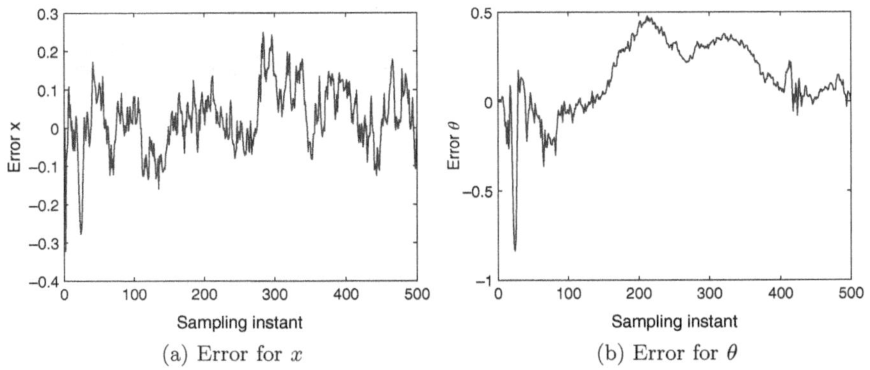

Figure 7.16 Estimation error for EKF (Example 7.8). $Q = 0.001I$ and $R = 0.1$.

The choices of Q and R matrices have more influence on the performance of the extended Kalman filter in comparison with their effects on the Kalman filter. Their optimal choices are the underlying variances of both process and measurement noises. When these are unknown, we will need to tune the parameters through trial-and-error practice.

This section is completed using the following example together with Tutorial 7.4.

Example 7.9

The dynamic model of two connected water tanks (see Figure 7.17) is described by the nonlinear differential equations:

$$\frac{dL_1(t)}{dt} = -\frac{a_s}{S_1}\sqrt{2g_0(L_1(t) - L_2(t))} + \frac{1}{S_1}u_1(t) \tag{7.102}$$

$$\frac{dL_2(t)}{dt} = \frac{a_s}{S_2}\sqrt{2g_0(L_1(t) - L_2(t))} - \frac{1}{S_2}u_2(t) \tag{7.103}$$

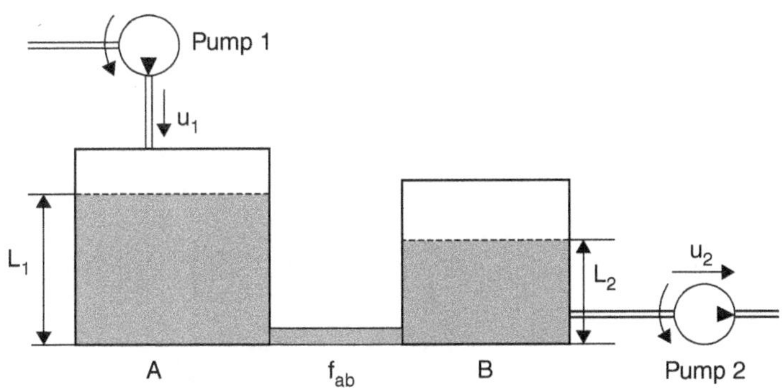

Figure 7.17 Schematic of a double tank.

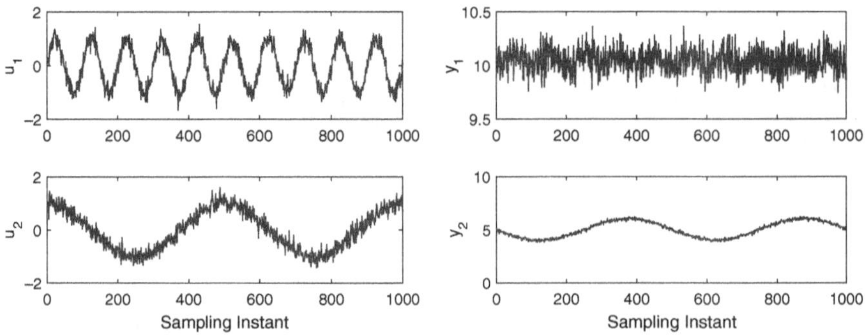

(a) Measured input signal (b) Measured output signal

Figure 7.18 Measured input and output signals (Example 7.9).

where water flows into the first tank and flows out from the second tank. A pump controls the water in-flow rate $u_1(t)$ (m^3/sec) to the first tank; and another pump controls the water out-flow rate $u_2(t)$ (m^3/sec) from the second tank. The variables $L_1(t)$ and $L_2(t)$ represent the water levels for tanks A and B respectively. For a given system, we have the following physical parameters: $a_s = 3.1416e-04$ (m^2), $S_1 = 3.1416$ (m^2), $S_2 = 0.7854$ (m^2) and $g_0 = 9.81$ (m/sec^2). This nonlinear system was used as an example to illustrate linearization in Wang [2020].

For this example, the sampling interval Δt is chosen to be 0.01 (sec). The input signals are the in-flow and out-flow rates that are measured and have measurement noise as shown in Figure 7.18(a). The water levels L_1 and L_2 are measured with low-cost sensors and as a result, they are severely corrupted with measurement noise, particularly with the measurement of L_1. Figure 7.18(b) shows the measured output signals. Design an extended Kalman filter to filter the noise and estimate the water tank levels.

Solution

With the sampling interval Δt, the derivatives $\frac{dL_1(t)}{dt}$ and $\frac{dL_2(t)}{dt}$ are approximated at the sampling instant k:

$$\frac{dL_1(t)}{dt} \approx \frac{L_1(k) - L_1(k-1)}{\Delta t} \tag{7.104}$$

$$\frac{dL_2(t)}{dt} \approx \frac{L_2(k) - L_2(k-1)}{\Delta t}. \tag{7.105}$$

Then, based on (7.102) and (7.103) we obtain the nonlinear difference equations:

$$\begin{aligned} L_1(k) &= f_1(L(k-1), u(k-1)) \\ &= L_1(k-1) + \Delta t \left(-\frac{a_s}{S_1}\sqrt{2g_0(L_1(k-1) - L_2(k-1))} + \frac{1}{S_1}u_1(k-1) \right) \end{aligned} \tag{7.106}$$

$$
\begin{aligned}
L_2(k) &= f_2(L(k-1), u(k-1)) \\
&= L_2(k-1) + \Delta t \left(\frac{a_s}{S_2} \sqrt{2g_0(L_1(k-1) - L_2(k-1))} - \frac{1}{S_2} u_2(k-1) \right)
\end{aligned} \tag{7.107}
$$

where $L(k) = \begin{bmatrix} L_1(k) & L_2(k) \end{bmatrix}^T$ *and* $u(k) = \begin{bmatrix} u_1(k) & u_2(k) \end{bmatrix}^T$.

The elements in the Jacobian matrix $A(k-1)$ *are computed based on (7.80) as:*

$$
\begin{aligned}
a_{11}(k-1) &= \frac{\partial f_1(L(k-1), u(k-1))}{\partial L_1(k-1)} |_{\hat{L}(k-1)^+} \\
&= 1 - 0.5\Delta t \frac{a_s}{S_1} \sqrt{2g_0} \frac{1}{\sqrt{\hat{L}_1(k-1)^+ - \hat{L}_2(k-1)^+}} \\
a_{12}(k-1) &= \frac{\partial f_1(L(k-1), u(k-1))}{\partial L_2(k-1)} |_{\hat{L}(k-1)^+} \\
&= 0.5\Delta t \frac{a_s}{S_1} \sqrt{2g_0} \frac{1}{\sqrt{\hat{L}_1(k-1)^+ - \hat{L}_2(k-1)^+}} \\
a_{21}(k-1) &= \frac{\partial f_2(L(k-1), u(k-1))}{\partial L_1(k-1)} |_{\hat{L}(k-1)^+} \\
&= 0.5\Delta t \frac{a_s}{S_2} \sqrt{2g_0} \frac{1}{\sqrt{\hat{L}_1(k-1)^+ - \hat{L}_2(k-1)^+}} \\
a_{22}(k-1) &= \frac{\partial f_2(L(k-1), u(k-1))}{\partial L_2(k-1)} |_{\hat{L}(k-1)^+} \\
&= 1 - 0.5\Delta t \frac{a_s}{S_2} \sqrt{2g_0} \frac{1}{\sqrt{\hat{L}_1(k-1)^+ - \hat{L}_2(k-1)^+}}.
\end{aligned}
$$

The output matrix C *is identity matrix because both water tank levels are measured with low-cost sensors. The role of the Kalman filter is to filter the measurement noise.*

To generate the data for simulation studies, we use the initial conditions $L_1(0) = 10$ *and* $L_2(0) = 5$*, the sampling interval* $\Delta t = 0.01$*. The input signals to the process are noise-free where the first input is* $\sin(2\pi k/100)$ *and the second input is* $\cos(2\pi k/500)$*. However, these noise-free signals are not available directly. The measured input signals are corrupted by noise, which are* $u_1(k) = \sin(2\pi k/100) + \epsilon_1(k)$ *and* $u_2(k) = \cos(2\pi k/500) + \epsilon_2(k)$ *where* $\epsilon_1(k)$ *and* $\epsilon_2(k)$ *are white noise and have variance* $\sigma_1^2 = \sigma_2^2 = 0.04$*. The measured input signals are used in the extended Kalman filter. The measurement noise* $v_1(k)$ *and* $v_2(k)$ *are white noise and have variance 0.01.*

In the next step, we will translate the measurement noise on the input signal to the process noise. From (7.106) it is seen that the process noise $w_1(k) = \frac{\Delta t}{S_1}\epsilon_1(k)$ *and* $w_2(k) = \frac{\Delta t}{S_2}\epsilon_2(k)$*. The covariance matrix* Q *is a diagonal matrix, where*

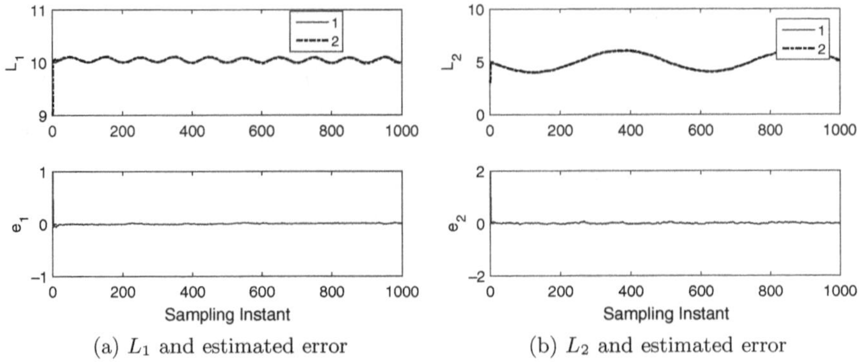

Figure 7.19 Comparison between the estimated and true water tank levels (Example 7.9). Key: line (1) true water level; line (2) estimated water level.

$Q(1,1) = \left(\frac{\Delta t}{S_1}\right)^2 \sigma_1^2 = 4 \times 10^{-7}$ *and* $Q(2,2) = \left(\frac{\Delta t}{S_2}\right)^2 \sigma_2^2 = 6 \times 10^{-6}$. *For the measurement noises, the variances are 0.001, thus we choose* $R = 0.001I$.

The initial conditions for the extended Kalman filter are chosen to be $P(0)^+ = I$, $\hat{L}_1(0)^+ = 9$, *and* $\hat{L}_2(0)^+ = 3$. *Since it is in a simulation environment, the true water tank levels are computed for comparison. Figures 7.19(a)–(b) evaluate the estimated water tank levels against the actual water tank levels, where the errors are calculated as* $e_1(k) = L_1(k) - \hat{L}_1(k)^+$ *and* $e_2(k) = L_2(k) - \hat{L}_2(k)^+$. *It is seen that the extended Kalman filter has removed most of the measurement noise and recovered the actual signals. We calculated the sum of mean squared error as*

$$\frac{1}{M}\sum_{k=1}^{M}\left((L_1(k) - \hat{L}_1(k)^+)^2 + (L_2(k) - \hat{L}_2(k)^+)^2\right) = 0.0055$$

where M is the data length. We can follow Tutorial 7.4 to verify the simulation results.

Tutorial 7.4 *This tutorial is to implement the extended Kalman filter given by Example 7.9.*

Step by Step

1. *Create a new file called* **Example4EKF.m.**
2. *Define the system parameters based on (7.102) and (7.103):*

```
rs=0.01;
rTank1=1;
rTank2=0.5;
as=pi*rs^2;
```

```
S1=pi*rTank1^2;
S2=pi*rTank2^2;
g=9.81;
gamma1=as*sqrt(2*g)/S1;
gamma2=as*sqrt(2*g)/S2;
Deltat=0.01;
```

3. *Generate the input signals and define the measurement matrix:*

```
k=1:1000;
u1=sin(2*pi*k/100);
u2=cos(2*pi*k/500);
C=[1 0;0 1];
B=[1/S1 0;0 1/S2];
```

4. *Generate the true state variables and the output signals for simulation study based on (7.102) and (7.103). The true state variables will be used in the evaluation of the Kalman filter. Continue entering the following program into the file:*

```
u=[u1;u2];
x1=10;
x2=5;
x=[x1;x2];
y1=C*x;
L=length(u1);
xmodel(:,1)=x;
y(:,1)=y1;
for kk=2:L;
    x1=x1+(-gamma1*sqrt(x1-x2)+1/S1*u1(kk-1))*Deltat;
    x2=x2+(gamma2*sqrt(x1-x2)-1/S2*u2(kk-1))*Deltat;
    x=[x1;x2];
    xmodel(:,kk)=x;
    y1=C*x;
    y(:,kk)=y1;
end
```

5. *Generate the measurement noises and add to the outputs. Enter the following program into the file:*

```
randn('seed',1);
noi1=0.1*randn(1,L);
```

```
noi2=0.1*randn(1,L);
y=y+[noi1;noi2];
```

6. *Generate the process noises to be used in the extended Kalman filter:*

```
[m,n]=size(C);
randn('seed',2);
w1=0.2*randn(1,L);
w2=0.2*randn(1,L);
```

7. *Define the Q and R matrices for the extended Kalman filter. Continue entering the following program into the file:*

```
Q=zeros(n);
Q(1,1)=(Deltat/S1)^2*0.04;
Q(2,2)=(Deltat/S2)^2*0.04;
R=0.001*eye(m);
```

8. *Choose initial conditions for the Kalman filter. Continue entering the following program into the file:*

```
P_plus=1*eye(n);
x1_plus=9;
x2_plus=3;
xhat(:,1)=[x1_plus;x2_plus];
```

9. *The following computation will be performed recursively. We will compute the* a priori *estimate of the state $\hat{x}(k)^-$ with the measured input signals by adding the noise $w_1(k)$ and $w_2(k)$. Continue entering the following program into the file:*

```
for k=2:L
u1n(k-1)=u1(k-1)+w1(k-1);
u2n(k-1)=u2(k-1)+w2(k-1);
x1_minus=x1_plus+(-gamma1*sqrt(x1_plus-x2_plus)+1/S1*
u1n(k-1))*Deltat;
x2_minus=x2_plus+(gamma2*sqrt(x1_plus-x2_plus)-1/S2*
u2n(k-1))*Deltat;
x_minus=[x1_minus;x2_minus];
```

10. *Compute the Jacobian matrix $A(k-1)$ and the covariance matrix $P(k)^-$:*

```
a11=-0.5*gamma1/sqrt(x1_plus-x2_plus);
a12=-a11;
a21=0.5*gamma2/sqrt(x1_plus-x2_plus);
a22=-a21;
```

```
A=eye(2)+[a11 a12;a21 a22]*Deltat;
P_minus=A*P_plus*A'+Q;
```

11. *Compute the Kalman filter gain, the* a posteriori *estimate of the state* $\hat{x}(k)^+$ *and the covariance matrix* $P(k)^+$*:*

```
K=P_minus*C'*inv(C*P_minus*C'+R);
x_plus=x_minus+K*(y(:,k)-C*x_minus);
x1_plus=x_plus(1,1);
x2_plus=x_plus(2,1);
P_plus=(eye(n)-K*C)*P_minus*(eye(n)-K*C)'+K*R*K';
```

12. *Save the estimated state variable and diagonal elements of* $P(k)^+$ *for performance evaluation, and complete one cycle of computation:*

```
xhat(:,k)=x_plus;
Pk(:,k)=diag(P_plus);
end
```

13. *Test the program based on Example 7.9.*

7.4.4 Food for Thought

1. Suppose that a continuous-time system is described by the following state space model:

$$\dot{x}(t) = ax(t)^2 + bx(t)u(t) + \overline{w}(t);\ \ y(t) = \sqrt{x(t)} + \overline{v}(t),$$

where a and b are constant. Assuming the sampling interval Δt, what is the discrete-time nonlinear state space model that is expressed as

$$x(k) = f(x(k-1), u(k-1)) + w(k-1); y(k) = h(x(k)) + v(k)?$$

2. What are the Jacobian matrices $A(k-1)$ and $C(k)$? What is the process noise $w(k)$ and the measurement noise $v(k)$?
3. What are the key equations for the extended Kalman filter algorithm for the continuous-time system given in Question 1?
4. Is it possible that for some applications, the extended Kalman filter estimate may diverge?

7.5 The Kalman Filter with Fading Memory

The idea of prescribed degree of stability in linear quadratic regulator can be carried over to Kalman filter design. This type of Kalman filter is called Kalman filter with fading memory.

7.5.1 The Algorithm for KF with Fading Memory

Assume that the model for the Kalman filter is described by the following difference equations:

$$x(k) = A(k-1)x(k-1) + B(k-1)u(k-1) + w(k-1) \tag{7.108}$$

$$y(k) = C(k)x(k) + v(k) \tag{7.109}$$

where $w(k)$ and $v(k)$ are zero mean noise vectors with normal distribution and have covariance matrices $Q(k)$ and $R(k)$.

The Kalman filter finds the sequence of the state estimates $\hat{x}(1)^-, \hat{x}(2)^-, \dots, \hat{x}(N)^-$ that minimizes the objective function $E(J_N)$, where J_N is given as

$$J_N = \sum_{k=1}^{N} [(y(k) - C(k)\hat{x}(k)^-)^T R(k)^{-1}(y(k) - C(k)\hat{x}(k)^-) + \hat{w}(k)^T Q(k)^{-1}\hat{w}(k)] \tag{7.110}$$

where $\hat{w}(k)$ is process noise with the estimated $\hat{x}(k)$. The idea of Kalman filter with fading memory is to include a time-varying weighting factor α^{-2k} in the objective function where J_N becomes

$$J_N = \sum_{k=1}^{N} (y(k) - C(k)\hat{x}(k)^-)^T \alpha^{-2k} R(k)^{-1}(y(k) - C(k)\hat{x}(k)^-) + \sum_{k=1}^{N} \hat{w}(k)^T \alpha^{-2(k+1)} Q(k)^{-1}\hat{w}(k). \tag{7.111}$$

The parameter $\alpha < 1$ is used to emphasize the weighted covariance of the residual at the recent time as factor α^{-2k} grows with k increases (see the first term of J_N). This means that the Kalman filter with fading memory will be forced to converge to the state estimates that discount old measurements and give a greater emphasis on the more current measurements. For those who are familiar with the idea of prescribed degree of stability in DLQR and in the observer design, a similarity is clearly at present.

To find the solution to the Kalman filter with fading memory, we re-write (7.111) as

$$\tilde{J}_N = \sum_{k=1}^{N} [(y(k) - C(k)\hat{x}(k)^-)^T \tilde{R}(k)^{-1}(y(k) - C(k)\hat{x}(k)^-) + \hat{w}(k)^T \tilde{Q}(k)^{-1}\hat{w}(k)] \tag{7.112}$$

where we have defined the following variables:

$$\tilde{R}(k)^{-1} = \alpha^{-2k} R(k)^{-1} \tag{7.113}$$

$$\tilde{Q}(k)^{-1} = \alpha^{-2(k+1)} Q(k)^{-1}. \tag{7.114}$$

For clarity, (7.113) and (7.114) imply that

$$\tilde{R}(k) = \alpha^{2k} R(k) \tag{7.115}$$

$$\tilde{Q}(k) = \alpha^{2(k+1)} Q(k). \tag{7.116}$$

Now, we can see that the minimization of $E(\tilde{J}_N)$ is identical to the minimization of the original objective function $E(J_N)$ except that the time-varying weighting functions $\tilde{R}(k)^{-1}$ and $\tilde{Q}(k)^{-1}$ are used. For this purpose, we present the two computational equations with the weighting matrices $\tilde{R}(k)$ and $\tilde{Q}(k)$, which are the Kalman filter gain $K(k)$ and the *a priori* covariance matrix $P(k)^-$:

$$K(k) = P(k)^- C(k)^T [C(k)P(k)^- C(k)^T + \tilde{R}(k)]^{-1} \tag{7.117}$$

$$P(k)^- = A(k-1)P(k-1)^+ A(k-1)^T + \tilde{Q}(k-1). \tag{7.118}$$

However, because the weighting matrices $\tilde{R}(k)$ and $\tilde{Q}(k-1)$ affect the *a posteriori* covariance matrix $P(k)^+$, the problem is more complex than the simple replacement of the weighting matrices. We need to find a way to update the Kalman filter gain, the *a priori* estimate and *a posteriori* estimate by incorporating fading memory.

Let us define the weighted *a priori* and *a posteriori* covariance matrices as,

$$\tilde{P}(k)^- = \alpha^{-2k} P(k)^-; \tag{7.119}$$

$$\tilde{P}(k)^+ = \alpha^{-2k} P(k)^+. \tag{7.120}$$

Multiplying both sides of (7.118) with the time-varying factor α^{-2k} gives:

$$\begin{aligned} \alpha^{-2k} P(k)^- &= \alpha^{-2k} A(k-1)P(k-1)^+ A(k-1)^T + \alpha^{-2k} \tilde{Q}(k-1) \\ &= \alpha^{-2} A(k-1)\alpha^{-2(k-1)} P(k-1)^+ A(k-1)^T + Q(k-1) \end{aligned} \tag{7.121}$$

where $\tilde{Q}(k-1) = \alpha^{2k} Q(k-1)$ is used on the basis of (7.116). By using the definitions of the variables $\tilde{P}(k)^-$ (see (7.119)) and $\tilde{P}(k)^+$ (see (7.120)), from (7.121), we have

$$\tilde{P}(k)^- = \alpha^{-2} A(k-1)\tilde{P}(k-1)^+ A(k-1)^T + Q(k-1). \tag{7.122}$$

This equation updates the weighted *a priori* covariance matrix.

Note that in (7.28) the simplified *a posteriori* covariance matrix is given by

$$P(k)^+ = P(k)^- - K(k)C(k)P(k)^- \tag{7.123}$$

Multiplying both sides of (7.123) with the time-varying factor α^{-2k} leads to:

$$\alpha^{-2k} P(k)^+ = \alpha^{-2k} P(k)^- - K(k)C(k)\alpha^{-2k} P(k)^- \tag{7.124}$$

which is, in terms of $\tilde{P}(k)^+$ and $\tilde{P}(k)^-$,

$$\tilde{P}(k)^+ = \tilde{P}(k)^- - K(k)C(k)\tilde{P}(k)^-. \tag{7.125}$$

This equation updates the weighted *a posteriori* covariance matrix.

The Kalman filter gain needs to be updated using the weighted *a priori* covariance matrix $\tilde{P}(k)^-$. Note that from (7.117), we write:

$$\begin{aligned} K(k) &= P(k)^- C(k)^T [C(k)P(k)^- C(k)^T + \alpha^{2k} R(k)]^{-1} \\ &= \alpha^{-2k} P(k)^- C(k)^T [C(k)\alpha^{-2k} P(k)^- C(k)^T + R(k)]^{-1} \\ &= \tilde{P}(k)^- C(k)^T [C(k)\tilde{P}(k)^- C(k)^T + R(k)]^{-1} \end{aligned} \tag{7.126}$$

where $\tilde{P}(k) = \alpha^{-2k} P(k)$.

In summary, from (7.122), (7.125) and (7.126), we obtain the Kalman filter with fading memory that will minimize that objective function with more emphasis on the current measurement, where

$$K(k) = \tilde{P}(k)^- C(k)^T [C(k)\tilde{P}(k)^- C(k)^T + R(k)]^{-1} \tag{7.127}$$

$$\tilde{P}(k)^- = \alpha^{-2} A(k-1)\tilde{P}(k-1)^+ A(k-1)^T + Q(k-1) \tag{7.128}$$

$$\tilde{P}(k)^+ = \tilde{P}(k)^- - K(k)C(k)\tilde{P}(k)^-. \tag{7.129}$$

Because at the initial conditions when $k = 0$, $\tilde{P}(k)^- = P(k)^-$ and $\tilde{P}(k)^+ = P(k)^+$ (see (7.119) and (7.120)), observing equations (7.127)–(7.129), the only difference between the original Kalman filter and the fading memory filter is the inclusion of the factor α^{-2} in (7.128). This means that with this small modification, the Kalman filter algorithms can be easily adopted to the Kalman filter with fading memory.

The following example illustrates the application of the Kalman filter with fading memory.

Example 7.10

A signal is composed of sequences of constants and ramp signals. This signal is corrupted with a measurement noise with standard deviation of 0.1 (see Figure 7.20(a)). Design a Kalman filter with fading memory to filter the noise.

Solution

We will find the dynamic model to describe the signal. Because it is not clear about the times when the signal switches from a ramp signal to a constant signal and vice versa, we will derive the dynamic model based on a ramp signal.

We choose $x_1(k)$ *to be the noise-free ramp signal* $y_1(k)$, *and* $x_2(k) = x_1(k) - x_1(k-1)$. *Then,* $x_2(k) - x_2(k-1) = 0$. *With the system knowledge, we find the state space model to describe the signal as*

$$\begin{bmatrix} x_1(k) \\ x_2(k) \end{bmatrix} = \begin{bmatrix} 1 & 1 \\ 0 & 1 \end{bmatrix} \begin{bmatrix} x_1(k-1) \\ x_2(k-1) \end{bmatrix}. \tag{7.130}$$

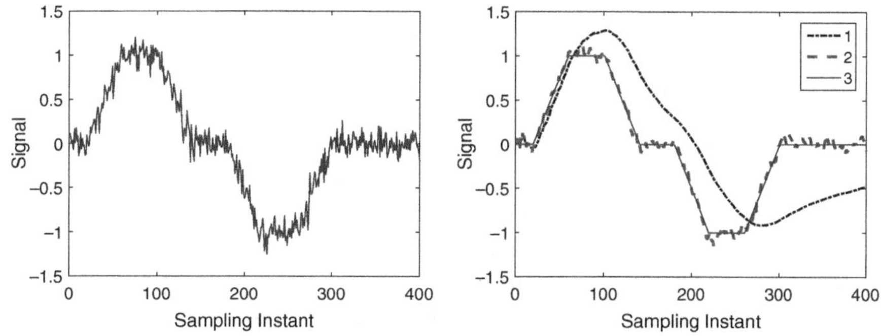

(a) Corrupted signal (b) Filtered signal

Figure 7.20 Application of Kalman filter with fading memory (Example 7.10). Key: line (1) $\alpha = 1$; line (2) $\alpha = 0.9$; line (3) noise-free signal.

The measurement equation is

$$y(k) = \begin{bmatrix} 1 & 0 \end{bmatrix} \begin{bmatrix} x_1(k) \\ x_2(k) \end{bmatrix} + v(k). \tag{7.131}$$

There are uncertainties in the state space equations (7.130), such as the switchings between ramp signals and constants, which could be captured by adding process noise $w(k)$ to (7.130). We could determine the covariance matrix Q through trial-and-error. As an alternative approach, we will use the Kalman filter with fading memory by selecting the parameter $\alpha < 1$.

The initial conditions for the Kalman filter with fading memory are $P(0)^+ = 100I$ (I is identity matrix) and $\hat{x}(0)^+ = [0\ 0]^T$. The covariance matrix Q is zero and $R = 0.001$. Figure 7.20(b) compares the cases of the estimated results for $\alpha = 1$ and $\alpha = 0.9$ against the true noise-free signal. Clearly it is seen that when $\alpha = 1$ corresponding to the case without fading memory, the estimated signal could not track the true signal (see line (1)), however, with $\alpha = 0.9$, the estimated signal tracks the true signal quite well with a small amount of variations due to the amplification of the noise (see line (2)). When the α value decreases further, the amplification of the noise becomes more severe.

7.5.2 Food for Thought

1. For the Kalman filter with fading memory, which step has used the parameter α?
2. When the parameter α is reduced, would you expect that the variances of the estimated state variables be increased? Why?

3. Suppose that a discrete-time system is time-invariant with the state space model:

$$x(k) = Ax(k-1) + Bu(k-1) + w(k-1); \; y(k) = Cx(k) + v(k).$$

With the choice of $\alpha = 0.99$, what can we say about the locations of the eigenvalues of the error system from the converged Kalman filter?
4. Do you think that there is a link between the observer designed using the prescribed degree of stability method and the fading memory Kalman filter? If so, could we simply scale the system matrices A and C by α like what we did previously to achieve the same results with the fading memory Kalman filter? Why not?
5. By studying the fading memory Kalman filter algorithm, can you find the equivalent $Q(k)$ matrix that will yield to the same original Kalman filter algorithm?
6. Is it possible that for some cases, when $\alpha > 1$ (say 1.1), the fading memory Kalman filter still works? What could it mean when this happened?

7.6 Relationship between Kalman Filter and Observer

The observer designed using the discrete-time linear quadratic regulator (see Sections 4.3 and 4.4) has a close relationship with the Kalman filter. Uncovering the relationship will help us further understand the Kalman filter.

7.6.1 One-step Kalman Filter Algorithm

Combining the *a priori* estimate and the *a posteriori* estimate of the state. Note that in the Kalman filter, the state estimate is obtained through the following two steps:

$$\hat{x}(k)^- = A(k-1)\hat{x}(k-1)^+ + B(k-1)u(k-1) \tag{7.132}$$

$$\hat{x}(k)^+ = \hat{x}(k)^- + K(k)(y(k) - C(k)\hat{x}(k)^-). \tag{7.133}$$

In order to match the state space model we used in Chapters 3-6, we re-write the *a priori* state estimate (7.132) using one-step ahead update:

$$\hat{x}(k+1)^- = A(k)\hat{x}(k)^+ + B(k)u(k) \tag{7.134}$$

and substituting the *a posteriori* state estimate (7.133) into (7.134) gives the one-step forward state estimate:

$$\hat{x}(k+1)^- = A(k)\hat{x}(k)^- + B(k)u(k) + A(k)K(k)(y(k) - C(k)\hat{x}(k)^-). \tag{7.135}$$

Combining the *a priori* and the *a posteriori* covariance matrices.
Note that in the Kalman filter, we compute the Kalman filter gain using the following three equations:

$$P(k)^- = A(k-1)P(k-1)^+A(k-1)^T + Q(k-1) \tag{7.136}$$

$$\begin{aligned} P(k)^+ &= (I - K(k)C(k))P(k)^-(I - K(k)C(k))^T + K(k)R(k)K(k)^T \\ &= (I - K(k)C(k))P(k)^- \end{aligned} \tag{7.137}$$

$$K(k) = P(k)^-C(k)^T[C(k)P(k)^-C(k)^T + R(k)]^{-1}. \tag{7.138}$$

Rewriting (7.136) in one-step ahead expression:

$$P(k+1)^- = A(k)P(k)^+A(k)^T + Q(k)$$

and substituting (7.137) into this equation leads to

$$P(k+1)^- = A(k)(I - K(k)C(k))P(k)^-A(k)^T + Q(k). \tag{7.139}$$

Replacing $K(k)$ in (7.139) with the expression from (7.138) gives us the discrete-time dynamic Riccati equation:

$$\begin{aligned} P(k+1)^- = A(k)\{&P(k)^- - P(k)^-C(k)^T[C(k)P(k)^-C(k)^T + R(k)]^{-1} \\ &C(k)P(k)^-\}A(k)^T + Q(k). \end{aligned} \tag{7.140}$$

If a Kalman filter is used for control applications, it is suited better to use the one-step forward form for state estimate (7.135) because it deploys the control signal $u(k)$ and measured output signal $y(k)$ for prediction of state at $k+1$. Both signals are available at the current sampling instant k. This implementation of Kalman filter will avoid the problem of algebraic loop with control system implementation. Certainly, one can adapt the two-step Kalman filter algorithm for control system applications.

7.6.2 Kalman Filter and Observer

Consider the observer equation used previously in Section 4.3, which has the following form:

$$\hat{x}(k+1) = A\hat{x}(k) + Bu(k) + K_{ob}(y(k) - C\hat{x}(k)). \tag{7.141}$$

By comparing (7.141) with the one-step Kalman filter update equation (7.135), we conclude that for a linear time- invariant system (i.e. the system matrices A, B, and C are constant), the state estimate equations of the observer and Kalman filter are identical when the observer gain K_{ob} is replaced by $A(k)K(k)$ from the Kalman filter.

To examine the relationship between the observer gain K_{ob} and Kalman filter gain $K(k)$, we examine the one-step estimate of the *a priori* covariance matrix given by (7.140). At the steady-state as $k \rightarrow \infty$, for a linear time invariant system, the dynamic Riccati equation (7.140) becomes the algebraic Riccati equation:

$$A[P - PC^T(R + CPC^T)^{-1}CP]A^T + Q - P = 0 \tag{7.142}$$

where P is the steady-state solution of $P(k)^-$ as $k \rightarrow \infty$. Note that the algebraic Riccati equation was used to find the observer gain K_{ob} as discussed in Section 4.4.

Once we establish that at the steady-state, the algebraic Riccati equation is identical to the dynamic Riccati equation, we turn to the relationship between the observer gain K_{ob} and the Kalman filter gain $K(k)$, for which we note that

$$K_{ob} = APC^T(CPC^T + R)^{-1}. \tag{7.143}$$

In comparison with the one-step Kalman filter gain given by (7.138), where

$$K(k) = P(k)^- C(k)^T [C(k)P(k)^- C(k)^T + R(k)]^{-1}$$

we conclude that at the steady-state, for a time-invariant system, we have,

$$K_{ob} = AK(k) \tag{7.144}$$

where $K(k)$ is the Kalman filter gain computed at the sampling instant k.

The question arises as under what conditions the dynamic Riccati equation (7.140) has a stabilizing steady-state solution which is also a solution of the corresponding algebraic Riccati equation (7.142). The necessary and sufficient conditions from the discrete-time Linear Quadratic Regulator state that the algebraic Riccati equation has a unique stabilizing, positive definite solution P that is the steady-state solution of the dynamic Riccati equation if and only if (A^T, C^T) is stabilizable and (A^T, Γ^T) is detectable where $Q = \Gamma^T\Gamma$.

A few comments follow.

1. For a time-invariant system, Kalman filter has strong similarities to the observer designed by solving the algebraic Riccati equation.
2. Each requires the solution of a Riccati equation together with the computation of the optimal gain. For the observer case, the observer gain is computed off-line and can be obtained using the MATLAB program **dlqr.m**. For the Kalman filter case, the Kalman filter gain is computed at each sampling instance and optimized by minimizing the *a posteriori* covariance matrix.
3. Kalman filter is computationally more demanding because of the online computation of the Kalman filter gain. However, it offers an optimal performance for each sampling instant and it is applicable to linear time-varying systems.

This section is completed with the following comparative studies.

Example 7.11

The state space model for the three springs and double mass system is described by the following differential equations:

$$\begin{bmatrix} \dot{x}_1(t) \\ \dot{x}_2(t) \\ \dot{x}_3(t) \\ \dot{x}_4(t) \end{bmatrix} = \begin{bmatrix} 0 & 1 & 0 & 0 \\ -\frac{\alpha_1+\alpha_2}{M_1} & 0 & \frac{\alpha_2}{M_1} & 0 \\ 0 & 0 & 0 & 1 \\ \frac{\alpha_2}{M_2} & 0 & -\frac{\alpha_1+\alpha_2}{M_2} & 0 \end{bmatrix} \begin{bmatrix} x_1(t) \\ x_2(t) \\ x_3(t) \\ x_4(t) \end{bmatrix} + \begin{bmatrix} 0 & 0 \\ \frac{1}{M_1} & 0 \\ 0 & 0 \\ 0 & \frac{1}{M_2} \end{bmatrix} \begin{bmatrix} u_1(t) \\ u_2(t) \end{bmatrix}$$

$$\begin{bmatrix} y_1(t) \\ y_2(t) \end{bmatrix} = \begin{bmatrix} 1 & 0 & 0 & 0 \\ 0 & 0 & 1 & 0 \end{bmatrix} \begin{bmatrix} x_1(t) \\ x_2(t) \\ x_3(t) \\ x_4(t) \end{bmatrix}$$

The system has two input variables that are the two applied forces u_1 and u_2 and two output variables that are the distances of the mass block movement y_1 corresponding to block one and y_2 corresponding to block two (see Example 1.1). As in Tongue [2002], the physical parameters are $M_1 = 2kg$, $M_2 = 4kg$, $\alpha_1 = 40N/m$ and $\alpha_2 = 100N/m$. We will use this system to illustrate the differences between the observer and the Kalman filter for the estimation of the states. There are measurement noises, however, no process noise.

Solution

The system is sampled to obtain a discrete time model using the sampling interval $\Delta t = 0.1$ (sec). With state initial conditions $x_1(0) = x_2(0) = x_3(0) = x_4(0) = 1$, and $u_1(k) = \sin \frac{2\pi k}{100}$, and $u_2(k) = 1$ for all k, the state variables are calculated using the discrete-time state space model. Two independent noise sequences with zero mean and unit variance are added to the output. Figure 7.21 shows the two output signals with noise corruptions.

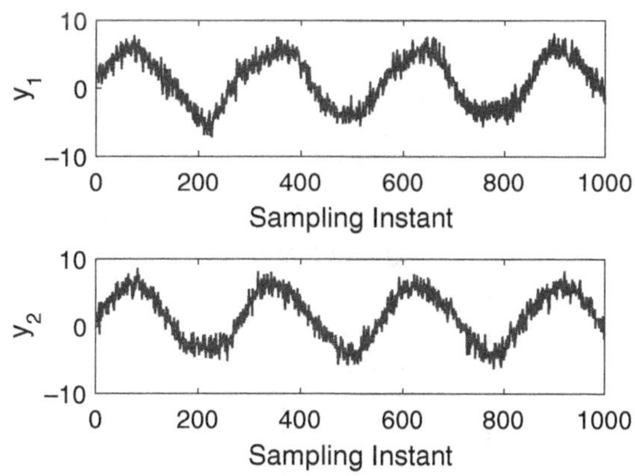

Figure 7.21 Output signals (Example 7.11).

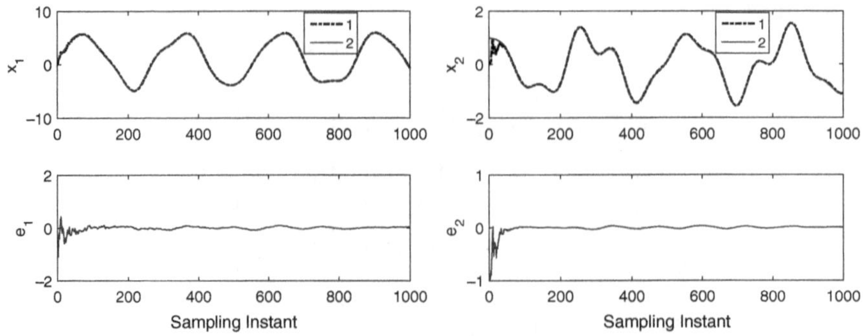

Figure 7.22 Estimation results from Kalman filter (Example 7.11). $Q = 0$ and $R = I$.

Estimation results from Kalman filter.
Because the system has no process noise, we choose Q to be zero matrix and R $= I$ *from the variances of the measurement noise. With* $P(0)^+ = 1I$ *and* $\hat{x}(0)^+$ *being a zero vector, the state variables are estimated using Kalman filter. Figures 7.22(a)–(b) show the comparison between the true and estimated state* (x_1 *and* x_2)*. With the chosen Q and R matrices, the estimated states are almost noise free. The closed-loop eigenvalues of the error system with the final Kalman filter gain are computed to obtain the following values:* $0.9977 \pm j0.0499$, $0.9988 \pm j0.0223$, *which have the magnitude of 0.9990 and are within the unit circle of the complex. This choice of Q and R matrices corresponds to the characteristics of the process and measurement noises, which is one of the key features of Kalman filter. We also calculated the sum of mean squared error:*

$$\frac{1}{M}\sum_{k=1}^{M}(x(k) - \hat{x}(k)^+)^T(x(k) - \hat{x}(k)^+) = 0.0333$$

where M is the data length.

When Q is a zero matrix, the pair (A^T, Γ^T) *is not detectable where* $Q = \Gamma^T\Gamma$ *because all four eigenvalues of A have the magnitude of one. This condition is required for the steady-state solution of the Riccati equation. Therefore, in order to compare with the observer design, we will choose* $Q = 0.001I$ *as the second case.*

For the choice of Q and R matrices, the Kalman filter gain converges to

$$K = \begin{bmatrix} 0.0693 & 0.0076 \\ 0.0204 & 0.0056 \\ 0.0076 & 0.0693 \\ 0.0056 & 0.0204 \end{bmatrix}$$

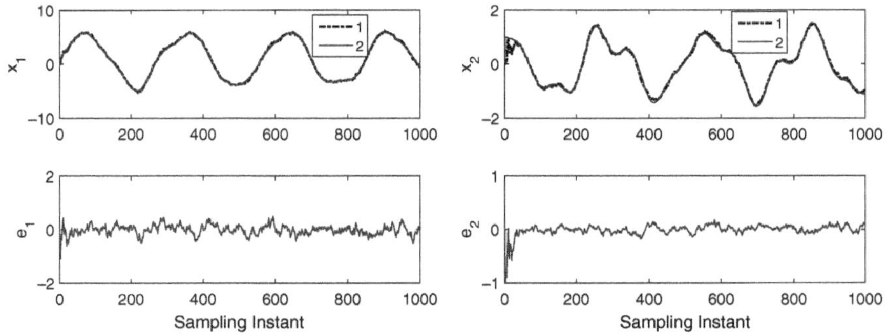

(a) x_1 and $\hat{x}_1$ (b) x_2 and $\hat{x}_2$

Figure 7.23 Estimation results from Kalman filter (Example 7.11). $Q = 0.001I$ and $R = I$.

Figures 7.23(a)–(b) show that the measurement noise has affected the estimated states, which is seen from the bottom plots of figures. The sum of mean squared error is calculated as

$$\frac{1}{M}\sum_{k=1}^{M}(x(k) - \hat{x}(k)^+)^T(x(k) - \hat{x}(k)^+) = 0.1112$$

where M is the data length.

Estimation results from the observer. *For the observer design, the choice of $Q = 0$ and $R = I$ does not lead to a steady-state solution of the Riccati equation, which is confirmed using the MATLAB function dlqr.m. Instead, we use $Q = 0.001I$ and $R = I$ for the observer design. The MATLAB program dlqr.m gives the following observer gain:*

$$K_{ob} = \begin{bmatrix} 0.0713 & 0.0082 \\ 0.0194 & 0.0062 \\ 0.0082 & 0.0713 \\ 0.0062 & 0.0194 \end{bmatrix},$$

and the four eigenvalues for the closed-loop observer error system: $0.9672 \pm j0.0532$, $0.9600 \pm j0.0384$. Here, K_{ob} is identical to the converged AK value. Figure 7.24 compares the four state errors obtained from using the KF and the observer where the solid line represents the error from the observer and dashed line represents the error from the Kalman filter. By visual inspection, we can conclude that in the steady-state, both Kalman filter and the observer have achieved the same estimation results, however, there are performance differences at the transient responses. For this initial choice of the a posteriori *covariance matrix $P(0)^+ = I$, the Kalman filter outperformed the observer during the transient period. This*

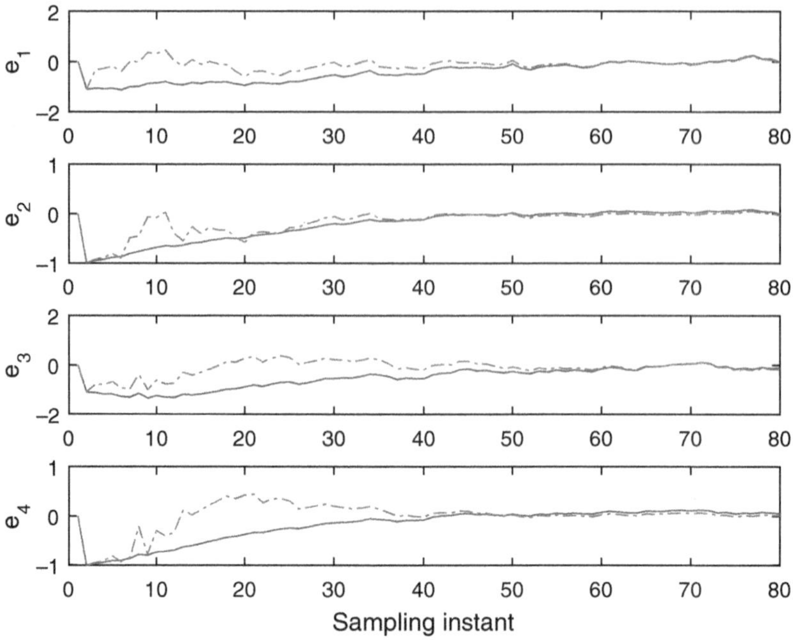

Figure 7.24 Comparative studies between KF and observer (Example 7.11). Key: solid-line observer error; dashed-line: Kalman filter error.

observation is confirmed by the calculation of the mean squared error:

$$\frac{1}{M}\sum_{k=1}^{M}(x(k) - \hat{x}(k))^T(x(k) - \hat{x}(k)) = 0.1665,$$

which is larger than the one obtained from using Kalman filter.

7.6.3 Food for Thought

1. Suppose that a system is described by the state space model:

$$x(k) = 0.8x(k-1) + 0.6u(k-1) + w(k-1); \;\; y(k) = x(k) + v(k),$$

   where $w(k)$ and $v(k)$ are independent white noise with variances σ_w^2 and σ_v^2. What are the equations that may be used for one-step Kalman filter algorithm?
2. If you used the two-steps Kalman filter algorithm for state estimation in feedback control, should you use $\hat{x}(k)^-$ or $\hat{x}(k)^+$ for the calculation of control signal $u(k)$? Or neither?

3. Continue from 1. Assume that the process noise $w(k)$ is not white noise, instead a random walk signal, which is

$$w(k) = \frac{\epsilon(k)}{1 - q^{-1}},$$

where $\epsilon(k)$ is white noise with variance σ_ϵ^2. From the knowledge accumulated from studying of observer, do you believe that the Kalman filter estimate $\hat{x}(k)^+$ would converge to true $x(k)$? If not, what would you propose to do?
4. In the event that we are not sure whether the random walk disturbance occurred at the input or output, do you think that it is critical that we modelled it correctly?
5. Would the answer to the question 4 depend on whether the system is time-varying or time-invariant?
6. Do you believe that checking the eigenvalues of closed-loop error system in Kalman filter provided us with an intuitive tool to access the information for convergence of Kalman filter?

7.7 Summary

This chapter has discussed various topics about the Kalman filter with MATLAB tutorials. More specifically, the important aspects of this chapter are summarized as follows.

1. The Kalman filter optimally estimates the unknown true state vector $x(k)$ using the knowledge of process dynamics and the measured output $y(k)$ based on the statistical properties of the process noise $w(k)$ and $v(k)$. This estimation is performed iteratively to produce *a posteriori* estimate $\hat{x}(k)^+$.
2. The Kalman filter gain $K(k)$ is obtained by minimizing the objective function $J(k)$ (see (7.17)) at the sampling instant k where $J(k)$ is chosen to be the sum of the variances of the errors between $x(k)$ and $\hat{x}(k)^+$.
3. The problems of sensor bias and load disturbance can be solved using the augmented state space models that include these unknown variables as part of the state vector.
4. With some small modifications, the Kalman filter algorithm has the capacity to handle missing data and to operate in a multi-rate sampling environment.
5. The extended Kalman filter is an effective tool for state estimation of nonlinear systems.
6. The Kalman filter with fading memory provides an additional means for tuning the filter's performance.

7.8 Further Reading

1. Kalman's original papers are Kalman [1960], Kalman and Bucy [1961].
2. Books for optimal filters include Anderson and Moore [1979], Kailath [1981], Goodwin and Sin [1984], Simon [2006], Brown and Huang [2012], Grimble and Johnson [1988a], Catlin [2012].
3. Tutorial papers on the Kalman filter include Rhodes [1971], Higgins [1975], Orderud [2005], Young [2011], Faragher [2012], Humpherys et al. [2012], Rhudy et al. [2017].
4. The Kalman filter has wide applications in navigation and localization of autonomous vehicles. The books for this area include Noureldin et al. [2013], Bekir [2007], Groves [2013]. A PhD thesis on localization is presented in Guan [2019].
5. An example of extended Kalman filter with low-cost sensors can be found in Driessen et al. [2018]. The Kalman filter with recursive covariance estimation is presented in Feng et al. [2014].

Problems

Problem 7.1 Assume that a continuous-time signal is corrupted by measurement noise where the noise-free signal $y_0(t)$ can be described by

$$y_0(t) = at + b$$

where the parameters a and b are unknown constant. The sampling interval for this continuous-time signal is Δt.

1. Choosing $x_1(t) = y_0(t)$ and $x_2(t) = \dot{y}_0(t)$, write down the equations associated with a discrete-time Kalman filter that will provide the best estimate for the noise-free signal $y_0(k)$ and the parameter a at the sampling instant k.
2. If we wish to provide the best estimate for the unknown parameters a and b, write down the equations associated with a discrete-time Kalman filter to achieve this objective.
3. Generate a set of test data with $a = 0.2$, $b = 0.6$ and $\Delta t = 0.1$ where the measurement noise is zero mean with variance 0.6.
4. Validate both Kalman filters using the same test data and compare the performance of the two filters in terms of the error signals $y(k) - \hat{y}(k)$ and $a(k) - \hat{a}(k)$. What are your observations?

Problem 7.2 Continue from Problem 7.1. In this exercise, we will test the Kalman filters in the scenario when the parameters a and b are piece-wise constant.

1. When generating the test data by doubling the simulation time from Problem 7.1 with the same Δt and noise variance, however, at one third of the simulation time, change the parameter a from 0.2 to −0.2 and at two third of the simulation time change parameter b from 0.6 to −0.6.
2. Explore the effect of Q matrix on the performance of the two filters. For simplicity, we choose $Q = q_0 I$ where I is the identity matrix and optimize the parameter q_0 with respect to the sum of squared errors $\sum_k |y(k) - \hat{y}(k)|^2$. Compare the performance of the two filters in terms of the error signals $y(k) - \hat{y}(k)$ and $a(k) - \hat{a}(k)$. What are your observations?

Problem 7.3 Continue from Problems 7.1 and 7.2. Using the same test data generated from Problem 7.2, in this exercise, we will apply fading memory Kalman filter to the time-varying system.

1. Choosing the weighting matrix Q to be zero, find the best α value in the fading memory Kalman filter that will minimize the sum of squared errors $\sum_k |y(k) - \hat{y}(k)|^2$ and apply the procedure for the two filters designed in Problem 7.1. Compare the performance of the two filters in terms of the error signals $y(k) - \hat{y}(k)$ and $a(k) - \hat{a}(k)$. What are your observations?
2. In comparison with the results obtained from Problem 7.2, which approach would you prefer? What are the advantages and disadvantages of each method?

Problem 7.4 A continuous time signal $y(t)$ is measured, which is severely corrupted by noise $\epsilon(t)$. The measured signal $y(t) = y_0(t) + \epsilon(t)$ where the noise-free $y_0(t)$ is described by the following continuous-time function

$$y_0(t) = a_m sin(\omega_0 t + \theta) + d$$

Here, the frequency ω_0 is known, but the amplitude a_m and the phase angle θ are unknown. $d \neq 0$ is an unknown constant.

1. Write down the continuous-time state space model that will describe the continuous-time function $y_0(t)$.
2. Generate a noise-free signal $y_0(t)$ where the parameters are $a_m = 1, \omega_0 = 1, \theta = 0$ and $d = 1$ and sampling interval $\Delta t = 0.05$ (sec), and add a noise signal which is generated using the MATLAB script:epsilon=0.2*randn(1,Nsim); where the number of samples Nsim=600.
3. Use Kalman filter to estimate the sinusoidal signal and the constant signal with your choice of initial conditions, Q and R matrices. Present the estimated states and compare them with true states.
4. Assuming that the frequency $\omega_0 = 0.2\cos(0.1t)$ (rad/sec), use Kalman filter to estimate the states and filter the noise. Present the estimated states and compare them with the true states.

Problem 7.5 A discrete-time system is described by the difference equations:

$$x(k) = 1.1x(k-1) - 2\gamma(k-1) + w(k-1)$$
$$\gamma(k) = \gamma_0 1.1^k$$

where γ_0 is an unknown constant, and $w(k)$ is a zero mean white noise with variance σ_w^2. The measurement $y(k)$ is known to have the following relationship with the variables $x(k)$ and $\gamma(k)$:

$$y(k) = 2x(k) + \gamma(k) + v(k)$$

where the measurement $v(k)$ is a zero-mean white noise with variance σ_v^2.

1. Write down the Kalman filter equations that will recursively estimate the state variables $x(k)$ and $\gamma(k)$.
2. Generate a set of test data with $x(0) = 0$, $\gamma_0 = 0.2$, $\sigma_w = 0.6$ and $\sigma_v = 0.9$.
3. Validate the Kalman filter with the test data. Plot the errors $y(k) - \hat{y}(k)$, $x(k) - \hat{x}(k)$ and $\gamma(k) - \hat{\gamma}(k)$ and calculate the mean squared errors. What are your observations?

Problem 7.6 A second order system with both actuator delay and sensor delay is described by the following state space model:

$$x(k) = Ax(k-1) + Bu(k-3) + w(k-1)$$
$$y(k) = Cx(k-4) + v(k)$$

where the system matrices are

$$A = \begin{bmatrix} 0.7 & 0.9 \\ 0 & 1 \end{bmatrix};\ B = \begin{bmatrix} 1 \\ 1 \end{bmatrix};C = \begin{bmatrix} 2 & 1 \end{bmatrix}.$$

1. Write down the equations associated with Kalman filter that will recursively estimate the state $x(k)$.
2. To generate the simulation data, the initial condition for $x(0)$ is chosen to be $[1\ 1]^T$ and $u(k) = \cos\frac{2\pi}{100}$, $w(k)$ and $v(k)$ are zero mean white noise with standard deviation of 0.1 and 0.3 respectively.
3. Evaluate the filter's performance using the simulation data. Plot the errors $y(k) - \hat{y}(k)$, $x_1(k) - \hat{x}_1(k)$ and $x_2(k) - \hat{x}_2(k)$ and calculate the mean squared errors.
4. Study the effect of mismatching time delay on the performance of the Kalman filter. In the model used for the Kalman filter, the output delay is reduced to 1 sample and the input delay is reduced to 0. Evaluate the filter's performance using the same set of simulation data. Plot the errors $y(k) - \hat{y}(k)$, $x_1(k) - \hat{x}_1(k)$ and $x_2(k) - \hat{x}_2(k)$ and calculate the mean squared errors. In comparison with

the results obtained from the previous step where correct time delay information was used in the model, what are your observations?

Problem 7.7 The dynamic model of a permanent magnetic synchronous machine is represented by,

$$\frac{di_d(t)}{dt} = \frac{1}{L_d}(v_d(t) - R_s i_d(t) + \omega_e(t) L_q i_q(t))$$

$$\frac{di_q(t)}{dt} = \frac{1}{L_q}(v_q(t) - R_s i_q(t) - \omega_e(t) L_d i_d(t) - \omega_e(t)\phi_{mg})$$

$$\frac{d\omega_e(t)}{dt} = \frac{Z_p}{J_m}(\frac{3}{2}Z_p \phi_{mg} i_q(t) - \frac{B_v}{Z_p}\omega_e(t) - T_L)$$

where L_d and L_q are the inductances, R_s is the resistance, ϕ_{mg} is the flux, Z_p is the number of pole pairs, J_m represents the inertia of the motor, B_v is viscous friction coefficient. All these physical parameters are known. However, the load torque T_L is unknown and is assumed to be a constant in the applications.

1. Assume that the input signals $v_d(t)$ and $v_q(t)$ are given, and the measured signal is $\omega_e(t)$ and the sampling interval is Δt. Write down the equations associated with an extended Kalman filter algorithm that will recursively estimate the current i_d, i_q and the load torque T_L.
2. Discuss the significance of including T_L in the filter design.

Problem 7.8 In Example 7.9, we have considered the case where both water tank levels are measured. In this exercise, we will continue from Example 7.9, however with measurement from one water level.

1. Assume that water level L_1 is measured with a low-cost sensor, but water level L_2 is not measured. Design an extended Kalman filter to estimate both water tank levels.
2. Assume that water level L_2 is measured with a low-cost sensor, but water level L_1 is not measured. Design an extended Kalman filter to estimate both water tank levels.
3. Compare the estimation results for the two cases and examine the effect of Q and R matrices on the estimation results.

8

Addressing Computational Issues in KF

8.1 Introduction

Computational issues in real-time applications when using the Kalman filter are very important. It sometimes happens that the Kalman filter is correct in theory and in simulation studies, however, they will not work in a real system. One of the main causes for this type of failures is the finite precision arithmetic existed in the computational devices such as micro-controllers or other low-cost computational devices.

The objective of this chapter is to address the real-time computational issues in the Kalman filter. There are two topics discussed. The first topic is related to the computation of Kalman gain in real-time. Because it requires the operation of matrix inversion, in the absence of a library, the implementation using a micro-controller presents a challenge in itself. To address these potential challenges for real-time applications, Section 8.2 introduces the sequential Kalman filter in order to avoid the matrix inversion in the computation of Kalman gain. In a digital processor, the arithmetic is finite precision, meaning that only a certain number of bits are used to represent the numbers from the Kalman filter equations. In Section 8.3 the Kalman filter using UDU^T factorization is discussed which may have the potential to double the accuracy and is relevant for real-time implementations. MATLAB tutorials are given in this chapter to illustrate real-time implementation of the Kalman filters.

8.2 The Sequential Kalman Filter

8.2.1 Basic Ideas about Sequential Kalman Filter

The basic idea behind the sequential Kalman filter is to avoid the matrix inversion required in the computation of Kalman filter gain. Note that from (7.24),

State Feedback Control and Kalman Filtering with MATLAB/Simulink Tutorials, First Edition.
Liuping Wang and Robin Ping Guan.

Companion website: www.wiley.com/go/feedback19

the inversion of the matrix $C(k)P(k)^{-}C(k)^{T} + R(k)$ is required in computing the Kalman filter gain. However, if the problem were to deal with one output measurement ($C(k)$ is a row vector and $R(k)$ is a scalar), then this matrix would degrade to a scalar, which would not involve a matrix inversion. Based on this observation, the sequential Kalman filter algorithm is to enter the output measurement one at a step and sequentially calculate the Kalman filter gain $K(k)$ column by column. The key assumption is that the covariance matrix $R(k)$ is diagonal, meaning that the measurement noise corrupted each output is independent. If $R(k)$ is not a diagonal matrix, a linear transformation will be used to achieve the diagonal realization.

We assume that the mathematical model is

$$x(k) = A(k-1)x(k-1) + B(k-1)u(k-1) + w(k-1) \tag{8.1}$$

$$y(k) = C(k)x(k) + v(k). \tag{8.2}$$

Supposing that the system has m output measurements, we denote the components in (8.2) as

$$y(k) = \begin{bmatrix} y_1(k) \\ \vdots \\ y_i(k) \\ \vdots \\ y_m(k) \end{bmatrix}; v(k) = \begin{bmatrix} v_1(k) \\ \vdots \\ v_i(k) \\ \vdots \\ v_m(k) \end{bmatrix}; C(k) = \begin{bmatrix} C_1(k) \\ \vdots \\ C_i(k) \\ \vdots \\ C_m(k) \end{bmatrix}$$

where $C_i(k)$ is the ith row in the output matrix $C(k)$. Accordingly, we denote the Kalman filter gain $K(k)$ as $K(k) = \begin{bmatrix} K_1(k) & K_2(k) & \dots & K_m(k) \end{bmatrix}$ where $K_i(k)$ is the ith column of the Kalman filter gain $K(k)$. We assume that the covariance matrix for the measurement noise $v(k)$ is $E(v(k)v(k)^T) = R$, and R is a diagonal matrix with diagonal elements $r_{11}, r_{22}, \dots, r_{mm}$.

Now, we will have a closer look on how the multiple measurement problem is solved in terms of a single measurement problem. Because the matrix inversion occurs at the *a posteriori* estimate step, there is no change for the *a priori* estimate and the covariance. We will enter the output measurement one by one to avoid the matrix inversion.

For the *a posteriori* estimate and the covariance, at the sampling time k, we choose

$$\hat{x}(k)_0^+ = \hat{x}(k)^- \tag{8.3}$$

$$P(k)_0^+ = P(k)^- \tag{8.4}$$

where the subscript 0 means that the initial conditions for the *a posteriori* estimate and covariance. To update the *a posteriori* estimate with the first output $y_1(k)$, we use the following equation:

$$\hat{x}(k)_1^+ = \hat{x}(k)_0^+ + K_1(k)(y_1(k) - C_1(k)\hat{x}(k)_0^+). \tag{8.5}$$

For the update from the first output, the *a posteriori* covariance matrix is expressed as

$$E[(x(k) - \hat{x}(k)_1^+)(x(k) - \hat{x}(k)_1^+)^T] = P(k)_1^+ \tag{8.6}$$

$$= (I - K_1(k)C_1(k))P(k)^-(I - K_1(k)C_1(k))^T + r_{11}K_1(k)K_1(k)^T, \tag{8.7}$$

where we have used the initial conditions (8.3) and (8.4). Based on (8.7), minimizing the following objective function

$$J_1(k) = trace\{E[(x(k) - \hat{x}(k)_1^+)(x(k) - \hat{x}(k)_1^+)^T]\}$$

gives the optimal solution of $K_1(k)$ as

$$K_1(k) = P(k)^-C_1(k)^T/(C_1(k)P(k)^-C_1(k)^T + r_{11}). \tag{8.8}$$

We then update the *a posteriori* covariance matrix $P(k)_1^+$ based on (8.7) with the optimal Kalman $K_1(k)$.

To update the *a posteriori* estimate with the second output $y_2(k)$, the *a posteriori* state estimate is written as

$$\hat{x}(k)_2^+ = \hat{x}(k)_1^+ + K_2(k)(y_2(k) - C_2(k)\hat{x}(k)_1^+) \tag{8.9}$$

where $\hat{x}(k)_1^+$ is the *a posteriori* estimate obtained with the first measurement $y_1(k)$ (see (8.5)). The *a posteriori* covariance matrix is calculated as

$$E[(x(k) - \hat{x}(k)_2^+)(x(k) - \hat{x}(k)_2^+)^T] = P(k)_2^+ \tag{8.10}$$

$$= (I - K_2(k)C_2(k))P(k)_1^+(I - K_2(k)C_2(k))^T + r_{22}K_2(k)K_2(k)^T. \tag{8.11}$$

Based on (8.11), minimization of the objective function $J_2(k)$:

$$J_2(k) = trace\{E[(x(k) - \hat{x}(k)_2^+)(x(k) - \hat{x}(k)_2^+)^T]\}$$

leads to the optimal solution of the Kalman filter gain:

$$K_2(k) = P(k)_1^+C_2(k)^T/(C_2(k)P(k)_1^+C_2(k)^T + r_{22}). \tag{8.12}$$

We update the *a posteriori* covariance matrix $P(k)_2^+$ using the Kalman filter gain $K_2(k)$ with equation (8.11).

We sequentially calculate the *a posteriori* estimates for all the outputs. For the last output (the mth), the update equations are

$$K_m(k) = P(k)_{m-1}^+C_m(k)^T/(C_m(k)P(k)_{m-1}^+C_m(k)^T + r_{mm})$$
$$\hat{x}(k)_m^+ = \hat{x}(k)_{m-1}^+ + K_m(k)(y_m(k) - C_m(k)\hat{x}(k)_{m-1}^+$$
$$P(k)_m^+ = (I - K_m(k)C_m(k))P(k)_{m-1}^+(I - K_m(k)C_m(k))^T + r_{mm}K_m(k)K_m(k)^T.$$

Upon completing the sequential measurement updating with all output measurements, the *a posteriori* estimate takes the calculations from the mth output (the final output entering the computation), which are given by

$$\hat{x}(k)^+ = \hat{x}(k)_m^+$$
$$P(k)^+ = P(k)_m^+.$$

For the sequential Kalman filter algorithm, $K_m(k)$ equals the last column of the original Kalman filter gain, but the rest of the gain vectors do not correspond to Kalman filter gain solved using matrix inversion.

To verify that this sequential measurement updating produces identical *a posteriori* estimate and covariance, we present the following simple numerical example.

Example 8.1

Consider that a system with two states and two measurements has the output matrix $C = \begin{bmatrix} 1 & 2 \\ 2 & 3 \end{bmatrix}$. *We also assume that at the sampling instant k, the measurement is* $y(k) = [1\ 2]^T$ *and* a priori $\hat{x}(k)^- = [1\ 2]^T$, *and the* a priori *covariance matrix at sampling instant k is* $P(k)^- = \begin{bmatrix} 3 & 0.1 \\ 0.1 & 2 \end{bmatrix}$. *With* $R = \begin{bmatrix} r_{11} & 0 \\ 0 & r_{22} \end{bmatrix}$ *where* $r_{11} = 1$ *and* $r_{22} = 2$, *verify that the* a posteriori *covariance with matrix inversion*

$$P(k)^+ = (I - K(k)C(k))P(k)^-(I - K(k)C(k))^T + K(k)R(k)K(k)^T$$

is identical to the sequentially calculated a posteriori *covariance*

$$P(k)_2^+ = (I - K_2(k)C_2(k))P(k)_1^+(I - K_2(k)C_2(k))^T + r_{22}K_2(k)K_2(k)^T$$

where

$$K_1(k) = P(k)^- C_1(k)^T / (C_1(k)P(k)^- C_1(k)^T + r_{11}),$$
$$K_2(k) = P(k)_1^+ C_2(k)^T / (C_2(k)P(k)_1^+ C_2(k)^T + r_{22}),$$
$$P(k)_1^+ = (I - K_1(k)C_1(k))P(k)^-(I - K_1(k)C_1(k))^T + r_{11}K_1(k)K_1(k)^T.$$

Additionally, show that K_1 *does not equal to the first column of K, however,* K_2 *equals to the second column of K.*

Solution

We first find Kalman filter gain and the a posteriori *covariance* $P(k)^+$ *using the matrix inversion:*

$$K(k) = \begin{bmatrix} -0.1866 & 0.2949 \\ 0.3255 & 0.0034 \end{bmatrix}$$
$$P(k)^+ = \begin{bmatrix} 1.7395 & -0.9631 \\ -0.9631 & 0.6443 \end{bmatrix}.$$

Then, we test the sequential Kalman filter approach to find $K_1(k)$ and $P(k)_1^+$, which are

$$K_1(k) = \begin{bmatrix} 0.2581 \\ 0.3306 \end{bmatrix}; P(k)_1^+ = \begin{bmatrix} 2.1742 & -0.9581 \\ -0.9581 & 0.6444 \end{bmatrix}; K_2(k) = \begin{bmatrix} 0.2949 \\ 0.0034 \end{bmatrix},$$

$$P(k)_2^+ = \begin{bmatrix} 1.7395 & -0.9631 \\ -0.9631 & 0.6443 \end{bmatrix}.$$

Furthermore, when we assume that at the sampling instant k, the measurement is $y(k) = [1\ 2]^T$ and the a priori $\hat{x}(k)^- = [1\ 2]^T$, *we can verify that the sequential Kalman filter estimate of the state $\hat{x}(k)_2^+ = [1.5114 - 0.3388]^T$, which equals to the one obtained using matrix inversion.*

The sequential Kalman filter algorithm is summarized below with the assumption R being a diagonal matrix with the ith diagonal element r_{ii}.

1. Compute the *a priori* estimate of the state and the covariance matrix:

$$\hat{x}(k)^- = A(k-1)\hat{x}(k-1)^+ + B(k-1)u(k-1) \tag{8.13}$$

$$P(k)^- = A(k-1)P(k-1)^+A(k-1)^T + Q(k-1). \tag{8.14}$$

2. Compute the Kalman filter gain sequentially with the following steps:
   a) Choose the initial *a posteriori* estimates:

$$\hat{x}(k)_0^+ = \hat{x}(k)^-$$
$$P(k)_0^+ = P(k)^-.$$

   b) For $1 \le i \le m$, perform the following computation for each output sequentially:

$$K_i(k) = P(k)_{i-1}^+ C_i(k)^T / (C_i(k)P(k)_{i-1}^+ C_i(k)^T + r_{ii}),$$
$$\hat{x}(k)_i^+ = \hat{x}(k)_{i-1}^+ + K_i(k)(y_i(k) - C_i(k)\hat{x}(k)_{i-1}^+),$$
$$P(k)_i^+ = (I - K_i(k)C_i(k))P(k)_{i-1}^+(I - K_i(k)C_i(k))^T + r_{ii}K_i(k)K_i(k)^T.$$

3. Update the *a posteriori* estimate of the state and the covariance:

$$\hat{x}(k)^+ = \hat{x}(k)_m^+$$
$$P(k)^+ = P(k)_m^+.$$

In the sequential updating step, we can choose to use the simpler equation:

$$P(k)_i^+ = (I - K_i(k)C_i(k))P(k)_{i-1}^+.$$

However, the *a posteriori* covariance may loss its symmetric property if one chooses to use this simpler form.

8.2.2 Non-diagonal *R*

The key assumption in the sequential Kalman filter algorithm is that the R matrix has a diagonal structure. For the cases that R is not a diagonal matrix, we will use singular value decomposition to write

$$R = VDV^T \tag{8.15}$$

where D is a diagonal matrix with all positive elements and $V^{-1} = V^T$.

We define a new variable $\overline{y}(k)$ as

$$\begin{aligned}\overline{y}(k) = V^T y(k) &= V^T Cx(k) + V^T v(k)\\ &= \overline{C}x(k) + \xi(k)\end{aligned} \tag{8.16}$$

where $\overline{C} = V^T C$ and $\xi(k) = V^T v(k)$. The covariance matrix for $\xi(k)$ is

$$\begin{aligned}E[\xi(k)\xi(k)^T] &= V^T E[v(k)v(k)^T]V\\ &= V^T VDV^T V = D = \overline{R}\end{aligned} \tag{8.17}$$

where we have used the property of singular value transformation: $V^T V$ is an identity matrix. Clearly, $\overline{R}$ is a diagonal matrix. After this transformation, the transformed measurement signal $\overline{y}(k)$ is used for the measurement update where $\overline{C}$ is to replace C, and $\overline{R}$ is to replace R in the sequential Kalman filter algorithm.

Note that the singular value decomposition of R matrix can be performed off-line using MATLAB function **svd.m** if R is a constant matrix. However, if R is time-varying for every sampling instant, then the sequential Kalman filter algorithm losses its computational advantage.

The following example is used to illustrate how the transformation works numerically.

Example 8.2

As an illustration, we will use the system described in Example 8.1, except that $R = \begin{bmatrix} 1 & 1 \\ 1 & 2 \end{bmatrix}$*, which is not diagonal matrix. We will show numerically that when R is not diagonal, the* a posteriori *estimate and covariance from the sequential Kalman filter are not identical to the solutions obtained using matrix inversion unless the linear transformation is used.*

Solution

When using matrix inversion, at sampling time k with the same $P(k)^-$ *and* $x(k)^-$ *as in Example 8.1, we obtain the* a posteriori *estimate:*

$$x(k)^+ = \begin{bmatrix} 1.4027 \\ -0.3095 \end{bmatrix} \tag{8.18}$$

and the covariance:

$$P(k)^{+} = \begin{bmatrix} 1.3968 & -0.7575 \\ -0.7575 & 0.5926 \end{bmatrix}. \tag{8.19}$$

Without singular value decomposition of the R matrix, the sequential Kalman filter gives

$$x(k)_2^{+} = \begin{bmatrix} 1.5114 \\ -0.3388 \end{bmatrix};\ P(k)_2^{+} = \begin{bmatrix} 1.7395 & -0.9631 \\ -0.9631 & 0.6443 \end{bmatrix},$$

which are different from the a posteriori *estimate (8.18) and covariance (8.19) when using the matrix inversion.*

The next step is to show that with the singular value transformation, the sequential Kalman filter works correctly. Using MATLAB function **svd.m**, *we decompose R matrix as*

$$R = \begin{bmatrix} 1 & 1 \\ 1 & 2 \end{bmatrix} = \begin{bmatrix} -0.5257 & -0.8507 \\ -0.8507 & 0.5257 \end{bmatrix} \begin{bmatrix} 2.6180 & 0 \\ 0 & 0.3820 \end{bmatrix} \begin{bmatrix} -0.5257 & -0.8507 \\ -0.8507 & 0.5257 \end{bmatrix}.$$

With the V^T matrix, we obtain the following transformed output matrix and covariance matrix of the measurement noise:

$$\overline{C} = V^T C = \begin{bmatrix} -0.5257 & -0.8507 \\ -0.8507 & 0.5257 \end{bmatrix} \begin{bmatrix} 1 & 2 \\ 2 & 3 \end{bmatrix} = \begin{bmatrix} -2.2270 & -3.6034 \\ 0.2008 & -0.1241 \end{bmatrix}$$

$$\overline{R} = \begin{bmatrix} 2.6180 & 0 \\ 0 & 0.3820 \end{bmatrix},$$

and the transformed measurement:

$$\overline{y} = V^T y = \begin{bmatrix} -2.2270 \\ 0.2008 \end{bmatrix}.$$

With the transformed variables, we calculate the a posteriori *estimates using the sequential Kalman filter to yield*

$$\hat{x}(k)_2^{+} = \begin{bmatrix} 1.4027 \\ -0.3095 \end{bmatrix};\ P(k)_2^{+} = \begin{bmatrix} 1.3968 & -0.7575 \\ -0.7575 & 0.5926 \end{bmatrix},$$

which are identical to those obtained from using the matrix inversion (see (8.18) and (8.19)).

8.2.3 MATLAB Tutorial for Sequential Kalman Filter

In this section, we will present an example followed by the MATLAB tutorial that could be used to verify it.

Example 8.3
Consider a system with the following state space model:

$$x(k) = Ax(k-1) + Bu(k-1); \; y(k) = Cx(k) + v(k), \tag{8.20}$$

where the system matrices are

$$A = \begin{bmatrix} 0.9 & -0.1 & 0 \\ -0.1 & 0.1 & 0.1 \\ 0 & 0.3 & 0.9 \end{bmatrix}; \; B = \begin{bmatrix} 1 & 0 \\ 0.1 & 0 \\ 0 & 0.1 \end{bmatrix}; \; C = \begin{bmatrix} 1 & 0 & 1 \\ 1 & 1 & 1 \end{bmatrix}$$

and the initial state $x(0) = [1\ 1\ 1]^T$, $u_1(k) = \sin\frac{2\pi k}{100}$ *and* $u_2(k)$ *is 1. The output measurement noise sequences,* $v_1(k)$ *and* $v_2(k)$*, are normally distributed white noise with standard deviation 1. Estimate the state variables using sequential Kalman filter.*

Solution
We generate the output data using (8.20) with 1000 sampled input data. Since the process noise $w(k)$ *is zero, we choose Q to be a zero matrix. R is chosen to be identity matrix. Although the initial state* $x(0)$ *is known from the simulation, for the majority of the applications, it is unknown. Thus, we choose* $\hat{x}(0)^+$ *to be zero.* $P(0)^+$ *is chosen to be* $100I$ *to start the estimation.*

Figure 8.1 compares the estimated state variables to the true state variables. We can see from this figure that the estimated variables have tracked the state variables closely after about 30 samples. The sum of the mean squared error is calculated as

$$\frac{1}{M}\sum_{k=1}^{M}(x(k) - \hat{x}(k)^+)^T(x(k) - \hat{x}(k)^+) = 0.0732.$$

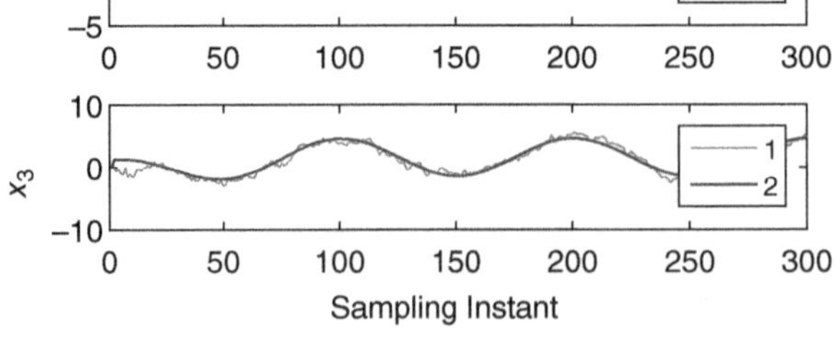

Figure 8.1 Comparison between $x(k)$ and $\hat{x}(k)^+$ (Example 8.3). Key: line (1) $\hat{x}(k)^+$; line (2) $x(k)$.

The following two tutorials are given to demonstrate the computational procedures of sequential Kalman filter by using Example 8.3.

Tutorial 8.1 *The objective of this tutorial is to write a MATLAB function* **KFSQ.m** *that will produce the estimated states and outputs for a multi-input and multi-output system without matrix inversion. The input variables to the function are:*

- *the system matrices A, B, C;*
- *the input and output data u ($p \times L$) y ($m \times L$), where p and m are the number of inputs and outputs respectively, L is the data length;*
- *the initial conditions for the Kalman filter $\hat{x}(0)^+$ and $P(0)^+$;*
- *the weighting matrices Q and R, where R is a diagonal matrix.*

The output variables from the function are $\hat{x}(k)^+$ and $\hat{y}(k)$.

Step by Step

1. *Create a new file called* **KFSQ.m**.
2. *Enter the following program into the file:*

```
function [xhat,yhat]=KFSQ(A,B,C,u,y,x_plus,P_plus,Q,R)
```

3. *Check the data length and the numbers of measurement and state. Continue entering the following program into the file:*

```
L=length(y);
[m,n]=size(C);
```

4. *The first estimates $\hat{x}(0)$ and $\hat{y}(0)$ are chosen to be the initial conditions $\hat{x}(0)^+$ and $C\hat{x}(0)^+$. Continue entering the following program into the file:*

```
xhat(:,1)=x_plus;
yhat(:,1)=C*x_plus;
```

5. *The computation will be performed recursively. Compute* a priori *estimate of the state $\hat{x}(k)^-$ and the covariance matrix $P(k)^-$. Continue entering the following program into the file:*

```
for k=2:L
x_minus=A*x_plus+B*u(:,k-1);
P_minus=A*P_plus*A'+Q;
```

6. *Compute the measurement update sequentially, which is different from the measurement update in a standard Kalman filter. Continue entering the following program into the file:*

```
for jj=1:m
K=P_minus*C(jj,:)'/(C(jj,:)*P_minus*C(jj,:)'+R(jj,jj));
```

```
x_plus=x_minus+K*(y(jj,k)-C(jj,:)*x_minus);
P_plus=(eye(n)-K*C(jj,:))*P_minus;
```

7. *Prepare for the update of the next output signal. Continue entering the following program into the file:*

```
P_minus=P_plus;
x_minus=x_plus;
end
```

8. *Save the estimated state variable and output variable, and complete one cycle of the computation. Continue entering the following program into the file:*

```
xhat(:,k)=x_plus;
yhat(:,k)=C*x_plus;
end
```

9. *Test this function using Example 8.3.*

Tutorial 8.2 *This tutorial is to illustrate how to validate the MATLAB function* **KFSQ.m**. *We will use Example 8.3 in this tutorial.*

Step by Step

1. *Create a new file called* **Ex4SKF.m**.
2. *Generate the output signals based on Example 8.3. We save the state variables for comparisons later on. Enter the following program into the file:*

```
A=[0.9 -0.1 0;-0.1 0.1 0.1;0 0.3 0.9];
B=[1 0;0.1 0; 0 0.1];
C=[1 0 1;1 1 1];
[m,n]=size(C);
kt=1:1000;
u1=sin(2*pi*kt/100);
u2=ones(1,1000);
u=[u1;u2];
x=[1;1;1];
y1=C*x;
L=length(u);
xmodel(:,1)=x;
for kk=2:L;
    x=A*x+B*u(:,kk-1);
    xmodel(:,kk)=x;
    y1=C*x;
    y(:,kk)=y1;
end
```

3. *Add the measurement noise to the output data. The MATLAB function* **randn.m** *is used to generate the measurement noise with standard deviation 1. By specifying the seed number, we know exactly which noise sequence used in the simulation. Continue entering the following program into the file:*

```
randn('seed',0);
noi1=1*randn(1,L);
randn('seed',1);
noi2=1*randn(1,L);
y=y+[noi1;noi2];
```

4. *Specify the Q and R matrices. Continue entering the following program into the file:*

```
Q=zeros(n);
R=eye(m);
```

5. *Specify the initial conditions for the Kalman filter algorithm. Continue entering the following program into the file:*

```
P_plus=100*eye(n);
x_plus=zeros(n,1);
```

6. *Call MATLAB function* **KFSQ.m** *written in Tutorial 8.1. Continue entering the following program into the file:*

```
[xhat,yhat]=KFSQ(A,B,C,u,y,x_plus,P_plus,Q,R);
```

7. *Test this program by comparing the estimated state variable vector xhat with the true state variable xmodel. Compute the error signals. Continue entering the following program into the file:*

```
E=xmodel-xhat;
S=E*E'/L;
trace(S)
```

8. *The answer for trace(S) is* 0.0732.

8.2.4 Food for Thought

1. In the sequential Kalman filter algorithm, what is the key assumption on the R matrix?
2. Do we make any changes to *a priori* estimates $\hat{x}(k)^-$ and $P(k)^-$?
3. If one of the diagonal elements in R matrix is zero, do you anticipate any problems from the sequential Kalman filter algorithm?
4. For the sequential Kalman filter algorithm, can we combine the *a priori* estimate with the *a posteriori* estimate into one-step Kalman filter?

5. How do we modify the sequential Kalman filter algorithm so that it also has the function of fading memory?
6. How do we write the extended Kalman filter algorithm using the idea of sequential updating on the estimates?

8.3 The Kalman Filter using UDU^T Factorization

A commonly encountered problem in real-time implementation of Kalman filters is degradation of accuracy when it is implemented in micro-controllers. There are several approaches that have been proposed in the literature, such as the square root Kalman filter and UDU^T factorization Kalman filter. These approaches typically offer double the accuracy of the original formulations, however with a higher computational demand. The UDU^T approach is considered here because it is relatively straightforward to implement once it is understood. Although the UDU^T filtering approach is derived based on a single measurement signal, this method can be easily extended to multiple measurement signals using the sequential Kalman filter algorithm. Therefore, throughout the whole section, we assume that the system has a single output variable.

8.3.1 Gram-Schmidt Orthogonalization Procedure

We will use the Gram-Schmidt orthogonalization procedure for the matrix decomposition. For those who are not familiar with matrix computation, the text books in this area include Horn and Johnson [2012], Moon and Stirling [2000], Watkins [2004], Demmel [1997], Trefethen and Bau III [1997], Lax [2007].

Assume that a matrix M is written as $M = W\Omega W^T$. The decomposition task is to find the U and V matrices such that

$$W\Omega W^T = UV\Omega V^T U^T,$$

where

$$V\Omega V^T = D$$

D is a diagonal matrix.

In the Gram-Schmidt orthogonalization procedure, the rows of W matrix are used to generate the row vectors in V matrix that are orthogonal with respect to the Ω inner product. The algorithm is summarized as follows.

Note that the matrix W has dimension of $n \times q$. The nth row vector v_n is chosen to be equal to the nth row of W. Then,

$$v_{n-1} = w_{n-1} - \frac{w_{n-1}\Omega v_n^T}{v_n \Omega v_n^T} v_n$$

$$v_{n-2} = w_{n-2} - \frac{w_{n-2}\Omega v_{n-1}^T}{v_{n-1}\Omega v_{n-1}^T} v_{n-1} - \frac{w_{n-2}\Omega v_n^T}{v_n \Omega v_n^T} v_n.$$

For the kth row vector in V, we have

$$v_k = w_k - \sum_{j=k+1}^{n} \frac{w_k \Omega v_j^T}{v_j \Omega v_j^T} v_j.$$

With the row vector v_k, we can find the diagonal matrix D as

$$D = V\Omega V^T,$$

where the kth diagonal element is

$$d_k = v_k \Omega v_k^T.$$

The element at the kth row and jth column in U become

$$U(k,j) = \frac{w_k \Omega v_j^T}{v_j \Omega v_j^T}$$

where $j, k = 1, \dots, n$.

The following tutorial is to write a function for Gram-Schmidt orthogonal decomposition.

Tutorial 8.3 *This tutorial is to illustrate how to produce UDU^T factorization for a matrix given by $W\Omega W^T$ using Gram-Schmidt orthogonal decomposition algorithm. We will use this function for the sequential Kalman filter implementation.*

Step by Step

1. *Create a new file called* **GSDecomp.m**. *The inputs to the function are W and Ω matrices, and the output variables from the function are U and D matrices. Enter the following program into the file:*

```
function [U,D]=GSDecomp(W,Omega)
```

2. *Check the dimension of W matrix. Continue entering the following program into the file:*

```
[n,q]=size(W);
```

3. *Initialize the U and D matrices. Continue entering the following program into the file:*

```
D=zeros(n);
U=eye(n);
```

4. *Compute the last row of V matrix and last diagonal element of D matrix. Continue entering the following program into the file:*

```
v(n,:)=W(n,:);
D(n,n)=v(n,:)*Omega*v(n,:)';
```

5. *Compute the rest rows of V matrix, the rest of the diagonal elements of D matrix and the elements in U matrix. Continue entering the following program into the file:*

```
for kk=n-1:-1:1;
    S=W(kk,:);
    for jj=kk+1:n;
        U(kk,jj)=W(kk,:)*Omega*v(jj,:)'/D(jj,jj);
       S=S-U(kk,jj)*v(jj,:);
    end
    v(kk,:)=S;
    D(kk,kk)=v(kk,:)*Omega*v(kk,:)';
end
```

6. *Test this function using the following matrices:*

$$W = \begin{bmatrix} 1 & 2 & 3 \\ 4 & 5 & 6 \end{bmatrix}; \ \Omega = \begin{bmatrix} 100 & 1 & 2 \\ 1 & 200 & 3 \\ 2 & 3 & 100 \end{bmatrix}$$

which gives the following U and D matrices:

$$U = \begin{bmatrix} 1 & 0.4118 \\ 0 & 1 \end{bmatrix}; \ D = 10^4 \begin{bmatrix} 0.0069 & 0 \\ 0 & 1.0516 \end{bmatrix}.$$

We can verify that $W\Omega W^T = UDU^T$.

8.3.2 Basic Ideas

We will first express the covariance $P(k)^-$ and $P(k)^+$ matrices using U^-, D^-, U^+ and D^+. For notational consistency, the *a priori* covariance matrix is denoted by

$$P(k)^- = U(k)^- D(k)^- (U(k)^-)^T$$

and the *a posteriori* covariance matrix is denoted by

$$P(k)^+ = U(k)^+ D(k)^+ (U(k)^+)^T$$

The central idea in Kalman filter using UDU^T factorization is to avoid the direct multiplications of badly scaled matrices. Instead of propagating $P(k)^-$ and $P(k)^+$

matrices, we propagate the upper triangular matrices U^-, U^+ and the diagonal matrices D^-, D^+.

Recall that for a single output system, at the sampling instant k, the key equations in the computation of the Kalman filter are the following three equations:

$$P(k)^- = A(k-1)P(k-1)^+A(k-1)^T + Q(k-1) \tag{8.21}$$

$$K(k) = \frac{P(k)^-C(k)^T}{C(k)P(k)^-C(k)^T + R(k)} \tag{8.22}$$

$$P(k)^+ = (I - K(k)C(k))P(k)^-. \tag{8.23}$$

Substituting the Kalman filter gain (8.22) into (8.23) gives the following *a posteriori* covariance matrix:

$$P(k)^+ = P(k)^- - \frac{P(k)^-C(k)^TC(k)P(k)^-}{C(k)P(k)^-C(k)^T + R(k)}. \tag{8.24}$$

To improve its numerical accuracy, the Kalman filter algorithm must avoid the direct manipulations of $P(k)^-$ and $P(k)^+$. Instead, all computations will be based on $U(k)^-$, $D(k)^-$, $U(k)^+$ and $D(k)^+$.

Expression of Kalman filter gain $K(k)$

$$\begin{aligned} K(k) &= \frac{P(k)^-C(k)^T}{C(k)P(k)^-C(k)^T + R(k)} \\ &= \beta(k)U(k)^-D(k)^-(C(k)U(k)^-)^T \end{aligned} \tag{8.25}$$

where the scalar $\beta(k)$ is calculated using:

$$\beta(k) = \frac{1}{C(k)U(k)^-D(k)^-(C(k)U(k)^-)^T + R(k)}. \tag{8.26}$$

Expression of the *a priori* covariance matrix $P(k)^-$

Assume that at the sampling instant $k-1$, the *a posteriori* covariance matrix $P(k-1)^+$ is decomposed in terms of $U(k-1)^+$ and $D(k-1)^+$ such that

$$P(k-1)^+ = U(k-1)^+D(k-1)^+(U(k-1)^+)^T.$$

At the sampling instant k, the *a priori* covariance matrix $P(k)^-$ is expressed as:

$$\begin{aligned} P(k)^- &= A(k-1)U(k-1)^+D(k-1)^+(U(k-1)^+)^TA(k-1)^T + Q(k-1) \\ &= W_1(k-1)\Omega_1(k-1)W_1(k-1)^T \end{aligned} \tag{8.27}$$

where $W_1(k-1)$ is given by

$$W_1(k-1) = \left[A(k-1)U(k-1)^+ \quad I_{n\times n}\right]$$

and $\Omega_1(k-1)$ is given by

$$\Omega_1(k-1) = \begin{bmatrix} D(k-1)^+ & 0_{n\times n} \\ 0_{n\times n} & Q(k-1) \end{bmatrix}$$

$0_{n\times n}$ is a zero matrix with dimension $n \times n$, and n is the number of state variables. The expression of $P(k)^-$ by (8.27) avoids the *a posteriori* covariance matrix $P(k-1)^+$ with the formulations of $W_1(k-1)$ and $\Omega_1(k-1)$, which will be used to find $U(k)^-$ and $D(k)^-$.

Expression of the *a posteriori* covariance matrix $P(k)^+$

With the expression of $\beta(k)$ from (8.26), and the computed $U(k)^-$ and $D(k)^-$ from the previous step, the *a posteriori* covariance matrix is written as

$$\begin{aligned} P(k)^+ &= P(k)^- - \beta(k)P(k)^- C(k)^T C(k)P(k)^- \\ &= U(k)^- \{D(k)^- - \beta(k)D(k)^-(U(k)^-)^T C(k)^T C(k)U(k)^- D(k)^-\}(U(k)^-)^T \\ &= U(k)^- \overline{P(k)}^+ (U(k)^-)^T \end{aligned} \tag{8.28}$$

where the variable $\overline{P(k)}^+$ is defined as

$$\overline{P(k)}^+ = D(k)^- - \beta(k)D(k)^-(U(k)^-)^T C(k)^T C(k)U(k)^- D(k)^-. \tag{8.29}$$

With the introduction of $\overline{P(k)}^+$, the task for finding the decomposition of $P(k)^+$ becomes the task to find the $\overline{U(k)}^+$ and $D(k)^+$ for the decomposition of $\overline{P(k)}^+$ defined by (8.29).

For this purpose, $\overline{P(k)}^+$ from (8.29) is re-written as

$$\overline{P(k)}^+ = W_2(k)\Omega_2(k)W_2(k)^T \tag{8.30}$$

where $W_2(k)$ is defined as

$$W_2(k) = \begin{bmatrix} I_{n\times n} & (C(k)U(k)^- D(k)^-)^T \end{bmatrix}$$

and $\Omega_2(k)$ is defined as

$$\Omega_2(k) = \begin{bmatrix} D(k)^- & 0_{n\times 1} \\ 0_{1\times n} & -\beta(k) \end{bmatrix}.$$

Here, $I_{n\times n}$ represents the identity matrix with the dimensions $n \times n$, $0_{p\times q}$ the zero vector with dimensions $p \times q$.

Note that the variable structures in $W_1(k-1)$ and $\Omega_1(k-1)$ represented by (8.27) are similar to those in $W_2(k)$ and $\Omega_2(k)$ (see (8.30)). The Gram-Schmidt orthogonalization procedure introduced in Section 8.3.1 is a suitable candidate to perform the orthogonal decomposition tasks required for both $P(k)^-$ and $\overline{P}(k)^+$. For this purpose, the MATLAB function **GSDecomp.m** written in Tutorial 8.3 will find the $U(k)^-$, $D(k)^-$, $U(k)^+$ and $D(k)^+$.

It is worthwhile to emphasis that $U(k)^-$ and $U(k)^+$ are triangular matrices with the diagonal elements being 1, $D(k)^-$ and $D(k)^+$ are diagonal matrices. These variables are to be stored in a computational device for each iteration and propagation in the Kalman filter algorithm.

8.3.3 Sequential Kalman Filter with UDU^T Decomposition

The sequential Kalman filter algorithm with Gram-Schmidt decomposition is summarized below with the assumption R being a diagonal matrix with the ith element r_{ii}. The initial covariance matrix $P(0)^+$ is decomposed to find $U(0)^+$ and $D(0)^+$ matrices.

1. Compute the *a priori* estimate of the state:

$$\hat{x}(k)^- = A(k-1)\hat{x}(k-1)^+ + B(k-1)u(k-1). \tag{8.31}$$

2. Based on (8.27), form $W_1(k-1)$ and $\Omega_1(k-1)$ matrices:

$$W_1(k-1) = \begin{bmatrix} A(k-1)U(k-1)^+ & I_{n\times n} \end{bmatrix}$$
$$\Omega_1(k-1) = \begin{bmatrix} D(k-1)^+ & 0_{n\times n} \\ 0_{n\times n} & Q(k-1) \end{bmatrix}.$$

3. Use the Gram-Schmidt orthogonalization procedure to find $U(k)^-$ and $D(k)^-$ matrices.
4. Compute the Kalman filter gain sequentially using the Gram-Schmidt orthogonalization procedure with the following steps:
   (a) Choose the initial *a posteriori* estimates:

$$\hat{x}(k)_0^+ = \hat{x}(k)^-$$
$$U(k)_0^+ = U(k)^-; D(k)_0^+ = D(k)^-.$$

   (b) For $1 \le i \le m$, compute the *a posteriori* estimates for each output sequentially. We drop the index i when there is no confusion. Form

$$\Gamma(k)_1 = C(k)U(k)^-$$
$$\beta(k) = \frac{1}{\Gamma(k)_1 D(k)^- \Gamma(k)_1^T + r_{ii}}.$$

   (c) Calculate the Kalman filter gain and update the estimate with the ith output:

$$K_i(k) = \beta(k)U(k)^- D(k)^- \Gamma(k)_1^T,$$
$$\hat{x}(k)_i^+ = \hat{x}(k)_{i-1}^+ + K_i(k)(y_i(k) - C_i(k)\hat{x}(k)_{i-1}^+).$$

(d) Base on (8.30), form $W_2(k)$ and $\Omega_2(k)$:

$$W_2(k) = \left[I_{n\times n} \quad (C(k)U(k)^- D(k)^-)^T\right],$$

$$\Omega_2(k) = \begin{bmatrix} D(k)^- & 0_{n\times 1} \\ 0_{1\times n} & -\beta(k) \end{bmatrix}.$$

(e) Use the Gram-Schmidt orthogonalization procedure to find $\overline{U(k)}^+$ and $D(k)^+$ matrices.

(f) Update $U(k)^+$ by calculating

$$U(k)^+ = U(k)^- \overline{U(k)}^+.$$

(g) Update

$$U(k)^- = U(k)^+$$
$$D(k)^- = D(k)^+$$

5. Using the values calculated from the mth output, update the *a posteriori* estimate of the state and $U(k)^+$,$D(k)^+$,

$$\hat{x}(k)^+ = \hat{x}(k)^+_m;$$
$$U(k)^+ = U(k)^+_m;$$
$$D(k)^+ = D(k)^+_m.$$

The following example illustrates the application of the algorithm.

Example 8.4

In Example 8.3, we have demonstrated the sequential Kalman filter algorithm. We can find that the final step of the sequential KF algorithm gives $P(1000)^-$ and $P(1000)^+$ as

$$P(1000)^- = 10^{-47} \begin{bmatrix} 0.0877 & -0.0414 & -0.2631 \\ -0.0414 & 0.0196 & 0.1242 \\ -0.2631 & 0.1242 & 0.7894 \end{bmatrix}$$

$$P(1000)^+ = 10^{-47} \begin{bmatrix} 0.0877 & -0.0414 & -0.2631 \\ -0.0414 & 0.0196 & 0.1242 \\ -0.2631 & 0.1242 & 0.7894 \end{bmatrix}$$

When the sequential Kalman filter algorithm converges, for this example, the matrices are identical. It is very difficult to propagate these matrices through the iterations with limited memories when using a micro-controller. With the identical conditions as in Example 8.3, find the U^-, D^-, U^+ and D^+ matrices for the final step of the Gram-Schmidt orthogonal decomposition algorithm.

Solution

The decomposition of the initial condition $P(0)^+$ gives,

$$U(0)^+ = \begin{bmatrix} 1 & 0 & 0 \\ 0 & 1 & 0 \\ 0 & 0 & 1 \end{bmatrix}; D(0)^+ = \begin{bmatrix} 100 & 0 & 0 \\ 0 & 100 & 0 \\ 0 & 0 & 100 \end{bmatrix}.$$

For the third iteration, we have:

$$U(3)^- = \begin{bmatrix} 1 & -7.3366 & -1.0469 \\ 0 & 1 & 0.2405 \\ 0 & 0 & 1 \end{bmatrix}; D(3)^- = \begin{bmatrix} 0.0386 & 0 & 0 \\ 0 & 0.0074 & 0 \\ 0 & 0 & 17.1315 \end{bmatrix}$$

$$U(3)^+ = \begin{bmatrix} 1 & -6.8463 & -1.0772 \\ 0 & 1 & 0.2442 \\ 0 & 0 & 1 \end{bmatrix}; D(3)^+ = \begin{bmatrix} 0.0359 & 0 & 0 \\ 0 & 0.0045 & 0 \\ 0 & 0 & 10.6008 \end{bmatrix}.$$

For the final iteration, we have:

$$U(1000)^- = U(1000)^+ = \begin{bmatrix} 1 & -8.8227 & -0.3333 \\ 0 & 1 & 0.1574 \\ 0 & 0 & 1 \end{bmatrix};$$

$$D(1000)^- = D(1000)^+ = \begin{bmatrix} d_{11} & 0 & 0 \\ 0 & d_{22} & 0 \\ 0 & 0 & d_{33} \end{bmatrix}$$

where $d_{11} = 6.8501 \times 10^{-64}$, $d_{22} = 8.7247 \times 10^{-66}$, $d_{33} = 7.8943 \times 10^{-48}$.

It is interesting to note that both U^- and U^+ are nicely scaled upper triangular matrices during the iterations. Only the diagonal matrices $D(k)^-$ and $D(k)^+$ decrease to near zero numbers.

8.3.4 MATLAB Tutorial

The following tutorial is to produce the Kalman filter with the UDU^T factorization.

Tutorial 8.4 *The objective of this tutorial is to write a MATLAB function* **KFUDU.m** *that will produce the estimated states and outputs for a multi-input and multi-output system using orthogonal decomposition. The input variables to the function are:*

- *the system matrices A, B, C;*
- *the input and output data u ($p \times L$) y ($m \times L$), where p and m are the number of inputs and outputs respectively, L is the data length;*
- *the initial conditions for the Kalman filter $x(0)^+$ and $P(0)^+$;*
- *the weighting matrices Q and R, where R is a diagonal matrix.*

The output variables from the function are $\hat{x}(k)^+$ and $\hat{y}(k)$.

Step by Step

1. *Create a new file called* **KFUDU.m**.
2. *Enter the following program into the file:*

```
function[xhat,yhat]=KFUDU(A,B,C,u,y,x_plus,P_plus,Q,R)
```

3. *Check the data length and the numbers of measurement and state. Continue entering the following program into the file:*

```
L=length(y);
[m,n]=size(C);
```

4. *The first estimates $\hat{x}(0)$ and $\hat{y}(0)$ are chosen to be the initial conditions $\hat{x}(0)^+$ and $C\hat{x}(0)^+$. Continue entering the following program into the file:*

```
xhat(:,1)=x_plus;
yhat(:,1)=C*x_plus;
```

5. *The initial values of $D(0)^+$ and $U(0)+$ are chosen according to $P(0)^+$, where we assume that $P(0)^+$ is a diagonal matrix. Continue entering the following program into the file:*

```
U_plus=eye(n);
D_plus=P_plus;
```

6. *The computation will be performed recursively. Compute* a priori *estimate of the state. Continue entering the following program into the file:*

```
for k=2:L
x_minus=A*x_plus+B*u(:,k-1);
```

7. *Form $W_1(k-1)$ and $\Omega_1(k-1)$ matrices. Continue entering the following program into the file:*

```
W1=[A*U_plus eye(n)];
Omega1=zeros(2*n,2*n);
Omega1(1:n,1:n)=D_plus;
Omega1(n+1:2*n,n+1:2*n)=Q;
```

8. *Compute $U(k)^-$ and $D(k)^-$ using the Gram-Schmidt orthogonal decomposition algorithm. Continue entering the following program into the file:*

```
[U_minus,D_minus]=GSDecomp(W1,Omega1);
```

9. *Perform the measurement update sequentially. Continue entering the following program into the file:*

```
for jj=1:m
Gamma1=C(jj,:)*U_minus;
beta=1/(Gamma1*D_minus*Gamma1'+R(jj,jj));
K=U_minus*D_minus*Gamma1'*beta;
x_plus=x_minus+K*(y(jj,k)-C(jj,:)*x_minus);
```

10. *Form $W_2(k)$ and $\Omega_2(k)$ matrices. Continue entering the following program into the file:*

```
Gamma2=Gamma1*D_minus;
W2=[eye(n) Gamma2'];
Omega2=zeros(n+1,n+1);
Omega2(1:n,1:n)=D_minus;
Omega2(n+1:n+1,n+1:n+1)=-beta;
```

11. *Compute $U(k)^+$ and $D(k)^+$ using the Gram-Schmidt orthogonal decomposition algorithm. Continue entering the following program into the file:*

```
[Ut,D_plus]=GSDecomp(W2,Omega2);
U_plus=U_minus*Ut;
```

12. *Prepare for the update of the next output signal. Continue entering the following program into the file:*

```
U_minus=U_plus;
D_minus=D_plus;
x_minus=x_plus;
end
```

13. *Save the estimated state variables and complete one cycle of computation. Continue entering the following program into the file:*

```
xhat(:,k)=x_plus;
yhat(:,k)=C*x_plus;
end
```

The MATLAB program **KFUDU.m** can be tested using the system data generated in Tutorial 8.2. With the data from Tutorial 8.2, the application of program **KFUDU.m** leads to the following errors:

```
E=xmodel-xhat;
S=E*E'/L;
trace(S)
```

The answer for *trace(S)* is 0.0732, which is identical to the result obtained from Tutorial 8.2.

8.3.5 Food for Thought

1. A matrix is given by $W\Omega W^T$, where the dimension of W is $n \times q$ and Ω is $q \times q$. Based on Gram-Schmidt orthogonal decomposition, what is the dimension of V matrix?
2. A matrix is given by $W\Omega W^T$, where

$$W = \begin{bmatrix} 1 & 2 & 3 \\ 4 & 2 & 1 \end{bmatrix}; \; \Omega = \begin{bmatrix} 2 & 1 \\ 1 & 2 \end{bmatrix}.$$

   Based on Gram-Schmidt orthogonal decomposition, what are the V, D and U matrices?
3. Is it correct to say that the Kalman filter with orthogonal decomposition algorithm is derived for a single measurement system, but can be applied to a multiple measurement system if R matrix is either constant or diagonal with time-varying elements?
4. What are the key equations we used to form $P(k)^-$ and $P(k)^+$ based on $U(k-1)^+, D(k-1)^+, U(k)^-$ and $D(k)^-$?
5. How do we adapt the Kalman filter with orthogonal decomposition to include the fading memory function?
6. What are the key steps do we need to modify in order to obtain the extended Kalman filter algorithm with the orthogonal decomposition?

8.4 Summary

This chapter addresses some real-time computational issues in the Kalman filter when it is implemented in an embedded system using a micro-controller or other low-cost computational devices. The important aspects are summarized as follows.

1. The sequential Kalman filter is proposed for computing the *a posteriori* state estimate $\hat{x}(k)^+$, the *a posteriori* covariance $P(k)^+$ and the Kalman filter gain $K(k)$. With the sequential Kalman filter algorithm, a system with multiple measurements is handled in the manner of a single output measurement and the computation of matrix inversion in the Kalman filter is avoided. As a result, the sequential Kalman filter eliminates the complicated issue of computing a matrix inversion when a library is not available for the development of an embedded system. This approach is particularly useful when the multiple measurements have a multiple sampling rate or encounter the problem of missing measurement data.
2. In order to improve numerical accuracy of the Kalman filter for its implementation in an embedded system with a low-cost computational device, the Kalman

filter with UDU^T decomposition is proposed. The algorithm has the potential to double its accuracy by avoiding the operation of multiplication of numerically ill-conditioned matrices.

3. Both sequential Kalman filter and Kalman filter with UDU^T decomposition are based on the assumption that the covariance matrix of the measurement noise R is a diagonal matrix. If R is not a diagonal matrix, as long as it is constant matrix, a singular value decomposition on R can be performed to transform it into a diagonal form so that the Kalman filter algorithms can be implemented successfully.

8.5 Further Reading

1. The book by Simon Dan Simon [2006] has extensively addressed the computational issues in the Kalman filter.
2. The sequential Kalman filter is introduced in Bucy and Joseph [1968] and Brown and Hwang [1992].
3. The sequential Kalman filter is discussed and applied in Willner et al. [1976], Kettner and Paolone [2017], Dong et al. [2020].
4. The Kalman filter with UDU^T decomposition is introduced in Bierman [1976], Thornton and Bierman [1978], Bierman and Thornton [1977].
5. Books about linear algebra and matrix computations include Watkins [2004], Demmel [1997], Trefethen and Bau III [1997], Lax [2007], Horn and Johnson [2012], Moon and Stirling [2000].

Problems

Problem 8.1 A discrete-time system is described by the following state space model:

$$x(k) = Ax(k-1) + w(k-1)$$
$$y(k) = Cx(k) + v(k)$$

where the system matrices are

$$A = \begin{bmatrix} 0.1 & 2 & 0 \\ 0 & 1 & 3 \\ 0 & 1 & 0.6 \end{bmatrix}; \; C = \begin{bmatrix} 1 & 2 & 0 \\ 0 & 1 & 1 \end{bmatrix},$$

the covariance of process noise is $E(w(k)w(k)^T) = Q$ and the measurement noise is $E(v(k)v(k))^T = R$.

1. Assuming that

$$R = \begin{bmatrix} 0.6 & 0 \\ 0 & 1 \end{bmatrix},$$

write down the equations associated with the sequential Kalman filter that will recursively estimate the state $x(k)$ without matrix inversion.

2. Assuming that

$$R = \begin{bmatrix} 0.6 & 0.2 \\ 0.2 & 1 \end{bmatrix},$$

what are the equations required for the sequential Kalman filter?

3. Assuming that R is a time-varying and diagonal matrix described by

$$R = \begin{bmatrix} r_{11}(k) & 0 \\ 0 & r_{22}(k) \end{bmatrix},$$

modify the sequential Kalman filter to accommodate the time-varying elements in R matrix.

Problem 8.2 Assuming that a discrete-time system is described by

$$\begin{bmatrix} x_1(k) \\ x_2(k) \end{bmatrix} = \begin{bmatrix} 0.3 & 1 \\ -0.1 & 0.6 \end{bmatrix} \begin{bmatrix} x_1(k-1) \\ x_2(k-1) \end{bmatrix} + \begin{bmatrix} 1 \\ 2 \end{bmatrix} u(k-1) + \begin{bmatrix} 2 \\ 1 \end{bmatrix} w(k-1)$$

$$y(k) = \begin{bmatrix} 1 & 2 \end{bmatrix} \begin{bmatrix} x_1(k) \\ x_2(k) \end{bmatrix} + v(k)$$

where $w(k)$ and $v(k)$ are zero mean white noise with variances σ_w^2 and σ_v^2. Write down the Kalman filter equations using UDU^T decompositions.

Problem 8.3 The state space model for the three springs and double mass system is described by the following differential equations:

$$\begin{bmatrix} \dot{x}_1(t) \\ \dot{x}_2(t) \\ \dot{x}_3(t) \\ \dot{x}_4(t) \end{bmatrix} = \begin{bmatrix} 0 & 1 & 0 & 0 \\ -\frac{\alpha_1+\alpha_2}{M_1} & 0 & \frac{\alpha_2}{M_1} & 0 \\ 0 & 0 & 0 & 1 \\ \frac{\alpha_2}{M_2} & 0 & -\frac{\alpha_1+\alpha_2}{M_2} & 0 \end{bmatrix} \begin{bmatrix} x_1(t) \\ x_2(t) \\ x_3(t) \\ x_4(t) \end{bmatrix} + \begin{bmatrix} 0 & 0 \\ \frac{1}{M_1} & 0 \\ 0 & 0 \\ 0 & \frac{1}{M_2} \end{bmatrix} \begin{bmatrix} u_1(t) \\ u_2(t) \end{bmatrix} \tag{8.32}$$

$$\begin{bmatrix} y_1(t) \\ y_2(t) \end{bmatrix} = \begin{bmatrix} 1 & 0 & 0 & 0 \\ 0 & 0 & 1 & 0 \end{bmatrix} \begin{bmatrix} x_1(t) \\ x_2(t) \\ x_3(t) \\ x_4(t) \end{bmatrix} \tag{8.33}$$

As in Tongue [2002], the physical parameters are $M_1 = 2kg$, $M_2 = 4kg$, $\alpha_1 = 40N/m$ and $\alpha_2 = 100N/m$.
The system has been used in comparative studies between Kalman filter and observer in Example 7.11 of Chapter 7. In this exercise, we will use this system to practice sequential Kalman filter with multi-rate sampled measurement data. We assume that there is no process noise, therefore, $Q = 0$, however, the measurement noise is described by

$$R = E(\upsilon(k)\upsilon(k)^T) = \begin{bmatrix} 0.8 & 0 \\ 0 & 1 \end{bmatrix}.$$

1. Assuming that the dynamic model (8.32) is sampled with $\Delta t = 0.01$ (sec), and the measurement model (8.33) is sampled with $\Delta t = 0.1$ (sec), write down the sequential Kalman filter equations that will estimate the state $x(k)$ corresponding to the faster sampling rate. One approach is to model the slower measurement data with a time-varying measurement matrix C.
2. Generate a set of simulation data with sampling interval $\Delta t = 0.01$ (sec), the initial condition $x(0) = \begin{bmatrix} 1 & 0 & 1 & 0 \end{bmatrix}^T$ and input signal

$$u(k) = \begin{bmatrix} \cos \frac{2\pi}{100}k & \sin \frac{2\pi}{50}k \end{bmatrix}^T$$

   The measurement noises for y_1 and y_2 are zero means with variances 0.8 and 1 respectively.
3. Validate the sequential Kalman filter algorithm using the simulation data. Present the error signals $x_i(k) - \hat{x}_i(k)$, where $i = 1,2,3,4$, and k corresponds to the faster measurement data.
4. Compare the estimation results with those obtained using a standard Kalman filter algorithm when the matrix inversion is used. What are your observations?

Problem 8.4 Continue from Problem 8.3. In this exercise, we will design a sequential Kalman filter for the application where each sensor has a different sampling rate.

1. Assuming that the sampling interval for the measurement $y_1(t)$ is 0.1 (sec) and for the measurement $y_2(t)$ is 0.5 (sec), write down the equations associated with the sequential Kalman filter algorithm that will recursively estimate the state $x(k)$ without matrix inversion.
2. Validate the sequential Kalman filter algorithm with the same set of testing data generated from Problem 8.3.
3. In comparison with the results obtained from Problem 8.3, what are your observations?

Bibliography

Vladislav Akhmatov, Hans Knudsen, and Arne Hejde Nielsen. Advanced simulation of windmills in the electric power supply. *International Journal of Electrical Power & Energy Systems*, 22(6):421–434, 2000.

Victor M Alfaro and Ramon Vilanova. *Model-reference robust tuning of PID controllers*. Springer, 2016.

B. D. O. Anderson and J. M. Moore. *Linear Optimal Control*. Prentice-Hall, Hemel Hempstead, 1971.

B. D. O. Anderson and J. M. Moore. *Optimal Filtering*. Prentice-Hall, Englewood Cliffs, New Jersey, 1979.

Brian DO Anderson and John B Moore. Linear system optimisation with prescribed degree of stability. In *Proceedings of the Institution of Electrical Engineers*, volume 116, pages 2083–2087. IET, 1969.

P. J. Antsaklis and A. N. Michel. *Linear systems*. New York: McGraw-Hill, 1997.

K. J. Astrom and T. Hagglund. *PID Controllers:Theory, Design, and Tuning*. Instrument Society of America, Research Triangle Park, NC, 1995.

K. J. Astrom and T. Hagglund. *Advanced PID control*. Instrument Society of America, Research Triangle Park, NC, 2006.

K. J. Astrom and B. Wittenmark. *Computer Controlled Systems:Theory and Design*. Prentice-Hall, Englewood Cliffs, N. J., 3rd edition, 1997.

Karl Johan Åström and Rodney D Bell. Drum-boiler dynamics. *Automatica*, 36(3):363–378, 2000.

J. S. Bay. *Fundamentals of Linear State Space Systems*. McGraw-Hill, 1999.

Esmat Bekir. *Introduction to modern navigation systems*. World scientific, 2007.

Gerald J Bierman. Measurement updating using the ud factorization. *Automatica*, 12(4):375–382, 1976.

Gerald J Bierman and Catherine L Thornton. Numerical comparison of Kalman filter algorithms: Orbit determination case study. *Automatica*, 13(1):23–35, 1977.

R. R. Bitmead and B.D.O. Anderson. Adaptive frequency sampling filters. *IEEE Trans. on Circuits and Systems*, Vol. 28:524–533, 1981.

State Feedback Control and Kalman Filtering with MATLAB/Simulink Tutorials, First Edition.
Liuping Wang and Robin Ping Guan.

Companion website: www.wiley.com/go/feedback19

Edgar Bristol. On a new measure of interaction for multivariable process control. *IEEE transactions on automatic control*, 11(1):133–134, 1966.

Robert Brown and P. C. Hwang. *Introduction to random signals and applied Kalman filtering*. New York, John Wiley & Sons, 1992.

Robert Grover Brown and Patrick Y.C. Huang. *Introduction to random signals and applied Kalman filtering with MATLAB exercises*. John Wiley and Sons, 2012.

A. E. Bryson and Y.C. Ho. *Applied optimal control*. Washington, DC: Hemisphere, 1975.

Richard S Bucy and Peter D Joseph. *Filtering for stochastic processes with applications to guidance*. John Wiley & Sons, New York, 1968.

R. L. Burden and J. D. Faires. *Numerical Analysis*. PWS-KENT Publishing Company, Boston, 1989.

Frank M Callier and Charles A Desoer. *Linear system theory*. Springer Science & Business Media, 2012.

Donald E Catlin. *Estimation, control, and the discrete Kalman filter*, volume 71. Springer Science & Business Media, 2012.

Chi-Tsong Chen. *Linear system theory and design*. Oxford University Press, Inc., 1998.

Wen-Hua Chen, Donald J Ballance, Peter J Gawthrop, and John O'Reilly. A nonlinear disturbance observer for robotic manipulators. *IEEE Transactions on industrial Electronics*, 47(4):932–938, 2000.

Wen-Hua Chen, Jun Yang, Lei Guo, and Shihua Li. Disturbance-observer-based control and related methods: an overview. *IEEE Transactions on Industrial Electronics*, 63(2):1083–1095, 2016.

Mahmoud Chilali and Pascal Gahinet. H_∞ design with pole placement constraints: an LMI approach. *IEEE Transactions on automatic control*, 41(3):358–367, 1996.

Min-Sen Chiu and Yaman Arkun. A methodology for sequential design of robust decentralized control systems. In *1991 American Control Conference*, pages 1836–1841. IEEE, 1991.

E Davison and Sa Wang. On pole assignment in linear multivariable systems using output feedback. *IEEE Transactions on Automatic Control*, 20(4): 516–518, 1975.

E Davison and W Wonham. On pole assignment in multivariable linear systems. *IEEE Transactions on Automatic Control*, 13(6):747–748, 1968.

James W Demmel. *Applied numerical linear algebra*. SIAM, 1997.

Yi Dong, Dingjie Wang, Liang Zhang, Qingsong Li, and Jie Wu. Tightly coupled gnss/ins integration with robust sequential Kalman filter for accurate vehicular navigation. *Sensors*, 20(2):561, 2020.

SPH Driessen, NHJ Janssen, Liuping Wang, JL Palmer, and Henk Nijmeijer. Experimentally validated extended Kalman filter for UAV state estimation using low-cost sensors. *IFAC-PapersOnLine*, 51(15):43–48, 2018.

Ramsey Faragher. Understanding the basis of the Kalman filter via a simple and intuitive derivation [lecture notes]. *IEEE Signal processing magazine*, 29 (5):128–132, 2012.

Bo Feng, Mengyin Fu, Hongbin Ma, Yuanqing Xia, and Bo Wang. Kalman filter with recursive covariance estimation-sequentially estimating process noise covariance. *IEEE Transactions on Industrial Electronics*, 61(11):6253–6263, 2014.

A. Feuer and G. C. Goodwin. *Sampling in Digital Signal Processing and Control.* Wiley, New York, 1997.

B.A. Francis and W.M. Wonham. The internal model principle of control theory. *Automatica*, 12:457–465, 1976.

G. F. Franklin, J. D. Powell, and A. Emami-Naeini. *Feedback Control of Dynamic Systems.* Addison-Welsley Publishing Company, Second Edition, 1991.

K Furuta and S Kim. Pole assignment in a specified disk. *IEEE Transactions on Automatic Control*, 32(5):423–427, 1987.

Sergio Galeani, Sophie Tarbouriech, Matthew Turner, and Luca Zaccarian. A tutorial on modern anti-windup design. In *Control Conference (ECC), 2009 European*, pages 306–323. IEEE, 2009.

Farid Golnaraghi and B. C. Kuo. *Automatic control systems.* Wiley and Sons, 2010.

G. C. Goodwin and K.S. Sin. *Adaptive Filtering Prediction and Control.* Prentice-Hall, Englewood Cliffs, NJ, 1984.

G. C. Goodwin, S. Graebe, and M. Salgado. *Control System Design.* Prentice-Hall, Englewood Cliffs, NJ, 2000.

Graham Goodwin, R Evans, R Leal, and R Feik. Sinusoidal disturbance rejection with application to helicopter flight data estimation. *IEEE transactions on acoustics, speech, and signal processing*, 34(3):479–484, 1986.

Graham C Goodwin, Mario E Salgado, and Eduardo I Silva. Time-domain performance limitations arising from decentralized architectures and their relationship to the rga. *International journal of control*, 78(13):1045–1062, 2005.

M. J. Grimble and M. A. Johnson. *Optimal Control and Stochastic Estimation: Theory and Applications, Volume 1.* John Wiley and Sons, Chichester, U. K., 1988a.

PD Groves. *Principles of GNSS, Inertial, and Multisensor Integrated Navigation Systems.* Artech House: Boston, MA, USA, 2013.

Robin Guan. *Contributions to mobile robot localisation in GPS-denied environments.* PhD thesis, RMIT University, 2019.

Tore Hägglund, Sumit Shinde, Alfred Theorin, and Ulf Thomsen. An industrial control loop decoupler for process control applications. *Control Engineering Practice*, 123:105138, 2022.

Jingqing Han. From PID to active disturbance rejection control. *IEEE transactions on Industrial Electronics*, 56(3):900–906, 2009.

S. Hara, Y. Yamamoto, T. Omata, and M. Nakando. Repetitive control system: a new type of servo system for periodic exogenous signals. *IEEE Transactions on Automatic Control*, 33:659–668, 1988.

Walter T Higgins. A comparison of complementary and Kalman filtering. *IEEE Transactions on Aerospace and Electronic Systems*, (3):321–325, 1975.

Roger A Horn and Charles R Johnson. *Matrix analysis*. Cambridge university press, 2012.

Morten Hovd and Sigurd Skogestad. Sequential design of decentralized controllers. *Automatica*, 30(10):1601–1607, 1994.

Jeffrey Humpherys, Preston Redd, and Jeremy West. A fresh look at the Kalman filter. *SIAM review*, 54(4):801–823, 2012.

T. Inoue, M. Nakano, T. Kubo, S. Matsumoto, and H. Baba. High accuracy control of a proton synchrotron magnet power supply. *Proceedings of the 8th IFAC World Congress*, pages 216–221, 1981.

Qing Wei Jia. Disturbance rejection through disturbance observer with adaptive frequency estimation. *IEEE Transactions on Magnetics*, 45(6):2675–2678, 2009.

Karl Henrik Johansson. The quadruple-tank process: A multivariable laboratory process with an adjustable zero. *IEEE Transactions on control systems technology*, 8(3):456–465, 2000.

M. A. Johnson and M. H. Moradi. *PID Control: New identification and design methods*. Springer Verlag, New York, USA, 2005.

T. Kailath. *Linear Systems*. Prentice Hall, Inc., Englewood Cliffs, N.J., 1980.

Thomas Kailath. Lectures on Wiener and Kalman filtering. In *Lectures on Wiener and Kalman Filtering*, pages 1–143. Springer, 1981.

RE Kalman and JE Bertram. A unified approach to the theory of sampling systems. *Journal of the Franklin Institute*, 267(5):405–436, 1959.

Rudolph E Kalman. A new approach to linear filtering and prediction problems. *Transactions of the ASME, Journal of Basic Engineering*, 82(1):35–45, 1960.

Rudolph E Kalman and Richard S Bucy. New results in linear filtering and prediction theory. *Transactions of the ASME, Journal of Basic Engineering*, pages 95–108, 1961.

Navneet Kapoor, Andrew R Teel, and Prodromos Daoutidis. An anti-windup design for linear systems with input saturation. *Automatica*, 34(5):559–574, 1998.

Andreas Martin Kettner and Mario Paolone. Sequential discrete Kalman filter for real-time state estimation in power distribution systems: Theory and implementation. *IEEE Transactions on Instrumentation and Measurement*, 66(9):2358–2370, 2017.

Hidenori Kimura. Pole assignment by gain output feedback. *IEEE Transactions on automatic Control*, 20(4):509–516, 1975.

Satoshi Komada, Muneaki Ishida, Kouhei Ohnishi, and Takamasa Hori. Disturbance observer-based motion control of direct drive motors. *IEEE Transactions on Energy Conversion*, 6(3):553–559, 1991.

Mayuresh V Kothare, Peter J Campo, Manfred Morari, and Carl N Nett. A unified framework for the study of anti-windup designs. *Automatica*, 30(12): 1869–1883, 1994.

E. Kreyszig. *Advanced Engineering Mathematics*. John Wiley and Sons, New York, 2006.

Vladimir Kucera. *Discrete linear control: the polynomial equation approach*. John Wiley & Sons, Inc., 1980.

Huibert Kwakernaak and Raphael Sivan. *Linear optimal control systems*. Wiley-interscience New York, 1972.

Peter D. Lax. *Linear algebra and its applications*. Wiley, 2007.

S. Li, J. Yang, W.H. Chen, and X. Chen. Generalized state observer based control for systems with mismatched uncertainties. *IEEE Transactions on Industrial Electronics*, Vol. 59:4792–4802, 2012.

S. Li, J. Yang, W.H. Chen, and X. Chen. *Disturbance Observer-based Control: Methods and Applications*. CRC Press, 2014.

Su Liu, Jinfeng Liu, Yiping Feng, and Gang Rong. Performance assessment of decentralized control systems: An iterative approach. *Control Engineering Practice*, 22:252–263, 2014.

L. Ljung. *System Identification: Theory for the User*. Prentice-Hall, Englewood Cliffs, New Jersey, 1999.

R. W. Longman. Iterative learning control and repetitive control for engineering practice. *International Journal of Control*, pages 930–954, 2000.

David Luenberger. Observers for multivariable systems. *IEEE Transactions on Automatic Control*, 11(2):190–197, 1966.

David Luenberger. An introduction to observers. *IEEE Transactions on automatic control*, 16(6):596–602, 1971.

David G Luenberger. Observing the state of a linear system. *IEEE transactions on military electronics*, 8(2):74–80, 1964.

Luke McNabb, Liuping Wang, and Brendan McGrath. Intrinsically stable realization of a resonant current regulator for a single phase inverter. In *Control Conference (ASCC), 2017 11th Asian*, pages 2256–2261. IEEE, 2017.

Wim Michiels, Koen Engelborghs, Patrick Vansevenant, and Dirk Roose. Continuous pole placement for delay equations. *Automatica*, 38(5):747–761, 2002.

Tsutomu Mita, Mitsuo Hirata, Kenichi Murata, and Hui Zhang. H_∞ control versus disturbance-observer-based control. *IEEE Transactions on Industrial Electronics*, 45(3):488–495, 1998.

Todd K Moon and Wynn C Stirling. *Mathematical methods and algorithms for signal processing*. Prentice-Hall, New Jersey, 2000.

Kevin L Moore. *Iterative Learning Control for Deterministic Systems*. Springer Science & Business Media, 2012.

N. Nalano and S. Hara. Microprocessor-based repetitive control. *in Microprocessor-based Control Systems. Editor*: N. K. Sinha, pages 279–296, 1986.

H Nicholson. Dynamic optimisation of a boiler. In *Proceedings of the Institution of Electrical Engineers*, volume 111, pages 1479–1499. IET, 1964.

Aboelmagd Noureldin, Tashfeen B Karamat, and Jacques Georgy. *Fundamentals of inertial navigation, satellite-based positioning and their integration*. Springer, 2013.

Katsuhiko Ogata. *Modern Control Engineering*. Prentice Hall, 2002.

Fredrik Orderud. Comparison of Kalman filter estimation approaches for state space models with nonlinear measurements. In *Proc. of Scandinavian Conference on Simulation and Modeling*, pages 1–8, 2005.

David H Owens. *Iterative Learning Control: an Optimization Paradigm*. Springer, 2015.

Youbin Peng, Damir Vrancic, and Raymond Hanus. Anti-windup, bumpless, and conditioned transfer techniques for PID controllers. *IEEE Control Systems*, 16(4):48–57, 1996.

Abram Perdana. *Dynamic Models of Wind Turbines*. Chalmers University of Technology, PhD thesis, 2008.

Sameer Ralhan and Thomas A Badgwell. Robust disturbance rejection for FIR systems with bounded parameters. *IFAC Proceedings Volumes*, 33(10): 815–820, 2000.

Ian Rhodes. A tutorial introduction to estimation and filtering. *IEEE Transactions on Automatic Control*, 16(6):688–706, 1971.

Matthew B Rhudy, Roger A Salguero, and Keaton Holappa. A Kalman filtering tutorial for undergraduate students. *International Journal of Computer Science & Engineering Survey*, 8(1):1–9, 2017.

Eric Rogers, Krzysztof Galkowski, and David H Owens. *Control Systems Theory and Applications for Linear Repetitive Processes*. Springer Science & Business Media, 2007.

Emre Sariyildiz and Kouhei Ohnishi. Stability and robustness of disturbance-observer-based motion control systems. *IEEE Transactions on Industrial Electronics*, 62(1):414–422, 2015.

E. Schrijver and J. van Dijk. Disturbance observers for rigid mechanical systems:equivalence, stability and design. *Journal of Dynamic Systems, Measurement, and Control*, Vol. 124:539–548, 2002.

H Seraji. A new method for pole placement using output feedback. *International Journal of Control*, 28(1):147–155, 1978.

J. H. She, X. Xin, and Y. Pan. Equivalent-input-disturbance approach: analysis and application to disturbance rejection in dual-stage feed drive control system. *IEEE/ASME Transactions on Mechatronics*, Vol. 16:330–340, 2011.

Dan Simon. *Optimal state estimation: Kalman filter,H infinity, and nonlinear approaches*. John Wiley and Sons, 2006.

Sigurd Skogestad and Ian Postlethwaite. *Multivariable feedback control: analysis and design*. Wiley, New York, 2007.

T. Soderstrom. *Errors-in-Variables Methods in System Identification*. Springer, 2018.

T. Soderstrom and P. Stoica. *System Identification*. Prentice-Hall, Englewood Cliffs, New Jersey, 1989.

Dennis D Sourias and Vasilios Manousiouthakis. Best achievable decentralized performance. *IEEE Transactions on Automatic Control*, 40(11):1858–1871, 1995.

K. K. Tan, Qing-Guo Wang, and Chang C Hang. *Advances in PID Control*. Springer Science & Business Media, 2012.

Wen Tan, Horacio J Marquez, and Tongwen Chen. Multivariable robust controller design for a boiler system. *IEEE Transactions on Control Systems Technology*, 10(5):735–742, 2002.

Sophie Tarbouriech and Matthew Turner. Anti-windup design: an overview of some recent advances and open problems. *IET control theory & applications*, 3(1):1–19, 2009.

C James Taylor, Peter C Young, and Arun Chotai. *True Digital Control: Statistical Modelling and Non-minimal State Space Design*. John Wiley & Sons, 2013.

C.J. Taylor, A. Chotai, and P.C. Young. State space control system design based on non-minimal state variable feedback: further generalization and unification results. *International Journal of Control*, Vol. 73:1329–1345, 2000b.

C Thornton and G Bierman. Filtering and error analysis via the udu^{T} covariance factorization. *IEEE Transactions on Automatic Control*, 23(5): 901–907, 1978.

Benson H Tongue. *Principles of Vibration*. Second edition, Oxford University Press New York, 2002.

Lloyd N Trefethen and David Bau III. *Numerical linear algebra*, volume 50. SIAM, 1997.

Daniel J Trudnowski, Andrew Gentile, Jawad M Khan, and Eric M Petritz. Fixed-speed wind-generator and wind-park modeling for transient stability studies. *IEEE Transactions on Power Systems*, 19(4):1911–1917, 2004.

E. T. van Donkelaar, O. H. Bosgra, and P.M.J. Van den Hof. Constrained model predictive control with on-line input paramterization. In *Proc. the 38th Conference on Decision and Control, Phoenix, Arizona*, pages 3718–3721, 1999.

Anastasia Veloni and Nikolaos I Miridakis. *Digital control systems: theoretical problems and simulation tools*. CRC Press, 2017.

Antonio Visioli. *Practical PID Control*. Springer Science & Business Media, 2006.

C. Wang and P. C. Young. Direct digital control by input-output, state variable feedback: theoretical background. *International Journal of Control*, Vol. 747: 97–109, 1988.

L. Wang. *Model Predictive Control System Design and Implementation using MATLAB*. Springer-Verlag, 2009.

L. Wang and W. R. Cluett. *From Plant Data to Process Control: Ideas for Process Identification and PID Design*. Taylor and Francis, London, 2000.

L. Wang, C. Freeman, and E. Rogers. Repetitive-predictive control system with constraints: design and implementation. *Journal of Process Control*, 23: 956–967, 2013.

Liuping Wang. *PID Control System Design and Automatic Tuning using MATLAB/Simulink*. Wiley-IEEE PRESS, 2020. ISBN 9781119469346.

Liuping Wang. Control design methods with anti-windup mechanism for disturbance rejection and reference following. *Transactions of the Institute of Measurement and Control*, 38(6):625–639, 2016a.

Liuping Wang. Tutorial review on repetitive control with anti-windup mechanisms. *Annual Reviews in Control*, 42:332–345, 2016b.

Liuping Wang, Peter C Young, Peter J Gawthrop, and C James Taylor. Non-minimal state-space model-based continuous-time model predictive control with constraints. *International Journal of Control*, 82(6):1122–1137, 2009.

Liuping Wang, Shan Chai, Eric Rogers, and Chris T Freeman. Multivariable repetitive-predictive controllers using frequency decomposition. *IEEE Transactions on Control Systems Technology*, 20(6):1597–1604, 2011a.

Liuping Wang, Chris T Freeman, Shan Chai, and Eric Rogers. Multivariable repetitive-predictive control of a robot arm with experimental results. *IFAC Proceedings Volumes*, 44(1):7672–7677, 2011b.

Liuping Wang, Chris Freeman, and Eric Rogers. Observer-based predictive repetitive control with experimental validation. *IFAC-PapersOnLine*, 50(1): 3674–3679, 2017.

Liuping Wang, Chris T Freeman, and Eric Rogers. Disturbance observer-based predictive repetitive control with constraints. *International Journal of Control*, pages 1–10, 2020.

Qing-Guo Wang, Zhen Ye, Wen-Jian Cai, and Chang-Chieh Hang. *PID control for multivariable processes*, volume 373. Springer Science & Business Media, 2008.

David S Watkins. *Fundamentals of matrix computations*, volume 64. John Wiley & Sons, 2004.

Dieter Willner, CB Chang, and KP Dunn. Kalman filter algorithms for a multi-sensor system. In *1976 IEEE conference on decision and control including the 15th symposium on adaptive processes*, pages 570–574. IEEE, 1976.

William A Wolovich. *Linear multivariable systems*. Springer Science & Business Media, 2012.

W Wonham. On pole assignment in multi-input controllable linear systems. *IEEE transactions on automatic control*, 12(6):660–665, 1967.

J. Yang, S. Li, and W.H. Chen. Nonlinear disturbance observer-based control for multi-input and multi-output nonlinear systems subject to mismatching conditions. *International Journal of Control*, Vol. 85:1071–1082, 2012.

Jun Yang, Wen-Hua Chen, Shihua Li, Lei Guo, and Yunda Yan. Disturbance/uncertainty estimation and attenuation techniques in PMSM drives: A survey. *IEEE Transactions on Industrial Electronics*, 64(4): 3273–3285, 2017.

Peter C Young. Gauss, Kalman and advances in recursive parameter estimation. *Journal of Forecasting*, 30(1):104–146, 2011.

Peter C Young. *Recursive Estimation and Time-series Analysis: an Introduction*. Springer Science & Business Media, 2012.

Cheng-Ching Yu. *Autotuning of PID Controllers*. Springer Science & Business Media, 2006.

Luca Zaccarian and Andrew R Teel. *Modern Anti-windup Synthesis: Control Augmentation for Actuator Saturation*. Princeton University Press, 2011.

Kemin Zhou and John C. Doyle. *Essentials of robust control*, volume 104. Prentice hall Upper Saddle River, NJ, 1998.

Index

State Feedback Control and Kalman Filtering with MATLAB/Simulink Tutorials, First Edition.
Liuping Wang and Robin Ping Guan.

Companion website: www.wiley.com/go/feedback19